Biochemistry
of Nickel

BIOCHEMISTRY OF THE ELEMENTS

Series Editor: Earl Frieden
Florida State University
Tallahassee, Florida

Recent volumes in this series:

A Continuation Order Plan is available for this series. A continuation order will bring delivery of each new volume immediately upon publication. Volumes are billed only upon actual shipment. For further information please contact the publisher.

Biochemistry of Nickel

Robert P. Hausinger

Departments of Microbiology and Biochemistry
Michigan State University
East Lansing, Michigan

PLENUM PRESS • NEW YORK AND LONDON

Library of Congress Cataloging-in-Publication Data

Hausinger, Robert P.
    Biochemistry of nickel / Robert P. Hausinger.
        p.   cm. -- (Biochemistry of the elements ; v. 12)
    Includes bibliographical references and index.
    ISBN 0-306-44541-7
    1. Nickel--Metabolism. 2. Nickel enzymes. 3. Nickel--Toxicology.
4. Nickel in the body. I. Title. II. Series.
    [DNLM: 1. Nickel--metabolism. 2. Biochemistry. Qu 130 B6144
1980 v. 12 1980]
    QP535.N6H38 1993
    574.19'214--dc20
    DNLM/DLC
    for Library of Congress   93-21371
                                                                        CIP

ISBN 0-306-44541-7

© 1993 Plenum Press, New York
A Division of Plenum Publishing Corporation
233 Spring Street, New York, N.Y. 10013

Printed in the United States of America

To Sue, Michael, and Melanie

Preface

This book represents an attempt to summarize the rapidly growing body of knowledge concerning the biochemistry of nickel. This metal ion is typically considered in terms of its harmful effects on biological systems. For example, numerous toxic effects of nickel salts on microbes, plants, and animals have been described, and the carcinogenic properties of nickel compounds in animals have recently been extensively studied. Nevertheless, it is now clear that nickel is essential for several metabolic processes, and four types of nickel-containing enzymes have been identified. Many microorganisms, plants, and animals possess mechanisms to transport this metal ion and incorporate it into specific proteins or organic complexes. As an extreme example, the plant *Sebertia acuminata* accumulates nickel ion as a complex with an organic acid to a concentration of 25% (dry weight) in its sap. Here, I attempt to provide a balanced discussion of nickel biochemistry that includes descriptions of both its harmful and its beneficial effects in microorganisms, plants, and animals.

This book divides the multiple facets of nickel biochemistry into nine chapters. Chapter 1 serves as a general introduction to the harmful and beneficial aspects of nickel from a historical perspective. In addition, it provides a summary of nickel concentrations, distributions, and fluxes in the environment. The coordination chemistry and spectroscopic properties of nickel are briefly summarized in Chapter 2. Chapters 3 through 6 describe the four distinct types of nickel-containing enzymes: urease, hydrogenase, carbon monoxide dehydrogenase, and methyl coenzyme M reductase. The properties of the nickel-dependent proteins and their nickel metallocenters are discussed in detail, along with related genetic information. Furthermore, the evidence for and properties of elaborate biosynthetic systems that function to specifically incorporate this metal ion into certain of these proteins are described. Chapter 7 focuses on other aspects of microbial nickel metabolism, including the mechanisms of nickel transport, toxicity, and resistance. The interactions between plants and nickel are described in Chapter 8. Topics discussed in this chapter include the evidence that nickel is necessary to plant life, the mechanisms of toxicity and resistance, and the special case of the nickel-hyperaccumulating plants. Finally, Chapter 9 describes animal nickel metabolism. The mechanisms for nickel uptake, transport, and elimination are described, the evidence for a nickel requirement in animals is evaluated, and the toxic and carcinogenic effects of nickel are summarized.

When Dr. Earl Frieden asked whether I would be willing to undertake writing a book on the biochemistry of nickel, I realized that no one person can be expert in all aspects of this field. Additionally, it was clear to me that any book on this topic would age rapidly because of the brisk pace of research in this area. Nevertheless, I felt that a critical and detailed summary of our current understanding of the biological interactions of this metal ion would be timely and of widespread interest. Furthermore, such a book would allow one to point out the key unanswered questions remaining in this field and perhaps help to define future research directions involving the biochemistry of nickel. I accepted his offer to write this book and have strived to present an up-to-date synopsis of this field. I hope that the reader will enjoy learning about this fascinating biological metal and be stimulated to pursue research in the challenging area of nickel biochemistry.

Robert P. Hausinger

Contents

6. Methyl Coenzyme M Reductase

Introduction

1

1.1 Historical Perspectives on Nickel

The element nickel was discovered by Cronstedt in 1751 (Cronstedt, 1770) and first purified as a metal by Berthier in the early 1800s (Berthier, 1820). Early German copper miners working with niccolite, a red-colored nickel arsenide that has the appearance of copper ore, are responsible for the name of this element. Not only were these workers unable to extract copper from the material, but the fumes derived from the reddish substance were toxic; hence, they called the ore kupfernickel, or Old Nick's copper, after the name of an evil spirit referred to as "Old Nick." Since the discovery of nickel, the evil spirit side of this element has received the most attention; for example, innumerable studies have characterized various aspects of nickel toxicity and carcinogenicity at a biochemical level. In addition to its many toxic effects, however, this metal recently has been shown to be required for certain metabolic processes. The following paragraphs will highlight key biological benchmarks in the history of nickel.

Historical milestones related to the potential hazards from exposure to nickel compounds are summarized in Table 1-1. Human health effects related to nickel were apparent already in the early 1500s, when the toxicity of kupfernickel ore was noted in miners [reviewed by Howard-White (1963)]. The first laboratory study of nickel toxicity was probably carried out by Gmelin (1826), who administered high doses of nickel sulfate to rabbits and dogs and observed toxic effects ranging from inflammation and generalized physical wasting to severe gastritis and fatal convulsions. Numerous other studies to assess the acute lethal dosages and other aspects of nickel toxicity for nickel salts in various animals were carried out in the 1880s (e.g., Stuart, 1883, 1884), as summarized by the National Research Council (1975). In 1891, a gaseous nickel compound, nickel carbonyl [$Ni(CO)_4$], was shown to be highly toxic to animals (McKendrick and Snodgrass, 1891). This substance, reported by Mond $et\ al.$ (1890) in the previous year, was immediately recognized as

1

Table 1-1. Historical Overview of the Harmful Effects of Nickel

Date	Topic	Reference
1500s	Toxicity of kupfernickel ore observed in miners	Howard-White (1963)
1826	First laboratory demonstration of nickel toxicity in animals	Gmelin (1826)
1880s	Assessment of acute lethal dose of nickel salts in various animals	Stuart (1883, 1884)
1891	Nickel carbonyl shown to be toxic to animals	McKendrick and Snodgrass (1891)
1893	First demonstration of nickel toxicity in plants	Haselhoff (1893)
1912	Nickel dermatitis reported	Herxheimer (1912)
1939–1958	Epidemiological evidence implicates nickel as a carcinogen	Doll (1958)
1943	First animal study to demonstrate the carcinogenicity of nickel compounds	Campbell (1943)
1967	Demonstration that a nickel compound (Ni_3S_2) can cause morphological transformation of tissue culture cells	Basrur and Gilman (1967)

being useful in the isolation of pure metallic nickel (Fig. 1-1) and became utilized on a large scale in the nickel industry. In the succeeding years, several hundred workers in the nickel industry have been exposed to $Ni(CO)_4$, resulting in a few deaths and numerous instances of severe health problems. At about this time, it also was reported that nickel compounds had biological effects outside of the animal kingdom. For example, Haselhoff (1893) provided the first demonstration that nickel salts were toxic to plants (corn and bean) grown in solution culture. Phytotoxic effects were subsequently noted and characterized in many species of plants (reviewed by Mishra and Kar (1974) and Hutchinson (1981); see Chapter 8). In addition to the toxic effects of nickel compounds at high concentrations in animals, plants, and microbes (Babich and Stotzky, 1983), other harmful effects of nickel became apparent.

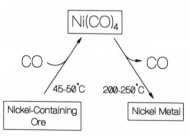

Figure 1-1. The Mond process for nickel purification. Nickel compounds and material containing metallic nickel are allowed to react with carbon monoxide at moderate temperatures and atmospheric pressure, resulting in the formation of gaseous nickel carbonyl, $Ni(CO)_4$. At elevated temperatures, nickel carbonyl decomposes to form metallic nickel and carbon monoxide. Nickel carbonyl is more toxic than hydrogen cyanide and has caused severe health effects and even death in victims of accidental exposure.

For example, Herxheimer (1912) described the skin disease known as nickel dermatitis in industrial workers in the nickel industry. Unlike the other toxic manifestations of nickel, this dermatitis subsequently was shown specifically to involve immunological responses. Finally, epidemiological studies such as that of Doll (1958) provided evidence implicating nickel in the development of pulmonary and nasal cancers. Again, this effect differs from general toxicity and involves cellular changes at the genetic level. Major advances in characterizing these changes were made possible by the ability to generate nickel-induced tumors in animals (e.g., Campbell, 1943) and to transform tissue culture cells (e.g., Basrur and Gilman, 1967). Studies such as those outlined above were important in identifying some of the harmful effects of nickel and laid the groundwork for detailed biochemical characterization of these phenomena, as described in Chapter 9.

Landmarks in the history of the biological benefits of nickel are summarized in Table 1-2. Several studies in the late 1800s reported potential therapeutic uses of nickel compounds in humans. For example, nickel salts

Table 1-2. Historical Overview of the Biological Benefits of Nickel

Date	Topic	Reference(s)
1850–1900	Therapeutic uses of nickel salts examined in humans	National Research Council (1975)
1915–1960	Enhanced plant growth and increased crop yield due to fungicidal action of nickel salts	Mishra and Kar (1974)
1965	Nickel-dependent chemolithotrophic growth observed for two *Hydrogenomonas* strains[a]	Bartha and Ordal (1965)
1967	Growth stimulation by nickel in the eucaryote *Chlorella vulgaris*	Bertrand and De Wolf (1967)
1970–1975	Evidence that nickel deficiency leads to ultrastructural changes and altered enzyme levels in animals	Nielsen and Ollerich (1974)
1974	Enhancement of microbial fixation of nitrogen gas by addition of nickel	Bertrand (1974)
1974	Nickel requirement for hydrogenase activity demonstrated using purified enzyme from *Nocardia opaca*	Aggagg and Schlegel (1974)
1975	Jack bean urease shown to contain nickel	Dixon *et al.* (1975)
1980	*Clostridium thermoaceticum* carbon monoxide dehydrogenase found to contain nickel	Drake *et al.* (1980)
1980	Methanogen F_{430} chromophore found to contain nickel	Diekert *et al.* (1980); Whitman and Wolfe (1980)
1982	F_{430} shown to be the coenzyme of methyl coenzyme M reductase	Ellefson *et al.* (1982)

[a] Currently *Alcaligenes.*

were found to serve as analgesics, sedatives, antidiarrheal agents, and antiepileptic drugs [summarized by National Research Council (1975)]. As the acute and chronic toxicity of nickel compounds became more widely known, however, the clinical interests were abandoned. From 1915 through at least the 1960s, many studies reported an apparent enhancement of plant growth or crop yield when low concentrations of nickel were applied to a wide variety of plant species [summarized by Mishra and Kar (1974)]. Most of these effects are now recognized as being artifactual, arising from the reduction in numbers of parasitic fungi due to the strong antifungal activity of nickel (see Chapters 7 and 8). The first clear demonstration of a bacterial requirement for nickel was the nickel-dependent chemolithotrophic growth of two *Alcaligenes* (formerly *Hydrogenomonas*) strains observed by Bartha and Ordal (1965). Similarly, Bertrand and De Wolf (1967) noted a 34% growth stimulation by nickel in an eucaryote, *Chlorella vulgaris*. In addition, the microbial fixation of dinitrogen into biological material was found by Bertrand (1974) to be enhanced by addition of nickel. In most cases, these early physiological results can now be explained by the functional requirement of a nickel-dependent hydrogenase (see Chapter 4). Not surprisingly, the first example of a nickel-dependent enzyme also was that of a hydrogenase. Aggagg and Schlegel (1974) demonstrated that purified hydrogenase from the gram-positive bacterium *Nocardia opaca* requires the presence of nickel ion in the assay buffer for activity. In addition to this exchangeable nickel ion, the *N. opaca* enzyme and many other hydrogenases were later found to possess tightly bound, catalytically active, nickel ion (Chapter 4). However the first description of a nickel-containing enzyme was that of Dixon *et al.* (1975), who showed that highly purified jack bean urease possesses tightly associated nickel that is required for activity. Ureases from other plants, numerous bacteria, some fungi, and even an invertebrate have subsequently been shown to contain nickel (Chapter 3). Two additional enzymes have since been found to contain nickel: carbon monoxide dehydrogenase (Drake *et al.*, 1980) and methyl coenzyme M reductase (Ellefson *et al.*, 1982). The former nickel-containing enzyme plays a key role in a recently described autotrophic pathway found in certain bacteria and functions in acetate degradation in other microbes, as described in Chapter 5. The latter enzyme, found only in methanogenic bacteria, possesses a unique chromophore, termed factor F_{430}, that earlier had been shown to contain nickel (Diekert *et al.*, 1980; Whitman and Wolfe, 1980). Detailed discussion of this chromophore and enzyme are provided in Chapter 6. Finally, evidence that certain animals possess a requirement for nickel began to mount starting early in the 1970s. For example, Nielsen and Ollerich (1974) summarized several changes in ultrastructural appearance and enzyme levels in nickel-deficient chicks. The universality of a nickel requirement in animals remains unestablished and the role(s) of this metal ion remains unclear, as described in Chapter 9.

1.2 Environmental Aspects of Nickel

Nickel is thought to comprise approximately 2% of the Earth; however, most of this element is located in the core or mantle regions of the planet, and only a small amount of nickel is biologically significant. For example, the core, represented by a sphere of $\sim$3500-km radius, contains 189×10^{25} g of nickel (at a concentration of 5.8%) and is surrounded by the mantle shell of $\sim$2900 km containing 407×10^{25} g of nickel (at a concentration of 0.22%) (Duke, 1980). In contrast, the crust of 10–50 km contains only 2.4×10^{25} g of nickel at a concentration of 0.0086%. At this concentration, nickel is the 24th most abundant element in the crust. The actual concentration of nickel in any particular soil sample may vary widely; for example, sandstone and granite may contain less than 0.0001% nickel, whereas the so-called ultramafic or ultrabasic rocks can contain substantially more than 0.3% nickel. This element is often found in nature associated with sulfur, antimony, and arsenic (e.g., NiS, $NiSb$, $NiAs_2$, $NiAsS$, $NiSbS$), and major deposits of magnesium–nickel silicates (garnierite) are known. The metal contents and other properties of various nickel-containing rocks have been summarized by Duke (1980) and Boyle and Robinson (1988). On average, soils contain nickel at a concentration of 16 ppm (Nriagu, 1980) to 40 ppm (Boyle and Robinson, 1988). The concentration of nickel in the oceans is 0.3–0.6 ppb, whereas somewhat higher levels are found in uncontaminated freshwater lakes and rivers (0.5–20 ppb). Human action, however, has led to high levels of nickel contamination in many sites, with potentially significant biological effects. Below, I briefly describe one example in which a historic method for enriching nickel from nickel-containing ores had a major environmental impact. In addition, I summarize the major components involved in global cycling of nickel and emphasize the importance of human activities as sources of nickel contamination.

The history of the Sudbury district of the Province of Ontario offers an extreme example of the types of environmental effects that have been associated with the nickel industry. This region, often called the nickel capital of the world, possesses nickel deposits that contain high concentrations of nickel (1–4%) as well as iron, copper, cobalt, and other metals in sulfide-rich ores. In 1886, the Canadian Copper Company began a primitive method of smelting in which nickel-containing ores were burned in open roastbeds to release gaseous sulfur dioxide and enrich the remaining material in nickel. By 1916, 600,000 tons of sulfur dioxide were emitted to the atmosphere annually by this process (Swift, 1977). The high levels of sulfur dioxide emissions caused a devastation of all plant and animal life in the area. In addition, acid rain from the SO_2 emissions caused an increase in soil and groundwater acidification that subsequently resulted in high levels of leaching of metal ions to further stunt plant growth. Moreover, the roasting process led to high airborne emissions of metal compounds, leading to nickel contamination of nearby soils and water systems. The metal emissions included the initial nickel sulfides in the ore and oxidation products such as various nickel oxides. Crystalline nickel sulfides and nickel oxides are now known to be carcinogenic to animal cells (see Chapter 9). Elimination of the open-pit roasting method and implementation of roasters and smelters with chimneys reduced the severity of the local problems, as did the completion of the so-called "superstack" in 1972, which allowed for more widespread dispersal of the emissions. However, these local improvements were somewhat negated by the potential negative impact on a larger area. Indeed, a 1974 study suggested that the annual price for environmental damage from sulfur dioxide in the Sudbury area was approximately $465,850,000 (Swift, 1977). Fortunately, numerous technological developments in the nickel industry (e.g., conversion of sulfur dioxide to sulfuric acid, a commercially viable product) combined with heightened environmental awareness and increased governmental regulation have led to successive drastic reductions of sulfur dioxide emissions over the years. Nevertheless, the area near Sudbury remained a wasteland for decades until recent collaborative efforts between the citizens of the area and the local nickel industries achieved remarkable progress in restoration of the area. Simply letting nature recover on its own after minimizing the emissions was not enough to rejuvenate the barren landscape; however, by liming the soils, fertilizing, and sowing seeds, the natural beauty of the area is rapidly being reestablished.

On a global scale, the anthropogenic sources of nickel in air, water, and soils have recently been assessed by Nriagu and Pacyna (1988), while Nriagu (1989) compared emissions by natural and human-derived sources. As shown in Table 1-3, wind-borne soil particles and volcanic activity account for the major natural sources of nickel emissions to the atmosphere. The ranges provided for the nickel emissions from these sources (1.8×10^9–20×10^9 and 0.93×10^9–28×10^9 g of nickel per year, respectively) are derived from yearly estimates of the total mass of soil particles taken up by the wind (0.6×10^{14}–5×10^{14} g), the typical soil concentration of nickel (30–40 $\mu g/g$ of soil), the total amount of volcanic sulfur released (15×10^{12}–50×10^{12} g), and the nickel/sulfur ratio in vent gas (0.62×10^{-4}–5.6×10^{-4}). The values for these and the other natural sources of nickel emissions account for only 35% of the total nickel dispersed into the atmosphere each year. Thus, human activities account for the major sources of atmospheric emissions of nickel via the processes enumerated in the table. Although the levels of nickel naturally released into aquatic systems and soils were not provided, the

Table 1-3. Worldwide Emissions of Nickel to the Atmosphere[a]

Source	Yearly nickel emissions (10^9 g)	
	Range	Median
Natural		
Wind-borne soil particles	1.8–20	11
Seasalt spray	0.01–2.6	1.3
Volcanoes	0.93–28	14
Wild forest fires	0.1–4.5	2.3
Biogenic	0.11–1.65	0.73
Total	3.0–57	30
Anthropogenic		
Coal combustion		
Electric utilities	1.395–9.3	5.35
Industry and domestic	1.98–14.85	8.42
Oil combustion		
Electric utilities	3.84–14.5	9.17
Industry and domestic	7.16–28.64	17.9
Non-ferrous metal production		
Mining		0.8
Pb production		0.331
Cu–Ni production		7.65
Steel and iron manufacturing	0.036–7.1	3.57
Refuse incineration		
Municipal	0.098–0.42	0.26
Sewage sludge	0.03–0.18	0.11
Phosphate fertilizers	0.137–0.685	0.41
Cement production	0.089–0.89	0.49
Wood combustion	0.6–1.8	1.2
Total	24.15–87.15	55.65

[a] From Nriagu and Pacyna (1988) and Nriagu (1989).

values for anthropogenic sources shown in Tables 1-4 and 1-5 suggest that industrial outputs greatly exceed the baseline burdens for this metal. Thus, it is clear that human activities play a major role in the global cycling of this element.

1.3 Overview

Subsequent chapters of this book bracket nickel biochemistry from the inorganic to the biological perspective. Chapter 2 is a nickel chemistry primer for biological scientists that briefly reviews the coordination chemistry and

Table 1-4. Anthropogenic Release of Nickel
into Aquatic Systems[a]

Source	Yearly nickel released (10^9 g)	
	Range	Median
Domestic wastewater	21–102	61.5
Steam electric	3.0–18	10.5
Mining	0.01–0.5	0.25
Smelting and refining	2.0–24	13
Manufacturing processes		
Metals	0.2–7.5	3.85
Chemicals	1.0–6.0	3.5
Pulp and paper	0–0.12	0.06
Petroleum products	0–0.06	0.03
Atmospheric fallout	4.6–16	10.3
Dumping of sewage sludge	1.3–20	10.15
Total	33–194	113

[a] From Nriagu and Pacyna (1988).

spectroscopy of this element. Chapters 3 through 6 describe the functions, structures, and properties of the four known nickel enzymes: urease, hydrogenase, carbon monoxide dehydrogenase, and methyl coenzyme M reductase. Microbial metabolism involving this metal ion is covered in Chapter 7. Topics include the mechanisms of nickel transport, toxicity, and resistance in microbes and the evidence for roles beyond those involving the four known nickel-containing enzymes. The interactions between plants and nickel are detailed in Chapter 8. In addition to discussions of the functions of nickel in plants, metal ion transport, and the mechanisms of toxicity and resistance, this chapter describes the unique features of nickel-hyperaccumulating plants. Finally, the relationships between animals and nickel are reviewed in Chapter 9. The three major themes in this final chapter relate to the flux of nickel in animals (uptake, systemic transport, distribution, and excretion), the evidence that nickel may be an essential element in animals, and the harmful influences of nickel in animals, including immunological effects, general toxicity, embryotoxicity, and carcinogenicity.

In closing this introductory chapter, I would like to mention the availability of several other books and excellent special-topic reviews related to nickel biochemistry. The reader is referred to these sources to obtain more details about certain of the topics presented here. For example, *Nickel in the Environment* (Nriagu, 1980) offers an early detailed assessment of the concentrations and fluxes of nickel in various types of environments and describes

Table 1-5. Anthropogenic Release of Nickel into Soils[a]

Source	Yearly nickel released (10^9 g)	
	Range	Median
Agricultural and food wastes	6–45	25.5
Animal wastes (manure)	3–36	19.5
Logging and wood wastes	2.2–23	12.6
Urban refuse	2.2–10	6.1
Municipal sewage sludge	5.0–22	13.5
Other organic wastes	0.17–3.2	1.7
Solid wastes (metal manufacturing)	0.84–2.5	1.7
Coal and bottom fly ash	56–279	167
Fertilizer	0.2–0.55	0.375
Peat	0.22–3.5	1.86
Wastage of commercial products	6.5–32	19.3
Atmospheric fallout	11–37	24
Mine tailings	22–64	43
Smelter slags and wastes	32–65	48.5
Total	160–673	417

[a] From Nriagu and Pacyna (1988).

some of the interactions between nickel and microbes, plants, and animals that were known at the time. More recent multiauthored compilations entitled *Nickel and Its Role in Biology* (Sigel and Sigel, 1988, Volume 23 of *Metal Ions in Biological Systems*) and *The Bioinorganic Chemistry of Nickel* (Lancaster, 1988) include contributions that focus on some of the inorganic chemistry and spectroscopy of nickel, the properties of nickel enzymes, and other biological features of nickel. Three noteworthy books that focus primarily on human health effects of nickel are the classic review simply called *Nickel* (National Research Council, 1975) and the proceedings from conferences on nickel metabolism and toxicity held in 1983 and 1988 called *Nickel in the Human Environment* (Sunderman *et al.,* 1984) and *Nickel and Human Health: Current Perspectives* (Nieboer and Nriagu, 1992). Finally, I would like to mention notable reviews of nickel enzymes (Thauer *et al.,* 1980; Hausinger, 1987; Walsh and Orme-Johnson, 1987; Cammack, 1992; Kolodziej, 1993) and health-related aspects of nickel (Coogan *et al.,* 1989; Sunderman, 1989; Christie and Katsifis, 1990; Snow, 1992). In the present book, I attempt to summarize our current understanding of each of these aspects of nickel biochemistry. As the reader will learn, this is a field that continues to progress at a rapid rate, yet many important questions remain to be answered.

References

Aggagg, M., and Schlegel, H. G., 1974. Studies on a gram-positive hydrogen bacterium, *Nocardia opaca* 1b. III. Purification, stability and some properties of the soluble hydrogen dehydrogenase, *Arch. Microbiol.* 100:25–39.

Babich, H., and Stotzky, G., 1983. Toxicity of nickel to microbes: Environmental aspects, *Adv. Appl. Microbiol.* 29:195–265.

Bartha, R., and Ordal, E. J., 1965. Nickel-dependent chemolithotrophic growth of two *Hydrogenomonas* strains, *J. Bacteriol.* 89:1015–1019.

Basrur, P. K., and Gilman, J. P. W., 1967. Morphological and synthetic response of normal and tumor muscle cultures to nickel subsulfide, *Cancer Res.* 27:1168–1177.

Berthier, M., 1820. *Ann. Chim. Phys.* 2:14–52.

Bertrand, D., 1974. Le nickel, oligo-élément dynamique pour les micro-organismes fixateurs de l'azote de l'air, *C. R. Acad. Sci.* 278:2231–2235.

Bertrand, D., and De Wolf, A., 1967. Le nickel, oligoélément dynamique pour les végétaux supèrieurs, *C. R. Acad. Sci.* 265:1053–1055.

Boyle, R. W., and Robinson, H. A., 1988. Nickel in the natural environment, in *Nickel and Its Role in Biology* (H. Sigel and A. Sigel, eds.), *Metal Ions in Biological Systems,* Vol. 23, Marcel Dekker, New York, pp. 1–29.

Cammack, R., 1992. Catalysis by nickel in biological systems, in *Bioinorganic Catalysis* (J. Reedijk, ed.), Marcel Dekker, New York, 189–225.

Campbell, J. A., 1943. Lung tumours in mice and man, *Br. Med. J.* 1:179–183.

Christie, N. T., and Katsifis, S. P., 1990. Nickel carcinogenesis, in *Biological Effects of Heavy Metals* (E. C. Foulkes, ed.), CRC Press, Boca Raton, Florida, pp. 95–128.

Coogan, T. P., Latta, D. M., Snow, E. T., and Costa, M., 1989. Toxicity and carcinogenicity of nickel compounds, *CRC Crit. Rev. Toxicol.* 19:341–384.

Cronstedt, A. F., 1770. *An Essay towards a System of Mineralogy,* E. & C. Dilly, London.

Diekert, G., Klee, B., and Thauer, R. K., 1980. Nickel, a component of factor F_{430} from *Methanobacterium thermoautotrophicum, Arch. Microbiol.* 124:103–106.

Dixon, N. E., Gazzola, C., Blakeley, R. L., and Zerner, B., 1975. Jack bean urease (EC 3.5.1.5). A metalloenzyme. A simple biological role for nickel?, *J. Am. Chem. Soc.* 97:4131–4133.

Doll, R., 1958. Cancer of the lung and nose in nickel workers, *Br. J. Ind. Med.* 15:217–223.

Drake, H. L., Hu, S.-I., and Wood, H. G., 1980. Purification of carbon monoxide dehydrogenase, a nickel enzyme from *Clostridium thermoaceticum, J. Biol. Chem.* 255:7174–7180.

Duke, J. M., 1980. Nickel in rocks and ores, in *Nickel in the Environment* (J. O. Nriagu, ed.), John Wiley & Sons, New York, pp. 27–50.

Ellefson, W. L., Whitman, W. B., and Wolfe, R. S., 1982. Nickel-containing factor F_{430}: Chromophore of the methylreductase of *Methanobacterium, Proc. Natl. Acad. Sci. USA* 79:3707–3710.

Gmelin, C. G., 1826. Expériences sur l'action de la baryte, de la strontiane, du chrôme, du molybdène, du tungstène, du tellure, de l'osmium, du platine, de l'iridium, du rhodium, du palladium, du nickel, du cobalt, de l'urane, du cérium, du fer et du manganèse sur l'organisme animale, *Bull Sci. Med.* 7:110–117.

Haselhoff, E., 1893. Versuche uber die schadlicke wirkung von nickel-haltigen wasser auf pflanzen, *Landwirtsch. Jahrb.* 22:1862–1868.

Hausinger, R. P., 1987. Nickel utilization by microorganisms, *Microbiol. Rev.* 51:22–42.

Herxheimer, K., 1912. Über die gewerblichen erkrankungen der haut, *Dtsch. Med. Wochenschr.* 38:18–22.

Howard-White, F. B., 1963. *Nickel—An Historical Review,* Longmans Canada, Toronto.

Hutchinson, T. C., 1981. Nickel, in *Effect of Heavy Metal Pollution on Plants* (N. W. Lepp, ed.), Applied Science Publishers, London, pp. 171–211.

Kolodziej, A. F., 1993. The chemistry of nickel-containing enzymes, *Prog. Inorg. Chem.* 41:493–597.

Lancaster, J. R., Jr., 1988. *The Bioinorganic Chemistry of Nickel,* VCH Publishers, New York.

McKendrick, J. G., and Snodgrass, S. W., 1891. On the physiological action of carbon monoxide on nickel, *Br. Med. J.* 1:1215–1217.

Mishra, D., and Kar, M., 1974. Nickel in plant growth and metabolism, *Bot. Rev.* 40:395–452.

Mond, L., Langer, C., and Quincke, F., 1890. The action of carbon monoxide on nickel, *J. Chem. Soc.* 57:749–753.

National Research Council, 1975. *Nickel,* National Academy of Sciences, Washington, D.C.

Nieboer, E., and Nriagu, J. O., 1992. *Nickel and Human Health: Current Perspectives,* John Wiley & Sons, New York.

Nielsen, F. H., and Ollerich, D. A., 1974. Nickel: A new essential trace element, *Fed. Proc.* 33:1767–1772.

Nriagu, J. O., 1980. *Nickel in the Environment,* John Wiley & Sons, New York.

Nriagu, J. O., 1989. A global assessment of natural sources of atmospheric trace metals, *Nature (London)* 338:47–49.

Nriagu, J. O., and Pacyna, J. M., 1988. Quantitative assessment of worldwide contamination of air, water and soils by trace metals, *Nature (London)* 333:134–139.

Sigel, H., and Sigel, A., 1988. *Nickel and Its Role in Biology, Metal Ions in Biological Systems,* Vol. 23, Marcel Dekker, New York.

Snow, E. T., 1992. Metal carcinogenesis: Mechanistic implications, *Pharmacol. Ther.* 53:31–65.

Stuart, T. P. A., 1883. Nickel and cobalt: Their physiological action on the animal organism. Part I. Toxicology, *J. Anat. Physiol.* 17:89–123.

Stuart, T. P. A., 1884. Über den einfluss der nickel und der kobaltwerbindungen auf den thierischen organisms, *Arch. Exp. Pathol. Pharmakol.* 18:151–173.

Sunderman, F. W., Jr., 1989. Mechanisms of nickel carcinogenesis, *Scand. J. Work Environ. Health* 15:1–12.

Sunderman, F. W., Jr., Aitio, A., Berlin, A., Bishop, C., Buringh, E., Davis, W., Gounar, M., Jacquignon, P. C., Mastromatteo, E., Rigaut, J. P., Rosenfeld, C., Saracci, R., and Sors, A., 1984. *Nickel in the Human Environment,* Oxford University Press, New York.

Swift, J., 1977. *The Big Nickel,* Between the Lines, Kitchener, Ontario.

Thauer, R. K., Diekert, G., and Schönheit, P., 1980. Biological role of nickel, *Trends Biochem. Sci.* 11:304–306.

Walsh, C. T., and Orme-Johnson, W. H., 1987. Nickel enzymes, *Biochemistry* 26:4901–4906.

Whitman, W. B., and Wolfe, R. S., 1980. Presence of nickel in the factor F_{430} from *Methanobacterium bryantii, Biochem. Biophys. Res. Commun.* 92:1196–1201.

Chemistry of Nickel   2

2.1 Introduction

This chapter on nickel chemistry is designed for readers with only a weak background in inorganic chemistry; hence, my treatment of this enormous field is necessarily superficial. Those interested in a more detailed discussion are referred to the excellent review by Saconni *et al.* (1987) and other references cited below.

Nickel, atomic number 28, has an atomic weight of 58.71 and is comprised of five stable isotopes: ^{58}Ni (68.274%), ^{60}Ni (26.095%), ^{61}Ni (1.134%), ^{62}Ni (3.593%), and ^{64}Ni (0.904%) (Weast, 1975). Whereas the two most abundant isotopes (^{58}Ni and ^{60}Ni) have no nuclear spin, ^{61}Ni has a nuclear spin of $\frac{3}{2}$ with a magnetic moment of -0.7487 Bohr magneton (BM). This nuclear spin can couple to the electronic spin of paramagnetic nickel (nickel oxidation states that possess free electrons; e.g., Ni^{1+}, Ni^{3+}), resulting in a characteristic splitting of the nickel-derived electron paramagnetic resonance (EPR) spectroscopic signal. Hyperfine spectra of this type have been used as a diagnostic signature for paramagnetic nickel species in protein samples isolated from ^{61}Ni-supplemented cultures (see Chapters 4–6). The unstable isotope ^{63}Ni also has been important in biological studies by acting as a radioactive tracer. This isotope decays by β emission with an energy of 0.067 MeV and a half-life of 92 years, making it an excellent tool in the laboratory. Several other nickel isotopes also are radioactive (^{56}Ni, ^{57}Ni, ^{59}Ni, ^{65}Ni, ^{66}Ni, and ^{67}Ni), but they have seen little use in biochemical studies. The positron and γ emitter $^{57}Ni(t_{1/2}$ 36 h), however, was the first nickel isotope used in nickel transport studies (Abelson and Aldous, 1950) and recently has been used successfully in toxicokinetic studies (Nielsen and Andersen, 1993).

The following sections briefly describe the coordination chemistry of nickel in its various oxidation states and summarize the spectroscopic properties of this element. With regard to the oxidation state, among the nickel compounds that are known the formal charge on the metal ion ranges from

1− to 4+; however, the most common state for nickel is the dication, and the properties of this state will be described in greatest detail. My treatment will focus on general characteristics of nickel complexes of various geometries and metal oxidation states; that is, an extensive discussion of the properties of specific nickel compounds will not be included here. It is important to note, however, that the properties of several model compounds related to the four nickel-containing enzymes and to mechanisms for nickel-dependent DNA damage are described in Chapters 3–6 and Chapter 9, respectively.

2.2 Coordination Chemistry of Nickel

An excellent and extensive discussion of the coordination chemistry of nickel has been presented by Sacconi *et al.* (1987), and more condensed summaries of nickel coordination chemistry include those of Coyle and Stiefel (1988), Tomlinson (1981), Nag and Chakravorty (1980), Cotton and Wilkinson (1980), and Jolly and Wilke (1974). Here, I present only an overview of this topic with an emphasis on biologically relevant chemistries. For the benefit of the reader, several common geometries of nickel are illustrated in Fig. 2-1.

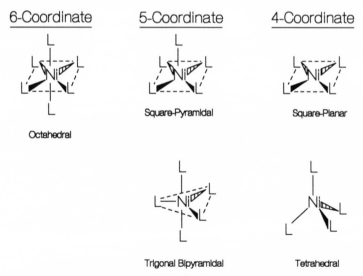

Figure 2-1. Coordination geometries accessible to nickel complexes. Most nickel complexes possess four, five, or six ligands (L) in geometries approaching those illustrated. In numerous cases, however, the geometry is distorted by compression or elongation along one or more axes.

2.2.1 High Oxidation States of Nickel: Ni(IV, III)

Complexes containing nickel(IV) or nickel(III) are generally unstable and rapidly oxidize a variety of organic compounds while forming the nickel(II) species. Partial stabilization of the higher oxidation states of the metal ion can be achieved by charge delocalization involving ligand-to-metal charge transfer from negatively charged ligands or ligands with high electron density. Examples of such complexes typically have nitrogen (tetraazamacrocycles, deprotonated amides, and oximes), oxygen (oxides), and fluorine donor atoms in the first coordination sphere and generally possess an octahedral or pseudooctahedral geometry, although square-planar and trigonal bipyramidal geometries are known.

From a biochemical perspective, nickel(IV) complexes have not been shown to be important. In contrast, nickel(III) species may play a biological role in genetic damage by nickel compounds, and nickel(III) states appear to be accessible to certain nickel-containing enzymes. Of particular importance to nickel carcinogenicity, Lappin *et al.* (1978) demonstrated that deprotonated peptides are capable of stabilizing the nickel(III) oxidation state [reviewed by Margerum and Anliker (1988)]; the postulated mechanisms of nickel-dependent oxidative damage to DNA by such nickel(III)–peptide complexes are discussed in Chapter 9. In the case of the nickel-dependent enzymes hydrogenase and carbon monoxide dehydrogenase, however, complexation by deprotonated peptides does not occur. Rather, the nickel in these enzymes is coordinated, at least in part, by sulfur ligands. Stabilization of nickel(III) by sulfur ligands is very difficult in synthetic complexes because of ligand oxidation giving a nickel(II)–sulfur ligand radical; however, nickel(III)-thiolate complexes recently have been established (Fox *et al.,* 1990; Krüger and Holm, 1987, 1990; Krüger *et al.,* 1991). The relevance of this oxidation state to the chemistry of these enzymes and the mechanisms by which these proteins can stabilize the nickel(III) species are discussed in Chapters 4 and 5. Despite the ability of tetraazamacrocycles to stabilize the nickel(III) state (Busch, 1978), there is presently no evidence for a biologically significant nickel(III) tetrapyrrolic species.

2.2.2 Nickel(II) Coordination Chemistry

Nickel(II) represents the most common oxidation state for nickel. Each of the coordination geometries shown in Fig. 2-1 is known for complexes of nickel in this oxidation state; however, a hallmark of many nickel(II) complexes is their propensity to form an equilibrium among several geometries. For example, the well-known Lifschitz salts are nickel(II) complexes of substituted ethylenediamines that, depending on the temperature, solvent, anions present, and other features, may vary widely in color and magnetic properties. These features perplexed scientists for years, but it is now clear that the anomalous results are explained simply by facile and reversible addition of ligands at the axial positions of square-planar complexes. The square-planar complexes of these ligands are yellow and have no free electron on the nickel (i.e., the nickel is diamagnetic), whereas octahedral complexes are blue and paramagnetic (2.9–3.4 BM). The equilibrium position depends on several factors including the temperature, the concentration, and the ligand strength.

A wide variety of ligands can complex with nickel(II). This element is considered to be at the borderline between hard and soft metals; thus, it can interact with ligands that are classified as hard (e.g., water, hydroxide, ammonia, chloride, phosphate, sulfate, nitrate, carbonate), intermediate (e.g., amines, azide, nitrite), and soft (e.g., thiols, thiolates, CO, cyanide). An especially important ligand with regard to nickel(II) coordination is dimethylglyoxime. This ligand was first used by Tschugaeff (1905) for determination of trace quantities of nickel. It forms a red square-planar bis(dimethylglyoximato)nickel(II) complex (Fig. 2-2) that allows for spectrophotometric or gravimetric determination of nickel in biological samples. A simple procedure to assay nickel in biological samples by using this compound was described by Dixon *et al.* (1980); however, most nickel analyses are carried out by atomic absorption or plasma emission spectroscopic methods.

From a biological perspective, nickel(II) is found as the sole oxidation state of urease (Chapter 3) and is one of several oxidation states available to other nickel-dependent enzymes (Chapters 4–6). In addition, organisms must take up nickel ion from the environment in this oxidation state, and, in the case of plants or animals, any transport form of nickel will also be as nickel(II). With particular reference to nickel uptake, several highly insoluble complexes of nickel(II) are found in the environment. For example, nickel sulfides, nickel carbonate, nickel oxide, and nickel phosphate exhibit no to little dissolution at pH 7. Furthermore, when nickel ion in aqueous buffer is adjusted to high pH, the $Ni(OH)_2$ complex precipitates. Soluble complexes of many common

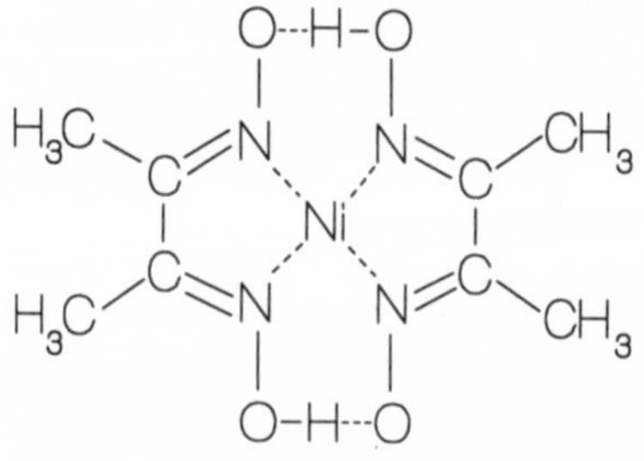

Figure 2-2. The bis(dimethylglyoximato)nickel(II) complex. Complexation by the dimethylglyoxime ligand has been used in spectrophotometric and gravimetric quantitation of nickel in biological samples.

organic compounds (e.g., citrate, histidine, and cysteine) are also highly stable and may interfere with transport processes.

2.2.3 Low Oxidation States of Nickel: Ni(I, 0, −I)

The best characterized synthetic nickel(I) complexes possess tetrahedral or trigonal bipyramidal coordination by phosphine ligands with π-acceptor capability. These types of ligands are unlikely to be significant in the biochemistry of this element. From a biological perspective, however, nickel(I) complexes of tetraazamacrocycles and thiol ligands are also well known (Sacconi *et al.*, 1987). Evidence for the presence of nickel(I) in a tetrapyrrole associated with methyl coenzyme M reductase is discussed in Chapter 6, and the possibility of nickel(I) in the sulfur-rich environment of the hydrogenase active site is examined in Chapter 4.

Nickel(0) stabilization similarly requires the presence of ligands with strong π-acceptor capacity (e.g., carbonyls, phosphines, phosphites). In general, these complexes are unstable toward oxygen and water. An exception is nickel carbonyl [$Ni(CO)_4$], the toxic intermediate in the Mond process for nickel purification (discussed in Chapter 1). This and other nickel(0) species appear to possess tetrahedral geometry. No clear examples of nickel(0) species that possess biologically relevant ligands have been described. A nickel(0) state has been hypothesized to occur in certain hydrogenases (Chapter 4); however, this state of the enzyme alternatively has been interpreted as possessing the element as nickel(II).

Nickel(−I) has been claimed to exist in several compounds, but this formal oxidation state of the metal probably has little physical meaning. Nickel(−I) is unlikely to have any biochemical significance.

2.3 Spectroscopic Properties of Nickel

This section provides a simplistic analysis of the spectroscopic properties of nickel compounds. I have purposely restricted the discussion to include only a few aspects of the electronic spectra and the magnetic properties of nickel species. The interested reader is encouraged to examine more detailed treatments such as those of Sacconi *et al.* (1987), Lever (1984), Salerno (1988), Nag and Chakravorty (1980), and Tomlinson (1981).

2.3.1 Electronic Spectroscopy of Nickel Compounds

As with other transition metals, the electronic (ultraviolet, visible, and near-infrared) spectra of nickel compounds can provide a wealth of information about the ligand type and geometry (Lever, 1984). A detailed discussion of the theory and interpretation of electronic spectra is beyond the scope of this book; however, I will attempt to provide the reader a sense of the type of information that may be available. The following paragraphs focus on two general types of transitions that give rise to the electronic spectra of nickel compounds: $d-d$ transitions and ligand-to-metal or metal-to-ligand charge-transfer transitions.

The electronic energy levels for nickel(II) in octahedral, square-pyramidal, trigonal bipyramidal, and tetrahedral geometries are illustrated in Fig. 2-3. Considering first the octahedral case, transitions between the $^3A_{2g}$ and the three other levels account for the three characteristic $d-d$ bands (at 20,000–29,000 cm^{-1}, 12,000–19,000 cm^{-1}, and 5,000–12,000 cm^{-1}) in the electronic spectra of six-coordinate nickel(II) complexes. For example, $[Ni(H_2O)_6]^{2+}$ is green with $d-d$ transitions at 25,000 cm^{-1} (400 nm), a doublet at 14,000 cm^{-1} ($\sim$700 nm), and 9000 cm^{-1} (1100 nm). In the presence of an alternative ligand such as ethylenediamine, however, the color shifts; $[Ni(ethylenediamine)_3]^{2+}$ is blue with $d-d$ transitions at 30,000 cm^{-1} ($\sim$350 nm), 18,500 cm^{-1} ($\sim$550 nm), and 11,000 cm^{-1} ($\sim$900 nm). The octahedral nickel(II) $d-d$ transitions are generally weak, with molar absorbances less than 10. Square-pyramidal nickel(II) complexes generally exhibit an intense transition at 19,000–29,000 cm^{-1} (molar absorbance of 100–800) along with three weak to moderate transitions at 4000–9000 cm^{-1}, 12,000–18,000 cm^{-1} (with a shoulder on the low-frequency side), and 17,000–25,000 cm^{-1}. Trigonal bipyramidal nickel(II) complexes also exhibit four bands: a high-intensity band at 17,000–22,000

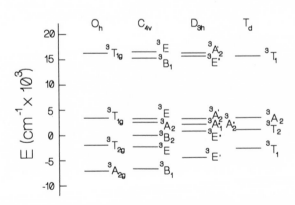

Figure 2-3. Electronic energy level diagram of nickel(II) in various geometries. This qualitative picture allows one to better define the electronic transitions that are accessible to the metal ion in octahedral (O_h), square-pyramidal (C_{4v}), trigonal bipyramidal (D_{3h}), and tetrahedral (T_d) complexes. Modified from Sacconi et al. (1987).

cm^{-1} with a shoulder at lower frequency and two transitions at 8000–14,000 cm^{-1} and one at 5000–8000 cm^{-1} of low intensity. Finally, tetrahedral complexes of nickel(II) exhibit three main bands: an intense band at 15,000–20,000 cm^{-1} (molar absorbance of 200–500) and weaker bands at 4000–7000 cm^{-1} and 7000–11,000 cm^{-1}. The levels giving rise to these transitions are all well characterized (Sacconi *et al.*, 1987).

In addition to the *d–d* bands, the electronic spectra of nickel complexes can include contributions from ligand-to-metal or metal-to-ligand charge-transfer transitions (referred to as LMCT and MLCT transitions, respectively). In some cases these charge-transfer transitions may dominate the spectrum of a species. For example, the interaction of urease with thiol compounds gives rise to an intense absorption that is readily explained by such a ligand-to-metal charge transfer transition (Chapter 3).

With a general understanding of molecular orbital and ligand field theories for analysis of electronic spectra and with the availability of electronic spectra of known model compounds, one can tentatively assign *d–d,* LMCT, and/or MLCT transitions and provide a reasonable guess of the geometry and types of ligands in an unknown nickel(II) species. In the case of hydrogenase and carbon monoxide dehydrogenase, however, the electronic spectra of the nickel metallocenters are obscured by the presence of iron–sulfur centers (Chapters 4 and 5). Furthermore, in the case of urease, the analysis of the spectrum is confounded by the presence of an apparent binuclear nickel metallocenter (Chapter 3). Nevertheless, electronic spectroscopy has been and will continue to be an important tool in the characterization of nickel in biological systems.

2.3.2 Magnetic Properties of Nickel Compounds

The magnetic properties of a nickel compound will depend on the metal ion oxidation state and the ligand geometry. Thus, by directly measuring the magnetic susceptibility of a complex (e.g., by saturation magnetization measurements) and by using a variety of magnetization-sensitive spectroscopic approaches, one can obtain valuable information relative to the structure and oxidation state of an unknown nickel center. One of the most widely used spectroscopic methods in this regard is EPR spectroscopy, and the following paragraphs will focus on this tool. Excellent books that focus on EPR spectroscopy of transition metals include those by Abragam and Bleaney (1970) and Pilbrow (1991).

Ni(III) complexes have a $3d^7$ electronic configuration and are always paramagnetic with magnetic moments ranging from 1.7 to 2.1 BM at room temperature. EPR spectroscopic studies are useful in distinguishing a true nickel(III) state from a nickel(II)-stabilized ligand radical complex: whereas nickel(II)-thiyl radical complexes exhibit a single, nearly isotropic g tensor near the free-electron value ($g = 2.0023$), true nickel(III) complexes exhibit anisotropic signals with significant deviations from the free-electron value. The EPR spectroscopic properties of nickel(III) in an octahedral environment are provided as a specific example. To facilitate this discussion, energy level diagrams for compressed and elongated octahedral nickel species are provided in Fig. 2-4. In the lowest energy state, the seven d electrons of nickel(III) include three pairs occupying the lowest three orbitals and a lone electron in the next highest orbital. Thus, in an elongated octahedral nickel(III) complex the electrons are found as $(d_{xz}d_{yz})^4(d_{xy})^2(d_{z^2})^1(d_{x^2-y^2})^0$, whereas for a compressed octahedral complex they populate these levels as $(d_{xz}d_{yz})^4(d_{xy})^2(d_{x^2-y^2})^1(d_{z^2})^0$. In the former case, microwave-induced transitions ($d_{xz} \rightarrow d_{z^2}$ and $d_{yz} \rightarrow d_{z^2}$) lead to the observation of $g_\perp > g_\| \approx 2.0$, whereas transitions in the compressed octahedral species ($d_{xz} \rightarrow d_{x^2-y^2}$ or $d_{yz} \rightarrow d_{x^2-y^2}$) exhibit $g_\| < g_\perp > 2.0$.

Nickel(II) complexes have a $3d^8$ configuration; that is, eight electrons occupy the five d orbitals either as four pairs or as three pairs and two unpaired electrons, depending on the energy levels of the two highest energy orbitals. The six-coordinate complexes are generally high spin ($S = 1$, two unpaired electrons), five-coordinate complexes can be either high spin (3.2–3.4 BM, found with electronegative donors such as O and N) or low spin ($S = 0$, found with donor atoms such as C, P, and S of low electronegativity), tetrahedral complexes have ligands of low donor strength and are high spin, and square-planar complexes have strong donor ligands and are low spin. In general, EPR spectra are not observed for octahedral nickel(II) complexes despite the presence of two free electrons. The absence of a spectrum can be explained by invoking spin–orbit coupling of orbitally nondegenerate ground states (Sacconi *et al.*, 1987).

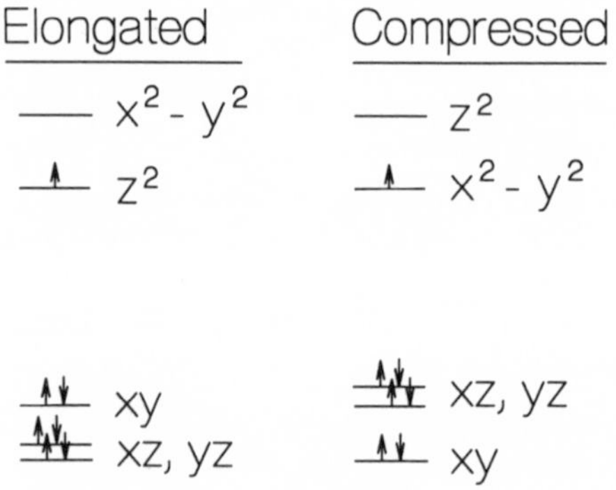

Figure 2-4. Comparison of the electron distributions among the five d orbitals of nickel(III) in compressed and elongated octahedral coordination. Electrons are denoted by arrows, shown as spin-up or spin-down.

Nickel(I) complexes possess a $3d^9$ configuration (nine d electrons) and are paramagnetic with a magnetic moment of 1.7–2.4 BM. The typical trigonal bipyramidal complexes of nickel(I) exhibit a pattern of $g_\perp > g_\parallel \approx 2.0$. Thus, it is not possible to distinguish nickel(I) from nickel(III) by EPR spectroscopy alone. This uncertainty has been observed in studies of several nickel enzymes (e.g., see Chapter 4), and alternative methods must be used to establish the actual oxidation state.

Additional methods to probe the oxidation state of nickel centers and to deduce the identity and geometry of nickel ligands are provided in individual chapters elsewhere in this book.

2.4 Perspective

This rather naive discussion of nickel chemistry should help the reader to better appreciate some of the detailed biophysical studies of nickel enzymes described in Chapters 3–6. Furthermore, knowledge of the fundamental properties of this element is essential for understanding how it interacts in beneficial and harmful ways with microbes, plants, and animals as described in Chapters 7–9.

References

Abelson, P. H., and Aldous, E., 1950. Ion antagonisms in microorganisms: Interference of normal magnesium metabolism by nickel, cobalt, cadmium, zinc, and manganese, *J. Bacteriol.* 60: 401–413.

Abragam, A., and Bleaney, B., 1970. *Electron Paramagnetic Resonance of Transition Metal Ions,* Oxford University Press, New York.

Busch, D. H., 1978. Distinctive coordination chemistry and biological significance of complexes with macrocyclic ligands, *Acc. Chem. Res.* 11:392–400.

Cotton, F. A., and Wilkinson, G., 1980. *Advanced Inorganic Chemistry: A Comprehensive Text,* 4th ed., John Wiley & Sons, New York, pp. 783–798.

Coyle, C. L., and Stiefel, E. I., 1988. The coordination chemistry of nickel: An introductory survey, in *The Bioinorganic Chemistry of Nickel* (J. R. Lancaster, Jr., ed.), VCH Publishers, New York, pp. 1–28.

Dixon, N. E., Blakeley, R. L., and Zerner, B., 1980. Jack bean urease (EC 3.5.1.5). I. A simple dry ashing procedure for the microdetermination of trace metals in proteins. The nickel content of urease, *Can. J. Biochem.* 58:469–473.

Fox, S., Wang, Y., Silver, A., and Miller, M., 1990. Viability of the $[Ni^{III}(SR)_4]^-$ unit in classical coordination compounds and in the nickel–sulfur center of hydrogenases, *J. Am. Chem. Soc.* 112:3218–3220.

Jolly, P. W., and Wilke, G., 1974. *The Organic Chemistry of Nickel, Vol. 1, Organonickel Complexes,* Academic Press, New York.

Krüger, H.-J., and Holm, R. H., 1987. Stabilization of nickel(III) in a classical N_2S_2 coordination environment containing anionic sulfur, *Inorg. Chem.* 26:3645–3647.

Krüger, H.-J., and Holm, R. H., 1990. Stabilization of trivalent nickel in tetragonal NiS_4N_2 and NiN_6 environments: Synthesis, structures, redox potentials, and observations related to [NiFe]-hydrogenases, *J. Am. Chem. Soc.* 112:2955–2963.

Krüger, H.-J., Peng, G., and Holm, R. H., 1991. Low-potential nickel(III, II) complexes: New systems based on tetradentate amidate-thiolate ligands and the influence of ligand structure on potentials in relation to the nickel site in [NiFe]-hydrogenases, *Inorg. Chem.* 30:734–742.

Lappin, A. G., Murray, C. K., and Margerum, D. W., 1978. Electron paramagnetic resonance studies of nickel(III)-oligopeptide complexes, *Inorg. Chem.* 17:1630–1634.

Lever, A. B. P., 1984. *Inorganic Electronic Spectroscopy,* 2nd ed., Elsevier Scientific Publishing Co., New York.

Margerum, D. W., and Anliker, S. L., 1988. Nickel(III) chemistry and properties of the peptide complexes of Ni(II) and Ni(III), in *The Bioinorganic Chemistry of Nickel* VCH Publishers, (J. R. Lancaster, Jr., ed.), New York, pp. 29–51.

Nag, K., and Chakravorty, A., 1980. Monovalent, trivalent, and tetravalent nickel, *Coord. Chem. Rev.* 33:87–147.

Nielsen, G. D., and Andersen, O., 1993. Application of ^{57}Ni in toxicokinetics, *Sci. Total Environ.* (in press).

Pilbrow, J., 1991. *Transition Ion Electron Paramagnetic Resonance,* Oxford University Press, New York.

Sacconi, L., Mani, F., and Bencini, A., 1987. Nickel, in *Comprehensive Coordination Chemistry: The Synthesis, Reactions, Properties & Applications of Coordination Compounds,* Vol. 5 (G. Wilkinson, R. D. Gillard, and J. A. McCleverty, eds.), Pergamon Press, New York, pp. 1–347.

Salerno, J. C., 1988. The EPR spectra of odd-electron nickel ions in biological systems: Theory for d^7 and d^9 ions, in *The Bioinorganic Chemistry of Nickel* (J. R. Lancaster, Jr., VCH Publishers, New York, pp. 53–71.

Tomlinson, A. A. G., 1981. Nickel, *Coord. Chem. Rev.* 37:221–296.

Tschugaeff, L., 1905. Determination of nickel, *Z. Anorg. Allg. Chem.* 44:144.

Weast, R. C., 1975. *Handbook of Chemistry and Physics,* 56th ed., CRC Press, Cleveland, Ohio.

Urease ③

3.1 Introduction

Urease catalyzes the hydrolysis of urea to yield ammonia and carbamate, which spontaneously decomposes to form carbonic acid and a second molecule of ammonia [reviewed by Andrews *et al.* (1984, 1988), Blakeley and Zerner (1984), Mobley and Hausinger (1989), and Zerner (1991)]:

$$H_2N\text{-}CO\text{-}NH_2 + H_2O \rightarrow NH_3 + H_2N\text{-}COOH$$

$$H_2N\text{-}COOH + H_2O \rightarrow NH_3 + H_2CO_3$$

The substrate in this reaction, urea, is constantly released into the environment through biological actions. For example, all mammals excrete urea in urine as a detoxification product (Visek, 1972). To provide a sense of the scale for urea excretion, human urine contains 0.4–0.5 M urea (Griffith *et al.*, 1976), resulting in an annual release of 10 kg of urea per adult (Visek, 1972). Urea is also formed by environmental catabolism of uric acid, the primary detoxification product excreted by birds, reptiles, and most terrestrial insects. Similarly, urea is a product of biodegradation of nitrogenous compounds including purines, arginine, agmatine, allantoin, and allantoic acid (Vogels and van der Drift, 1976). The urea generated by these reactions is rapidly degraded by ureases found in a wide range of bacteria, several fungi, a few invertebrates, and a variety of plants. The significance of urease in these various organisms is summarized below.

Bacterial ureases are important in environmental nitrogen transformations, in ruminant metabolism, and in the development of certain human and animal pathological states [reviewed by Mobley and Hausinger (1989)]. In many soil and aquatic bacteria, the function of the enzyme is simply to supply cells with ammonia as a nitrogen source for growth (Bremner and Mulvaney, 1978). A similar function is ascribed to the enzyme associated

23

with certain bacteria that inhabit the rumen of cattle, sheep, and other animals that contain a forestomach (Huntington, 1986). A nitrogen cycle appears to exist in these ruminants: the animal produces copious amounts of urea, substantial quantities of this urea diffuse into the rumen, where growth of the endemic microbial population is limited by nitrogen, ureolytic bacteria utilize this nitrogen source to enhance their growth, and the increased microbial biomass is subsequently used as a nutrient by the ruminant. In contrast to these cases where urease is used to provide a nitrogen source, the enzyme is a virulence factor in other microbes. Infection-induced urinary stones (accounting for 15–20% of all urinary stones) arise from microbial urease-induced alkalinization of urine that leads to supersaturation of polyvalent anions and results in their precipitation as struvite or carbonate apatite (Griffith *et al.,* 1976). Colonization of catheterized urinary tracts by urease-producing microbes can result in similar precipitation events within the catheter, leading to obstruction (Mobley and Warren, 1987). Urease-induced toxicity contributes to tissue damage, inflammation, and cell invasion during acute pyelonephritis, causing damage to the liver (Braude and Siemienski, 1960). Hyperammonemia can result from excessive urease-generated ammonia production in the intestinal tract; the ammonia overload can lead to hepatic encephalopathy (Samtoy and DeBreukelaer, 1980) or even hepatic coma (Sabbaj *et al.,* 1970). Finally, peptic ulceration recently has been demonstrated to arise from colonization of the stomach lining by *Helicobacter pylori* (formerly *Campylobacter pyloridis*) (Goodwin *et al.,* 1986); this microbe possesses a potent urease which may be a virulence factor in development of gastrointestinal inflammatory lesions (Hazell and Lee, 1986; Smoot *et al.,* 1990; Eaton *et al.,* 1991; Mégraud *et al.,* 1992; Segal *et al.,* 1992; Pérez-Pérez *et al.,* 1992). The regulation of bacterial urease varies with the source, but in general those microbes that utilize urease for provision of nitrogen for growth repress urease activity in the presence of a good nitrogen source, whereas many of the medically important ureolytic microorganisms regulate urease activity by urea induction [reviewed by Mobley and Hausinger (1989)]. Other microbes constitutively express very high levels of urease consistent with another, still unknown, role for the enzyme; for example, urease comprises 1% of the total cell dry weight of *Bacillus pasteurii* (Larson and Kallio, 1954). The enzyme was already known to be present in more than 200 species of bacteria by 1947 (Sumner and Somers, 1953), and the list has increased tremendously since then. The activity is found in a wide range of bacterial divisions, but within any division some species have urease and some do not. Because of the ecological and medical importance of bacterial ureases and the ease of manipulation of bacterial genes, the enzyme from these microbes has been very well characterized.

Fungal ureases are clearly involved in environmental transformations of nitrogenous compounds to provide ammonia to the cells for growth. In addition, certain species of fungi may utilize urease for other less well characterized roles. For example, the enzyme accounts for 8.5% of the total soluble protein of *Aspergillus tamarii,* where it was suggested to play a role as a storage protein (Zawada and Sutcliffe, 1981). The genetics of fungal ureases have been studied for several microbes, but very little biochemical analysis of the enzyme has been reported [reviewed by Mobley and Hausinger (1989)]. Even less is known about invertebrate urease, as not even the physiological function of the enzyme is certain.

Ureases are found in every division of plant phylogeny (Granick, 1937); however, the most extensive biochemical studies have been conducted with the enzyme from jack bean (*Canavalia ensiformis*), and the genetics are best characterized for soybean (*Glycine max*). The plant enzyme was formerly thought to be involved in degradation of allantoin and allantoic acid (Polacco *et al.,* 1985), two ureide nitrogen transport compounds, but Winkler *et al.* (1987) provided convincing evidence that ureide degradation in soybean does *not* include urea as an intermediate. Nevertheless, urease does have an essential role in soybean and cowpeas as shown by the accumulation of toxic urea concentrations in plants grown in nickel-deficient nutrient solutions (Eskew *et al.,* 1983, 1984; Walker *et al.,* 1985). Alternative roles for plant ureases include their participation in the degradation of canavanine (Loyola-Vargas *et al.,* 1988) or arginine (Thompson, 1980), major nitrogen storage forms in certain plants, and their possible role as a toxic defense protein (Stebbins *et al.,* 1991).

Biochemical analysis of urease has a long and fascinating history. Jack bean urease was the first enzyme ever crystallized (Sumner, 1926), an effort that won J. B. Sumner the Nobel prize. This experiment was considered to be very important in establishing the proteinaceous nature of enzymes; that is, enzymes were demonstrated to be more than simple inorganic catalytic complexes as they had previously been considered. However, nearly 50 years after the crystalline protein became available, the jack bean enzyme was shown to possess stoichiometric levels of tightly bound nickel ions (Dixon *et al.,* 1975b). In this chapter, I compare the general properties of nickel-containing ureases, summarize the characteristics of the nickel active site, compare the properties of the enzyme to chemical models, and describe the current understanding of how nickel is specifically incorporated into this protein.

3.2 General Properties of Nickel-Containing Ureases

Nickel has been demonstrated to be a component of purified ureases from two plants, one invertebrate, and several bacteria, as shown in Table 3-

Table 3-1. Compositions of Purified Nickel-Containing Ureases

Source	Apparent subunit M_r[a]	Apparent stoichiometry[b]	Ni content	Reference(s)
Eucaryotes				
Canavalia ensiformis (jack bean)	96,600 (α)	α6	2.0 Ni/α	Dixon et al. (1975b, 1980a)
Glycine max (soybean)	93,500 (α)	α6	2 Ni/93,500	Polacco and Havir (1979)
Otala lactea (land snail)	NR[c]	NR	Present[d]	McDonald et al. (1980)
Procaryotes				
Arthrobacter oxydans	NR	NR	0.3 Ni/242,000	Schneider and Kaltwasser (1984)
Bacillus pasteurii	65,500 (α)	α4	0.82 Ni/α	Christians and Kaltwasser (1986)
Brevibacterium ammoniagenes	67,000	α3	0.8 Ni/α	Nakano et al. (1984)
Helicobacter pylori[e]	66,000 (α), 29,500 (β)	α6β6	NR	Hu and Mobley (1990)
	62,000 (α), 30,000 (β)	α4β4	NR	Dunn et al. (1990)
	61,000 (α), 28,000 (β)	(αβ)[f]	Present	Hawtin et al. (1991)
	66,000 (α), 31,000 (β)	α6β6	Present	Evans et al. (1991)
Klebsiella aerogenes	72,000 (α), 11,000 (β), 9,000 (γ)	α2β2γ4	2.1 Ni/αβ2γ2	Todd and Hausinger (1987)
Lactobacillus fermentum	67,000 (α), 16,800 (β), 8,600 (γ)	α2β4γ2	1.9 Ni/αβ2γ	Kakimoto et al. (1990)
Lactobacillus reuteri	68,000 (α), 16,100 (β), 8,800 (γ)	α2β4γ2	1.8/αβ2γ	Kakimoto et al. (1989)
Providencia stuartii[g]	73,000 (α), 10,000 (β), 8,000 (γ)	α2β4γ4	1.9 Ni/αβ2γ2	Mulrooney et al. (1988)
Selenomonas ruminantium	70,000 (α), 8,000 (β), 8,000 (γ)	NR	2.1 Ni/α	Hausinger (1986); Todd and Hausinger (1987)
Sporosarcina ureae	63,100 (α), 14,500 (β), 8,500 (γ)	(αβγ2)[f]	2.1 Ni/αβγ2	McCoy et al. (1992)
Staphylococcus xylosus	64,000 (α), 17,800 (β), 16,300 (γ)	NR	3.91 Ni/300,000	Christians et al. (1991)
Streptococcus mitior	66,000 (α), 15,600 (β), 8,600 (γ)	α2β4γ2	2 Ni/αβ2γ	Yamazaki et al. (1990)
Ureaplasma urealyticum	72,000 (α), 14,000 (β), 11,000 (γ)	α2β2γ2	Present	Thirkell et al. (1989)

[a] Subunit sizes were estimated by denaturing gel electrophoresis. For several microbes the urease structural genes have been sequenced, and the calculated M_r values were found to differ considerably from those obtained by gel analysis (see text).

[b] Apparent subunit stoichiometry for the native urease. For heteropolymeric ureases, the subunit stoichiometry is based, in part, on denaturing gel electrophoretic band staining intensity and must be considered tentative. Furthermore, there is uncertainty in the native size estimates that were based primarily on gel filtration chromatography results. Thus, the overall stoichiometries must be considered with caution.

[c] NR, not reported.

[d] Although not quantitated, nickel was shown to be present.

[e] Formerly Campylobacter pyloridis.

[f] The ratio of subunits was determined, but the multiplicity of the basic unit was not reported.

[g] Formerly Proteus stuartii. The P. stuartii protein was isolated from Escherichia coli cells expressing the heterologous genes.

1. In addition to its presence in these purified proteins, nickel is known to be required for urease activity in other plants, fungi, algae, and bacteria, as noted in Table 3-2. Indeed, nickel has been found in all ureases that have been examined for this metal ion.

The nickel content and subunit stoichiometry of jack bean urease form the benchmark upon which other ureases are compared. Dixon *et al.* (1975b,

Table 3-2. Additional Ureases That Exhibit a Nickel Requirement for Activity

Source	Reference(s)
Plants	
Alnus rubra (red alder)	Dalton *et al.* (1988)
Lemna paucicostata (duckweed)	Gordon *et al.* (1978)
Nicotiana tabacum (tobacco)	Polacco (1977)
Oryza sativa (cotton)	Polacco (1977)
Robinia pseudoacacia (black locust)	Benchemsi-Bekkari and Pizelle (1992)
Vigna unguiculata (cowpea)	Walker *et al.* (1985)
Fungi	
Aspergillus nidulans[a]	Mackay and Pateman (1980)
Filobasidiella neoformans	Booth and Vishniac (1987)
Penicillium sp.	Polacco (1977)
Algae	
Cyclotella cryptica	Oliveira and Antia (1984)
Phaeodactylum tricornutum	Rees and Bekheet (1982)
Tetraselmis subcordiformis	Rees and Bekheet (1982)
Thalassiosira weissflogii	Price and Morel (1991)
Lichen	
Evernia prunastri	Pérezurria *et al.* (1986)
Bacteria	
Anabaena cylindrica[b]	Mackerras and Smith (1986)
Anabaena doliolum	Singh (1990)
Anacystis nidulans	Singh (1991)
Chromatium vinosum	Bast (1988)
Proteus mirabilis[c]	Rando *et al.* (1990)
Streptococcus aureus	Jose *et al.* (1991)
Thiocapsa roseopersicina	Bast (1988)
Endemic soil microbes	Dalton *et al.* (1985)
Mixed rumen population	Spears *et al.* (1977); Spears and Hatfield (1978)

[a] Urease from this fungus has been purified, but the nickel content of the protein was not assessed (Creaser and Porter, 1985).

[b] Urease from this cyanobacterium has been highly enriched, but the metal content was not determined (Argall *et al.,* 1992).

[c] *P. mirabilis* urease has been highly purified, but nickel quantitation was not carried out (Breitenbach and Hausinger, 1988).

1980a) demonstrated that two nickel ions are associated with each jack bean urease subunit of apparent M_r 96,600. The subunits appear to associate predominantly into a homohexamer of apparent M_r 590,000 (Dixon et al., 1980d), although a homotrimeric structure has also been observed (Blattler et al., 1967; Contaxis and Reithel, 1971; Fishbein et al., 1973). A similar situation appears to be present in the soybean seed enzyme (Polacco and Havir, 1979), and a two-Ni/subunit homopolymeric structure may be representative of all eucaryotic ureases. In contrast, the metal and subunit compositions of bacterial ureases appear to vary greatly with the source (Table 3-1). In the following paragraphs, I discuss aspects of the subunit number, stoichiometry, and nickel content for ureases isolated from procaryotes.

A cursory examination of bacterial urease properties reveals that the number of distinct subunits and their sizes appear to be highly species variable and clearly distinct from the single apparent M_r 96,600 subunit found in the jack bean enzyme. DNA sequence analysis, however, has provided compelling evidence that these proteins are highly homologous. An illustration of the sequence relatedness among several of these proteins is provided in Fig. 3-1. Both protein (Takishima et al., 1988) and DNA (Riddles et al., 1991) sequencing have shown the jack bean urease to be comprised of 840 amino acids with M_r 90,770. DNA sequences encoding the three-subunit enzymes from *Klebsiella aerogenes* (Mulrooney and Hausinger, 1990), *Proteus mirabilis* (Jones and Mobley, 1989), *Proteus vulgaris* (Mörsdorf and Kaltwasser, 1990), and *Ureaplasma urealyticum* (Blanchard, 1990) reveal extensive homology (i.e., greater than 50% identity) between the three structural genes and portions of the single-subunit jack bean urease. For example, the γ, β, and α subunits

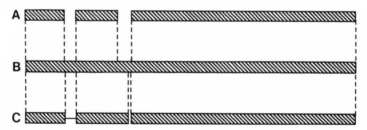

Figure 3-1. Sequence relationships among urease proteins. Jack bean urease possesses a single type of subunit comprised of 840 amino acids with M_r 90,770 (Riddles et al., 1991; Takishima et al., 1988), shown by the bar labeled B. As indicated by the three bars labeled A, DNA sequence analyses of urease genes from *Klebsiella aerogenes* (Mulrooney and Hausinger, 1990), *Proteus mirabilis* (Jones and Mobley, 1989), *Proteus vulgaris* (Mörsdorf and Kaltwasser, 1990), and *Ureaplasma urealyticum* (Blanchard, 1990) reveal that the three subunits in these enzymes are related to portions of the jack bean enzyme. Similarly, DNA sequence analyses of the *Helicobacter pylori* urease genes (Clayton et al., 1990; Labigne et al., 1991) indicate that the two-subunit enzyme (bars labeled C) is related to portions of the jack bean protein.

(encoded by the *ureA, ureB,* and *ureC* genes) of *K. aerogenes* urease correspond to residues 1–101, 132–237, and 271–840 of the jack bean protein. Similarly, genes encoding the two-subunit *Helicobacter pylori* urease have been sequenced (Clayton *et al.,* 1990; Labigne *et al.,* 1991), and the predicted protein sequences show homology to two stretches (although one is briefly interrupted) of the plant protein. These results are consistent with gene fusion or gene disruption events in the evolution of the urease proteins. No sequence data have appeared, however, for the apparent single-subunit *Bacillus pasteurii* or *Brevibacterium ammoniagenes* ureases. These enzymes may represent a distinct class of enzymes; however, it is also possible that small subunits were overlooked during denaturing gel electrophoretic analysis of these preparations. For example, the ureases from *Selenomonas ruminantium* (Hausinger, 1986) and *U. urealyticum* (e.g., Precious *et al.,* 1987) were initially suggested to possess a single subunit type, but subsequent characterization revealed three distinct subunits in each of these enzymes (Thirkell *et al.,* 1989; Todd and Hausinger, 1987). Consistent with this view, a three-subunit urease was isolated from *Sporosarcina ureae* (McCoy *et al.,* 1992), a microbe that is closely related to *B. pasteurii* (Pechman *et al.,* 1976), and from several other gram-positive microbes. Hence, it appears that all ureases are highly related in sequence despite differences in the precise number or size of subunits. Another point that can be made regarding the results from DNA sequence analysis concerns the discrepancies in calculated subunit sizes versus the apparent M_r values estimated by denaturing gel electrophoresis. In some cases, these differences are significant. For example, the calculated M_r values for the three *K. aerogenes* urease subunits are 60,304, 11,695, and 11,086, whereas the gel-derived apparent values were 72,000, 11,000, and 9,000. This example underscores the caution that must be used in estimation of subunit size by denaturing gel electrophoretic methods.

Values for the bacterial urease subunit stoichiometries shown in Table 3-1 must be considered very tentative. The stoichiometry of the subunits has been estimated by scanning the intensity of stained subunit bands after denaturing gel electrophoresis. This method assumes equivalent dye binding per unit mass by each peptide band, an assumption that is clearly incorrect for many proteins. Near-integer values have been reported for the subunit stoichiometry of various ureases; however, the ratio of $\alpha{:}\beta{:}\gamma$ subunits has varied among the different enzymes. For example, a 1:1:1 ratio was reported for *U. ureaplasma* (Thirkell *et al.,* 1989) [similarly, a 1:1 ratio for the two subunit protein from *Helicobacter pylori* (Dunn *et al.,* 1990; Hu and Mobley, 1990)], a 1:1:2 ratio for *S. ureae* (McCoy *et al.,* 1992), a 1:2:1 ratio for *Lactobacillus fermentum* (Kakimoto *et al.,* 1990), *Lactobacillus reuteri* (Kakimoto *et al.,* 1989), and *Streptococcus mitior* (Yamazaki *et al.,* 1990), and a 1:2:2 ratio for *Providencia stuartii* (Mulrooney *et al.,* 1988) and *K. aerogenes* (Todd and Hausinger, 1987). Support for the 1:2:2 stoichiometry in the case of *K. aerogenes* urease is available from chemical modification studies in which the enzyme was alkylated with radioactively labeled iodoacetamide, and the small subunits were separated from the larger subunits (Todd and Hausinger, 1991b). Sequence analysis had previously indicated that the α, β, and γ subunits possess eight, one, and no cysteines, respectively (Mulrooney and Hausinger, 1990), so the observed 4.8:1 ratio of radioactivity in the α and $\beta + \tau$ pools is most consistent with a 1:2 ratio of α:β subunits (no information could be obtained regarding the γ subunit by this experiment). Despite this corroborative evidence, the true ratio of subunits remains open to question, and the extensive sequence similarities between the bacterial and plant enzymes are most consistent with equivalent subunit levels among the procaryotic ureases.

The subunit stoichiometries proposed for the native enzymes are based on the staining intensity of the subunit ratios, as described above, and on the apparent size as determined by gel filtration chromatography or other methods. As an illustration of the problems inherent in deducing the true stoichiometry, the *K. aerogenes* enzyme was estimated to possess an $\alpha_2\beta_4\gamma_4$ structure (Table 3-1) by these methods. However, preliminary X-ray crystallographic analysis has shown that the native enzyme is a trimer of the catalytic unit ($\alpha\beta\gamma$ or $\alpha\beta_2\gamma_2$) (see Section 3.3.1).

The nickel content of purified bacterial ureases generally agrees with the presence of two nickel ions per minimal structural unit. This value nicely coincides with that observed for the jack bean urease of two nickel ions per single subunit. The three exceptions to this generality involve *Arthrobacter oxydans* (Schneider and Kaltwasser, 1984), where the nickel incorporation was clearly very low, and *B. ammoniagenes* (Nakano *et al.*, 1984) and *B. pasteurii* (Christians and Kaltwasser, 1986), which appear to possess a single nickel ion per subunit. In the latter two cases, no nickel ion (or only trace amounts of radioactively labeled nickel ion) was added to the medium; thus, it is possible that the purified enzyme included partial apoprotein and that the fully active enzyme would possess a higher nickel content. Overall, it is reasonable to conclude that the peptide structures and nickel contents are similar in all ureases.

3.3 Characteristics of the Urease Nickel Active Site

This section describes several types of experimental evidence related to the structure and mechanism of the urease active site. Enzyme-active-site characterization is greatly facilitated by the availability of a three-dimensional structure. Given that jack bean urease was the first enzyme ever to be crystallized (Sumner, 1926), one might expect that the crystallographic structure has been determined; however, this assumption would be in error. Despite the absence of a three-dimensional structure, many aspects of the urease active site are beginning to be clarified by using kinetic inhibition techniques, biophysical and spectroscopic methods, chemical modification approaches, and site-directed mutagenesis studies. Each of these topics is discussed separately below.

3.3.1 Crystallographic Analysis of Urease

The tetragonal bipyramidal jack bean urease crystals initially described by Sumner (1926) have recently been reproduced and characterized by Jabri et al. (1992). Preliminary diffraction studies show that the crystals belong to the cubic space group $F4_132$ and contain one or two subunits per asymmetric unit ($a = b = c = 364$ Å). By using a synchrotron source, the jack bean urease crystals were shown to diffract to near 3.5 Å; hence, limited three-dimensional structural information may become available from further study of this protein.

The researchers who obtained the jack bean urease crystals also obtained crystals of *Klebsiella aerogenes* urease (Jabri *et al.*, 1992). These crystals belong to the cubic space group 123 or 12_13 and contain a single catalytic site per asymmetric unit ($a = b = c = 170.8$ Å). The high-quality crystals of the native bacterial enzyme diffract to better than 2 Å, which should provide a detailed picture of the enzyme active site. Furthermore, several heavy-metal-atom derivatives have been obtained, and crystals of *K. aerogenes* urease apoenzyme were generated and appear to crystallize in the same space group (E. Jabri, M. H. Lee, R. P. Hausinger, and P. A. Karplus, unpublished observations). A preliminary 6-Å-resolution structure clearly shows the enzyme to be a trimer of the catalytic unit. A threefold rotational symmetry has also been observed in *Helicobacter pylori* urease by using electron microscopic methods (Austin *et al.*, 1992). Interestingly, electron microscopic analysis of negatively stained jack bean urease is consistent with the native hexameric structure being a dimer of trimers (Fishbein *et al.*, 1977). Furthermore, the homohexameric jack bean enzyme can be dissociated into trimers under selected conditions (Blattler *et al.*, 1967; Contaxis and Reithel, 1971; Fishbein *et al.*, 1973). Thus, the quaternary structures of bacterial and plant ureases are likely to be highly related.

3.3.2 Inhibition Studies

Inhibition studies have been extremely important in developing a model for the urease active site. As described in this section, urease inhibitor analyses furnish evidence for the presence of two nickel ions per catalytic unit, demonstrate that nickel is at the active site rather than playing a structural role, and allow some insight into the ionic environment of the substrate binding site. In each case, studies involving the plant enzyme are compared to those carried out with bacterial urease in order to highlight the similarity between these proteins.

In order to deduce the size of the jack bean urease catalytic unit, Dixon *et al.* (1975a) carried out a series of studies with ^{14}C-labeled acetohydroxamic acid and ^{32}P-labeled phosphoramidate. Because these inhibitors exhibit very slow rates of dissociation from the enzyme, it is possible to prepare partially inhibited samples, separate the protein from excess inhibitor by size exclusion chromatography, and assess the amount of inhibitor bound to the protein. By correlating the residual specific activity to the level of bound inhibitor, they were able to demonstrate that each catalytic unit (equivalent to the inhibitor binding site) possesses an equivalent weight of approximately 105,000. This value agrees quite well with the apparent subunit M_r of 96,600. Subsequent nickel quantitation demonstrating two moles of nickel ion per mole of subunit was used to suggest the presence of two nickel ions associated with each jack bean urease catalytic unit (Dixon *et al.*, 1975b). Using a different kinetic procedure, Todd and Hausinger (1989) exploited the slow-binding inhibitor phenyl phosphorodiamidate to quantitate the number of active sites per native *Klebsiella aerogenes* urease. Assuming that one mole of inhibitor bound per active site, they found that each catalytic unit possesses an equivalent weight of 102,000. This value, when combined with previous nickel analysis, is consistent with the presence of two nickel ions per bacterial urease catalytic unit (Todd and Hausinger, 1989).

The interaction of urease with thiol inhibitors was important in establishing that nickel is associated with the active site and not just playing a structural role. In the case of jack bean urease, the binding of β-mercaptoethanol leads to increases in absorbance at $\sim$320, $\sim$380, and 425 nm, and the value of the dissociation constant, K_d, of 0.95 mM associated with these spectral changes correlates very well to the value of the inhibitor constant, K_i, of 0.72 mM determined for this competitive inhibitor (Dixon *et al.*, 1980c). Similarly, *K. aerogenes* urease exhibits increases in absorbance at 322, 374, and 432 nm that provide a dissociation constant of 0.38 mM for β-mercaptoethanol, again consistent with the kinetically determined K_i value of 0.55 mM (Todd and Hausinger, 1989). The spectral differences were shown to be due to thiolate anion $\rightarrow$ nickel(II) charge-transfer transitions (Blakeley *et al.*, 1983). The spectroscopic demonstration of a nickel–sulfur interaction in urease upon binding of this competitive inhibitor, where the K_d value matches the K_i value, provides strong support for a mechanism where urea also binds to nickel.

Inhibitor studies have provided some evidence for the presence of a negatively charged residue, likely a carboxylate group, at the active site of bacterial urease. Using the *K. aerogenes* enzyme, Todd and Hausinger (1989) determined the K_i values for a series of thiol competitive inhibitors. They found that thiolates containing a negatively charged carboxylate group are significantly poorer inhibitors than uncharged or positively charged thiolates. For example, cysteine possesses a higher K_i value than does cysteine methyl ester. These results are consistent with repulsion of the carboxylated inhibitors by a negative charge in the active site.

3.3.3 Biophysical and Spectroscopic Analysis of the Urease Active Site

This section describes attempts to characterize the urease metallocenter by using a variety of biophysical and spectroscopic techniques, each of which has its advantages and its limitations. These results are not internally consistent among the various studies; nevertheless, several important conclusions concerning the unique urease metallocenter are highlighted.

A detailed analysis of the optical absorption spectrum of jack bean urease has been reported (Blakeley *et al.*, 1983). The native enzyme exhibits a major absorption at 278.5 nm due to protein tryptophan and tyrosine residues; however, a weak tail extends into the visible and infrared regions with a shoulder at ~407 nm. By contrast, in the presence of thiols the enzyme displays new shoulders at 316, 376, and 425 nm, along with prominent peaks at 745 and 1060 nm. Very similar results have been observed for the *Klebsiella aerogenes* enzyme (Todd and Hausinger, 1989). As discussed in Section 3.3.2, the first three features were interpreted as arising from a thiolate anion → nickel(II) charge-transfer transition. Absorption features near 400, 750, and 1000 nm are consistent with six-coordinate octahedral nickel(II) in the thiol-bound species; however, Blakeley *et al.* (1983) were careful to note the hazards involved in extrapolating the spectroscopic results from well-characterized mononickel model compounds to the uncharacterized, possibly binickel, urease active site. Furthermore, additional weak features at 630, 820, and 910 nm were noted in the thiol-bound form of the jack bean enzyme. The situation in the thiol-free form of urease is less clear. Although quantitative differences between the two groups were observed, both Blakeley *et al.* (1983) and Clark and Wilcox (1989) reported the presence of weak, long-wavelength absorption maxima for the jack bean enzyme that are distinct from those of the thiol-bound form of the protein. The spectra were interpreted as arising from octahedral coordination; however, Day *et al.* (1993), using saturation magnetization measurements, described below, demonstrated that much of the nickel in thiol-free enzyme could *not* be present in this geometry. Nevertheless, the spectroscopic results indicate that at least a portion of the nickel in urease appears to possess pseudooctahedral geometry. Furthermore, the optical spectrum of native jack bean urease (Blakeley *et al.*, 1983), and subsequently that of the bacterial enzyme (Todd and Hausinger, 1989), is important in precluding the presence of thiol ligands bound to the nickel.

X-ray absorption spectroscopy (XAS) is a powerful tool that can provide information on metallocenter symmetry, radial distances to ligands, coordination numbers, types of ligands, and cluster nuclearity. Hasnain and Piggott (1983) were the first to perform XAS on urease. Preliminary comparison of the jack bean enzyme spectrum with those of three model compounds led to the very tentative proposal that the urease nickel ions may be present in a histidine-type environment. An XAS fine-structure analysis by this group (Alagna *et al.*, 1984) concluded that the nickel is associated with approximately three oxygen/nitrogen scatterers at 2.04 Å, approximately two more O/N ligands at 2.06 Å, and another O/N at 2.25 Å (it was not possible to distinguish oxygen from nitrogen in their analysis). Higher quality XAS results were obtained for both the native and β-mercaptoethanol-bound forms of jack bean urease by Clark *et al.* (1990). These investigators found that the nickel in the native enzyme possesses five or six oxygen or nitrogen ligands at an average distance of 2.06 Å in a pseudooctahedral environment. Upon addition of β-mercaptoethanol, the best fit corresponds to the presence of one sulfur atom at 2.29 Å and approximately five oxygen/nitrogen ligands at 2.07 Å. These data confirm the absence of cysteine ligands bound to the nickel center in the native enzyme and verify that exogenous thiols bind to the nickel ion. However, it was not possible to distinguish whether each nickel ion binds a separate thiol or whether the thiol bridges the two active-site nickel ions. More detailed analysis of the native and thiol-bound jack bean enzyme as well as the native and thiol-inhibited *K. aerogenes* urease is consistent with the presence of five-coordinate geometry for nickel in all cases (Wang *et al.*, 1993). Using refined analysis programs that allow improved fitting of second-shell atoms, the fits for all enzyme forms are consistent with the presence of several imidazole ligands and additional oxygen or nitrogen donors to each nickel center, where one nitrogen/oxygen atom is replaced by sulfur in the inhibited cases. Importantly, the thiol-bound samples exhibited nickel–nickel scattering interactions with a distance of ~3.5 Å. In the thiol-free samples, by contrast, any Ni–Ni scattering is much weaker, and the metal–metal distance could not be determined.

Magnetic susceptibility measurements for three states of the jack bean urease metallocenter were reported by Clark and Wilcox (1989). They found that the thiol-bound form of the enzyme is primarily diamagnetic [compatible with strong antiferromagnetic coupling between two high-spin ($S = 1$) nickel ions], the acetohydroxamic acid-bound form is consistent with the presence of noncoupled high-spin nickel, and the native enzyme is a mixture of two states: 78% of the nickel was suggested to be present in a binuclear complex exhibiting weak antiferromagnetic exchange coupling, the remaining 22% of the nickel being mononuclear. Day *et al.* (1993) criticized the latter interpretation because it was based on results from two separate single-field measurements and the previous workers had assumed that all of the nickel would be paramagnetic. Examination of both jack bean and *K. aerogenes* ureases by multifield saturation magnetization methods revealed no evidence for exchange coupling between the nickel ions in either enzyme (Day *et al.*, 1993). Rather, these investigators reported that inhibitor-free enzyme possesses a mixed population of high-spin ($S = 1$) and low-spin ($S = 0$) nickel centers. Further, they showed that the relative population of the two species is pH dependent, with the diamagnetic contribution increasing with increasing pH (Day *et al.*, 1993).

Variable-temperature magnetic circular dichroism spectroscopy has provided additional insight into the properties of the native and inhibitor-treated urease active site. Finnegan *et al.* (1991) reported that analysis of native jack bean enzyme reveals bands at 420 and 745 nm that increase in intensity with decreasing temperature, consistent with octahedral nickel(II). Importantly, these spectra are incompatible with an antiferromagnetically coupled binuclear center in the inhibitor-free enzyme, whereas the data could be fit to a ferromagnetically exchange-coupled center. A possible danger in making these interpretations, however, is that the dominating spectra may actually arise from a minor contaminating species in the sample. In the case of the β-mercaptoethanol-treated jack bean enzyme, the same investigators found bands at 324, 380, 422, and 750 nm that increase in intensity with increasing temperature up to 150 K. This behavior was attributed to an antiferromagnetically exchange-coupled binuclear center in the thiol-bound urease. Such a result is consistent with the thiol acting to bridge the two nickel ions at the active site. Analogous results have been observed for *K. aerogenes* urease (M. G. Finnegan, K. L. Kiick, R. P. Hausinger, and M. K. Johnson, unpublished observations).

In summary, the urease active site possesses two nickel atoms that probably function in substrate binding. The metal ions are bound, at least in part, by histidinyl ligands. The two nickel atoms are located very close together and appear to be magnetically coupled under some conditions.

3.3.4 Chemical Modification Studies of the Urease Active Site

Protein chemical modification approaches have been used to provide information regarding the types of residues that serve as ligands to the metallocenter and to identify other residues that function at the active site. Below, I summarize chemical evidence that ligands to the nickel include histidinyl groups and that the active site contains additional histidinyl, cysteinyl, and carboxyl residues.

Support for the presence of histidinyl ligands to the urease nickel metallocenter is provided by comparison of the chemical reactivity for holoenzyme and apoprotein and by photooxidation studies. Lee *et al.* (1990) isolated urease from *Klebsiella aerogenes* that was grown in a medium depleted of nickel ion and showed that the inactive protein resembles the native enzyme except for being free of metal ions. The reactivity of apoprotein toward diethyl pyrocarbonate, a histidine-selective reagent, greatly exceeds that of holoenzyme. These results are consistent with reagent inaccessibility to several histidine residues in the native enzyme because they participate in metal ligation. Additional evidence for histidine ligation comes from photooxidation studies of jack bean urease (Sakaguchi *et al.,* 1983). These investigators observed a destruction of histidine residues when enzyme was subjected to intense light in the presence of methylene blue, whereas photooxidation is suppressed in the presence of the tight-binding urease inhibitor acetohydroxamic acid. As in all chemical modification studies, these results must be interpreted with caution; nevertheless, these studies provide consistent evidence that histidinyl residues participate in urease metal ion binding.

A histidinyl residue may also serve as a general base in urease catalysis. The pH dependence of catalysis for both jack bean (Dixon *et al.,* 1980e) and *K. aerogenes* (Todd and Hausinger, 1987) ureases is consistent with the existence of a general base with a pK_a of 6.5, a value that is appropriate for a histidine residue. Furthermore, Park and Hausinger (1993a) have shown that the native *K. aerogenes* enzyme is rapidly inactivated by diethyl pyrocarbonate and that the inactivation rate is reduced in the presence of substrate or active-site inhibitors. Inactivation is associated with a residue of pK_a 6.5 and spectroscopic changes were observed at 242 nm, consistent with histidine being the reactive residue. These results are compatible with the presence of an essential active-site histidine residue that functions as a general base in catalysis.

Several chemical modification studies have implicated the presence of an essential cysteine residue in urease. For example, Norris and Brocklehurst (1976) demonstrated that the jack bean enzyme is inactivated by 2,2'-dipyridyl disulfide, a reagent that is specific for cysteine groups. Enzyme inactivation correlates with the modification of six essential cysteine residues per protein (one per active site). Similarly, Todd and Hausinger (1991a) showed that both alkylating and disulfide reagents inactivate *K. aerogenes* urease, and reactivity difference studies in the presence and absence of active-site inhibitors demonstrate that a single essential cysteine is associated with each catalytic unit. The pH dependence of inactivation for the jack bean enzyme was interpreted to suggest that the essential cysteine possesses a pK_a of 9.15 (Norris and Brocklehurst, 1976); however, studies of the bacterial enzyme demonstrate that the cysteine is tightly associated with a second ionizable residue, and the coupled system possesses macroscopic pK_a values of less than 5 and $\sim$12 (Todd and Hausinger, 1991a).

The essential cysteine residue has been identified for both the jack bean and *K. aerogenes* ureases. Takishima *et al.* (1983) modified the nonessential cysteine residues of the jack bean enzyme with *N*-ethylmaleimide and then derivatized the essential thiol with the chromophore *N*-(4-dimethylamino-dinitrophenyl)maleimide. They isolated the specifically modified peptide and showed alkylation of a single residue, Cys-592. In contrast, Sakaguchi *et al.* (1984) carried out similar studies using the reagent diazonium-($1H$)-tetrazole and found that two peptides were labeled. The only peptide that was further characterized possessed Cys-207. More recently, Takishima *et al.* (1988) showed that enzyme activity loss correlates with modification of both Cys-592 and Cys-207, but that the latter residue could be completely modified with only 50% loss of activity; thus, the authors concluded that Cys-592 is the essential cysteine residue in jack bean urease. Peptide mapping studies of *K. aerogenes* enzyme, differentially modified with radioactively labeled iodoacetamide, demonstrate that a single cysteine residue is protected from modification at pH 6.3 in the presence of phosphate, an active-site inhibitor that prevents enzyme inactivation (Todd and Hausinger, 1991b). Additionally, modification of the same bacterial urease cysteine parallels activity loss at pH 8.5. The cysteine modified, Cys-319 in the large subunit of the heteropolymeric bacterial enzyme (Todd and Hausinger, 1991b), corresponds in sequence to Cys-592 in the homopolymeric plant urease (Takishima *et al.*, 1988).

Finally, chemical modification approaches have provided some evidence for the presence of a carboxyl group at the urease active site. Zerner and colleagues [quoted by Dixon *et al.* (1980e)] have found that the jack bean enzyme is irreversibly inhibited by triethyloxonium ion, consistent with the modification of a carboxylate ion. In addition, Todd and Hausinger (1991a) compared the reactivities of a series of alkylating and disulfide reagents; those reagents possessing negative charges are less accessible to the essential cysteine residue, consistent with charge repulsion by a nearby anionic group.

3.3.5 Site-Directed Mutagenesis of the Urease Active Site

Site-directed mutagenesis can be used as a selective method to test the importance of residues suspected to be at the active site and to examine the validity of proposals regarding the enzyme mechanism. Mutagenesis studies have been carried out regarding the essential cysteine and conserved histidine residues in the protein.

Martin and Hausinger (1992) used site-directed mutagenesis to change Cys-319 in the large subunit of *Klebsiella aerogenes* urease, previously identified as being essential (Todd and Hausinger, 1991b), to alanine, serine, aspartic acid, and tyrosine (C319A, C319S, C319D, and C319Y, respectively). None of these mutations affects the size or level of synthesis of the urease subunits. Each of the proteins was purified, and their properties were compared. Whereas the C319Y protein is inactive, the C319A, C319S, and C319D proteins possess 48, 4.5, and 0.03% of the wild-type activity levels. Thus, the "essential" cysteine is *not* essential. All active mutants exhibit a small increase in K_m compared to that of the wild-type protein. The mutant proteins display a greatly reduced rate of inactivation by iodoacetamide compared to wild-type protein, confirming the identity of the cysteine associated with alkylation-dependent activity loss. The pH dependence of activity is substantially altered in the mutants: C319S and C319D show optima near pH 5.2 and C319A near 6.7, compared to the pH 7.75 optimum of wild-type protein. These data are consistent with Cys-319, in association with a second ionically coupled unidentified residue (X), facilitating catalysis by participating as a general acid. In the mutant proteins lacking this cysteine, the requisite proton may be donated by the protonated residue X. Thus, Cys-319 may only be required for catalysis at neutral or basic pH values.

As already described, urease histidine residues may function as ligands to the nickel metallocenter and as a general base in catalysis. Furthermore, a histidine group is a reasonable possibility for residue X, the ionizable residue coupled to the active-site cysteine. In order to begin to characterize the roles for histidine groups in urease, Park and Hausinger (1993b) have changed each of ten histidines individually to alanine in *K. aerogenes* urease. These ten residues are conserved among all six ureases that have been sequenced. Urease proteins were purified from each of the ten mutants and characterized for activity and nickel contents. Five of the conserved histidines (residue 96 in the γ subunit, residues 39 and 41 in the β subunit, and residues 312 and 321 in the α subunit) have no catalytic function as shown by the lack of any effect when they were changed to alanine. Evidence for a role in nickel ligation was obtained for three histidines (residues 134, 136, and 246 in the α subunit): when these histidines were substituted by alanine, the proteins were inactive and possessed approximately 1, 0, and 0 nickel ions bound per catalytic unit. Finally, two additional histidines (residues 219 and 320 in the α subunit) appear to have essential roles in catalysis as shown by the near abolishment of activity in the alanine mutants, but the nickel contents were not affected compared to that of the wild-type protein (2 nickel ions per catalytic unit). The resistance to inactivation by diethyl pyrocarbonate of the H320A protein is consistent with a functional role of His-320 as the general base in catalysis. In contrast, the role of His-219 is associated, at least in part, with substrate binding, as shown by a dramatic increase in the urea K_m value for the H320A mutant. Further mutagenesis studies combined with three-dimensional structure determination will provide a more complete description of the residues at the active site.

3.3.6 Structure of the Urease Active Site and the Mechanism of Catalysis

The results from the above spectroscopic, chemical modification, and other studies have been assimilated into the model of the urease active site illustrated in Fig. 3-2. The model indicates that each active site of urease possesses two nickel ions that are liganded by histidine residues as well as other nitrogen or oxygen donors. The two metal ions are not exchange-coupled in the native enzyme, but they are near enough to provide for bridging by thiols, perhaps other inhibitors (e.g., see Todd and Hausinger, 1989), and possibly substrate. An essential histidine residue (possibly His-320 in *Klebsiella aerogenes* urease) functions as a general base (pK_a 6.5). A cysteine residue (Cys-319 in the large subunit of the bacterial enzyme or Cys-592 in jack bean urease) associates with X, a second ionizable residue, and together they can serve as a general acid in catalysis. A nearby carboxyl group is also indicated.

A reasonable mechanism of urease catalysis, modified from that originally proposed by Dixon *et al.* (1980e), can be developed from the above model as illustrated in Fig. 3-3. In the resting enzyme, one nickel ion is proposed to coordinate a water molecule, and a second coordinates hydroxide or a second water molecule. Urea is thought to displace the water molecule, and the illustrated nickel-coordinated urea resonance structure is proposed to be stabilized by interaction with the nearby carboxylate anion. The general base could activate the nickel-coordinated hydroxyl group (or water molecule) for

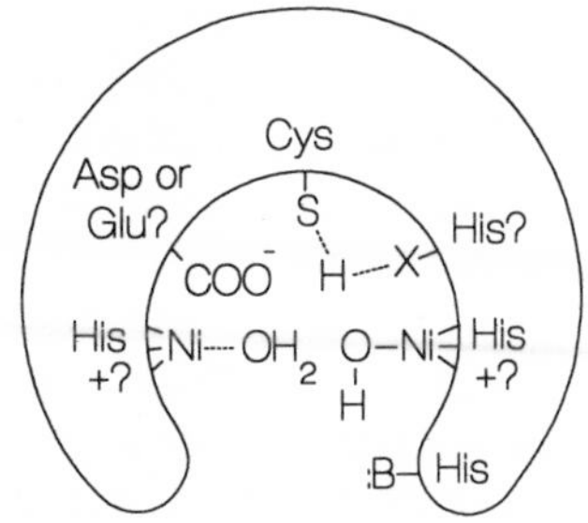

Figure 3-2. Model of the urease active site. The current model of the urease active site assigns two nickel ions per catalytic center, where the metal ions are coordinated by histidinyl and other N or O donors. In the native enzyme, the nickel ions are not exchange-coupled in a binuclear center. A nearby histidine residue functions as a general base in catalysis. Furthermore, a cysteine residue is ionically coupled to an unidentified residue, and together they function as a proton donor during urea hydrolysis. Finally, a negatively charged group suspected to be at the active site is assigned to a carboxylate residue.

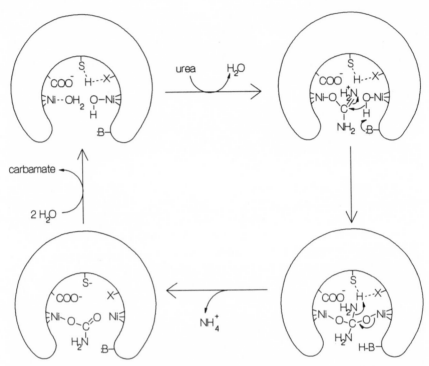

Figure 3-3. Mechanism for urea hydrolysis. In the first step of the reaction, urea displaces a water molecule and coordinates to one nickel ion. The nickel ion acts as a Lewis acid, and the O-coordinated resonance form is proposed to be charge-stabilized by interaction with the carboxylate group. General base-facilitated activation of the nickel-coordinated hydroxide ion leads to nucleophilic attack on the urea carbon atom to form the tetrahedral intermediate shown in the lower right-hand portion of the figure. Decomposition of this species is facilitated by general acid catalysis by the cysteine–X couple, resulting in release of ammonia. Finally, carbamate is released and the resting enzyme form restored.

nucleophilic attack on the urea carbon atom to form a tetrahedral intermediate. Decomposition of this intermediate and release of ammonia is facilitated by general acid catalysis involving the cysteine–X couple. Finally, carbamate is released from the enzyme with regeneration of the initial state. The rate of the reaction catalyzed by this enzyme represents a greater than 10^{14} rate enhancement over the uncatalyzed rate of urea degradation (Dixon *et al.*, 1980e). Furthermore, this mechanism is distinct from the uncatalyzed reaction, which involves elimination of ammonia to yield cyanic acid (Blakeley *et al.*, 1982).

3.4 Chemical Models of Urease

Examination of synthetic model compounds for their capacity to carry out urease-like chemistry or their ability to mimic aspects of the metallocenter could be important in assessing whether the amide nitrogen or the carbonyl oxygen of urea binds to the nickel, in examining potential roles for the two nickel ions at each active site, and perhaps in understanding why all ureases require nickel. Unfortunately, no functional synthetic model has yet been demonstrated. In the following, I describe several mononickel and binickel compounds that have been characterized as potential models of the urease active site.

Blakeley *et al.* (1982) reported that *N*-(2-pyridylmethyl)urea undergoes ethanolysis to give ethyl *N*-(2-pyridylmethyl)carbonate and hydrolysis to give (2-pyridylmethyl)amine in the presence of excess nickel chloride. Less effective as a catalyst are the cobalt and manganese salts, and no reaction is observed with magnesium or calcium salts, or in the absence of any metal ion. The authors proposed a mechanism for these reactions in which nickel coordination of the carbonyl oxygen facilitates solvent attack on the carbonyl group (Scheme 3-1).

This evidence was thought to support the proposal of these same authors for urea carbonyl coordination by nickel in urease (Dixon *et al.,* 1980e); however, Curtis *et al.* (1983) have pointed out that the same products would be generated by the N-coordinated intermediates in the reaction shown in Scheme 3-2.

Structural studies have shown that metal complexes of uncharged amides and ureas show almost exclusive coordination to the carbonyl oxygen; hence, this feature is retained in the mechanism of Fig. 3-3. However a nitrogen-coordinated nickel complex cannot yet be eliminated as a possible intermediate. For example, the copper complex shown in Scheme 3-3 has been characterized (Maslak *et al.,* 1991). Possible geometric limitations of the ligand structure cannot explain the observed coordination, because the zinc complex

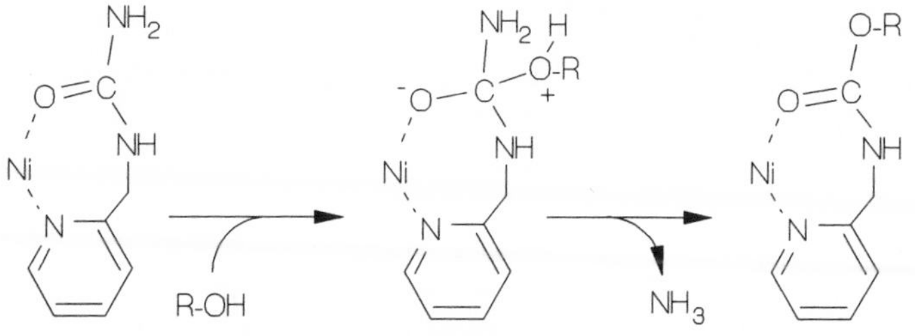

Scheme 3-1

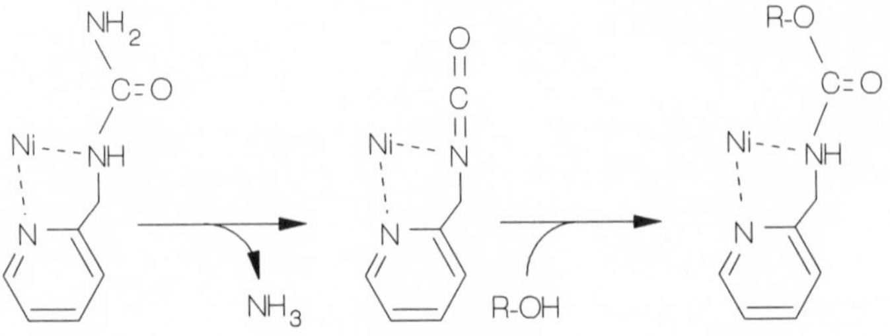

Scheme 3-2

of the same ligand exhibits metal binding to oxygen (Maslak *et al.,* 1991). For the nitrogen-coordinated copper complex, the infrared spectral band associated with the carbonyl group exhibits a shift of 50–100 cm^{-1} to higher frequency with respect to its position in the infrared spectrum of the free ligand. The same trend is observed for the nickel derivative whereas the zinc compound exhibits a shift in the opposite direction. Thus, it is likely that the nickel complex of this ligand also possesses nitrogen coordination. Both the nickel and copper species catalyze hydrolysis of the substituted urea ligands, whereas the zinc derivative does not (Maslak *et al.,* 1991).

In addition to the mononickel compounds described above, several binuclear complexes have been examined. Buchanan *et al.* (1989) characterized the structure of the triply bridged μ-phenoxo di-μ-acetato binickel complex shown in Scheme 3-4. The light green crystals of this compound display absorption bands at 373, 622, and 1037 nm in acetonitrile. In addition, evidence was obtained for weak antiferromagnetic exchange coupling in the complex.

A yellow binickel complex in which the nickel ions are bridged by an imidazole group has been reported by Salata *et al.* (1989). Although crystals of the nickel complex were not obtained, the analogous copper compound does crystallize, and its structure is represented in Scheme 3-5. Electron para-

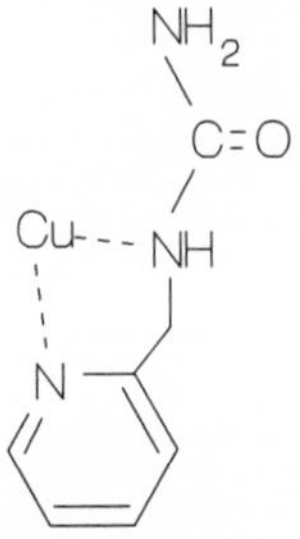

Scheme 3-3

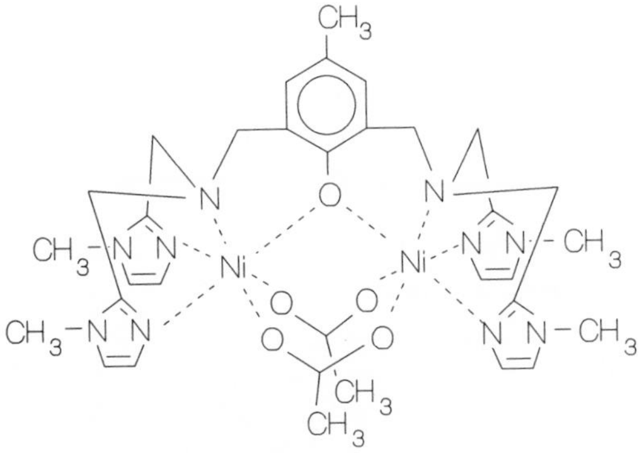

Scheme 3-4

magnetic resonance spectroscopic analysis of the green copper compound provides clear evidence for antiferromagnetic exchange coupling. A series of derivatives of the same complex was examined in which the copper or nickel ions are bridged by various ligands (Salata *et al.,* 1991). Interestingly, one of the nickel derivatives exhibits some, albeit modest, degradative activity toward *p*-nitrophenyl acetate. Because the nickel complex is destroyed in this chemical process, the relevance of this reaction to urea hydrolysis is unclear.

3.5 Mechanism of Urease Metallocenter Assembly

Recent evidence has demonstrated that assembly of the urease metallocenter is a complex process that requires participation by several accessory

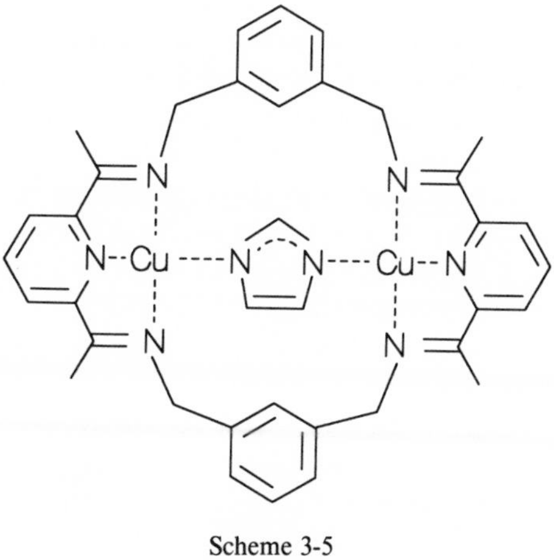

Scheme 3-5

proteins. Below, I briefly comment on urease metallocenter stability, describe how the apoenzyme can be reconstituted only by an energy-dependent *in vivo* process, and summarize evidence for the presence of auxiliary proteins that facilitate nickel incorporation into ureases from both procaryotes and eucaryotes. Speculative roles for three of these accessory proteins are provided.

3.5.1 Urease Metallocenter Stability

The nickel ions in urease are very tightly associated with the enzyme as evidenced by retention of the metal ion during protein isolation in buffers containing 1 mM EDTA. Furthermore, dialysis experiments have shown that nickel is not easily removed from the native enzyme at neutral pH. For example, Martin and Hausinger (1992) reported essentially no loss of nickel from *Klebsiella aerogenes* urease over 20 days of dialysis in 1 mM EDTA-containing buffers at pH 7.0. Similarly, Zerner (1991) reported that jack bean urease retains over 50% of its nickel ion when dialyzed for 60 days against pH 7.0 buffers containing sulfite.

In contrast to the urease metallocenter stability observed under the gentle conditions described above, nickel can be released from the enzyme by denaturants or acidic conditions, leading to irreversible loss of activity. For example, denaturation of jack bean urease in 2.5 M guanidinium chloride at pH 7.6 leads to the loss of nickel ion and concomitant loss of activity (Dixon *et al.*, 1980b). Surprisingly, Thirkell *et al.* (1989) and Hawtin *et al.* (1991) claimed that tracer amounts of nickel remain associated with the large subunits of *Ureaplasma urealyticum* and *Helicobacter pylori* ureases following polyacrylamide gel electrophoresis for samples denatured in sodium dodecyl sulfate. The latter results, however, are open to dispute because of possible adventitious binding of the metal ion to the major peptide. At low pH, EDTA was found to promote dissociation of nickel from jack bean urease (Dixon *et al.*, 1980b). As in the guanidinium chloride-treated sample, urease activity cannot be restored upon neutralization and nickel addition. Irreversible loss of nickel under acidic conditions was also reported for the bacterial enzyme purified from *Arthrobacter oxydans* (Schneider and Kaltwasser, 1984).

The stability and chelator inaccessibility of the urease metallocenter in the native enzyme may be due to the nickel being buried in the protein, perhaps with each metal ion possessing a single ligand position for catalysis. Any model proposing the presence of a highly sequestered enzyme metallocenter immediately raises questions concerning the metal incorporation mechanisms utilized during protein biosynthesis. As described below, nickel incorporation into urease apoprotein is not simply an associative type interaction.

3.5.2 Reconstitution of Urease Apoenzyme

Urease apoenzyme is synthesized in both procaryotic and eucaryotic cells when they are grown in the absence of nickel ion (e.g., Mulrooney *et al.,* 1989; Winkler *et al.,* 1983). Thus, nickel ion is not involved in transcriptional regulation of urease. The following paragraphs describe properties of the apoenzyme purified from *Klebsiella aerogenes* and detail the results of experiments involving *in vivo* reconstitution of this and other ureases.

Lee *et al.* (1990) purified the *K. aerogenes* urease apoenzyme from cells that were grown on nickel-depleted medium that contained the normal concentrations of other trace metal ions. The inactive protein possesses no significant levels of any metal ions as measured by plasma emission spectroscopy, but nevertheless it behaves identically to the holoprotein in terms of subunit stoichiometry and native size. The authors were unable to activate the protein by providing nickel ion in buffers containing no other additive or in buffers containing thiols, EDTA, guanidine hydrochloride, glycerol, or KCl or mixtures of these compounds. These results clearly demonstrate the inability of nickel ion to form a functional active site by simple association with the apoprotein. Furthermore, nickel addition to disrupted cells failed to generate active enzyme even upon addition of various concentrations of Mg-ATP, thiols, EDTA, glucose, or mixtures of these compounds. Similar negative results were reported for soybean apoenzyme that was treated with 1 mM nickel in the presence of KCN, dimethyl sulfoxide, or urea (Winkler *et al.,* 1983). Indeed, *in vitro* activation of urease apoprotein has not been reported for any organism. In unpublished work, however, significant levels of activation of *K. aerogenes* urease apoenzyme were obtained by adding 1 mM nickel chloride to highly concentrated and freshly prepared cell extracts prepared from cultures that were grown in the absence of nickel (M. B. Carr and R. P. Hausinger, unpublished observations).

Urease apoenzyme can be activated under *in vivo* conditions. For example, preformed urease apoprotein was activated by nickel addition to intact nickel-depleted *K. aerogenes* cells that had been treated with an inhibitor of protein synthesis (Lee *et al.,* 1990). The observed activation was slow: there was a delay of $\sim$30 min after nickel addition, followed by a continuous increase in activity for at least 4 h. Furthermore, activation appears to be an energy-dependent process as shown by the inability to generate active enzyme in cells previously treated with dinitrophenol (a proton uncoupler) or *N,N'*-dicyclohexylcarbodiimide (DCCD; an ATP synthase inhibitor). *In vivo* reconstitution of urease apoenzyme also has been reported for another enteric bacterium, *Proteus mirabilis* (Rando *et al.,* 1990), as well as for a purple sulfur bacterium (Bast, 1988), a cyanobacterium (Mackerras and Smith, 1986), several algae (Rees and Bekheet, 1982), and soybean (Winkler *et al.,* 1983).

3.5.3 Accessory Genes Facilitating Nickel Incorporation into Bacterial Urease

In addition to providing primary structural information related to genes encoding the urease subunits, DNA sequence analysis has revealed the presence of several additional genes that are part of the urease gene cluster. For example, in *Klebsiella aerogenes* the *ureA*, *ureB*, and *ureC* genes encoding the urease subunits are immediately preceded by the *ureD* gene and followed by the *ureE*, *ureF*, and *ureG* genes (Lee *et al.*, 1992; Mulrooney and Hausinger, 1990), as illustrated in Fig. 3-4. Sequence analysis of other bacterial urease genes reveals that many of these accessory genes are conserved. *Proteus mirabilis* possesses highly similar counterparts to at least the *ureD*, *ureE*, and *ureF* genes (Jones and Mobley, 1989). *Helicobacter pylori* possesses analogs to each of the four *K. aerogenes* accessory genes (Cussac *et al.*, 1992), although the counterpart to *ureD* is located after *ureG* and was termed *ureH* by the authors. Portions of *ureD* and *ureE* genes can be observed in the sequences immediately surrounding the *P. vulgaris* urease subunit genes (Mörsdorf and Kaltwasser, 1990). Finally, all of these adjacent genes appear to be present in a ureolytic strain of *Escherichia coli* (Collins and Gutman, 1992; Carleen Collins, personal communication).

The nonsubunit auxiliary genes recently have been shown to be required for urease metallocenter assembly. Transposon mutagenesis and other inactivation methods have demonstrated that *ureD*, *ureF*, and *ureG* (or unidentified upstream or downstream regions that are likely to correspond to these genes) are essential for obtaining a functional urease (Collins and Falkow, 1988; Collins and Gutman, 1992; Gerlach *et al.*, 1988; Mulrooney *et al.*, 1988; Mulrooney and Hausinger, 1990; Walz *et al.*, 1988). To begin to assess

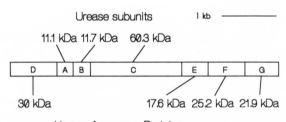

Urease Accessory Proteins

Figure 3-4. Structure of the bacterial urease gene cluster. In the best characterized example, seven genes are associated with the urease gene cluster in *Klebsiella aerogenes* (Lee *et al.*, 1992; Mulrooney and Hausinger, 1990). The genes all appear to be present in a single transcriptional unit that is transcribed from left to right. The *ureA*, *ureB*, and *ureC* genes encode the three urease subunits, whereas the *ureD*, *ureE*, *ureF*, and *ureG* genes are required for the functional incorporation of the urease metallocenter.

the roles for the accessory genes, Lee *et al.* (1992) examined the *K. aerogenes* urease properties in recombinant *E. coli* cells containing plasmids with the intact urease gene cluster or deletion mutants in each of *ureD, ureE, ureF,* and *ureG* genes. In the deletions involving *ureD, ureF,* and *ureG,* the urease protein is synthesized in an inactive form that is devoid of nickel. Furthermore, mutants in *ureE* possess a reduced urease activity, and the nickel contents of the purified urease enzymes are correspondingly reduced (Lee *et al.,* 1992). Each of the four genes was shown to function via a *trans*-acting factor. Thus, all four accessory genes encode proteins that appear to be necessary for functional incorporation of nickel into urease.

Although specific functions cannot be assigned definitively to any of the four accessory proteins that are required for nickel incorporation into urease, speculative roles for the action of three of these proteins can be hypothesized. Consistent with sequence analysis of the translated *K. aerogenes ureE* gene that reveals several potential metal binding sites (Mulrooney and Hausinger, 1990), purified UreE protein has been shown to bind approximately six nickel ions per dimer with a K_d of ~ 10 μM (Lee *et al.,* 1993). It is tempting to speculate that UreE may bind nickel ion and act as the nickel donor to the urease apoprotein. In the absence of UreE, other nickel donors may be used, but with decreased efficiency. Because this protein is synthesized at high levels in the urease clone, large quantities of it are available, and numerous studies of this new nickel-binding protein have been carried out (Lee *et al.,* 1993), including XAS and variable-temperature magnetic circular dichroism spectrascopic analyses of the nickel metallocenter.

The UreG protein is also synthesized at high levels; however, the protein has not been purified to homogeneity at this time. Sequence analysis of the translated *ureG* gene reveals a P-loop motif (Saraste *et al.,* 1990) that is found in a variety of ATP- and GTP-binding proteins. Combining this feature with the observed energy dependence for *in vivo* nickel ion incorporation described above, one can hypothesize that UreG binds ATP or GTP and couples its hydrolysis to the nickel incorporation event. Sequence analysis also revealed (Wu, 1992) a relationship ($\sim 25\%$ identity) between the *K. aerogenes* UreG protein (Mulrooney and Hausinger, 1990) and a portion of the *E. coli hypB* gene product (Lutz *et al.,* 1991). The *hyp* operon encodes a *hy*drogenase *p*leiotropic operon that is required for functional activation of each of the three hydrogenases in *E. coli* (see Chapter 4). A mutation in the *hypB* gene had previously been shown to be complemented, in part, by the addition of high levels of nickel (Waugh and Boxer, 1986), consistent with a role for this protein in nickel processing. Thus, at least one of the urease accessory genes and one of the hydrogenase accessory genes may have diverged from a common ancestral nickel incorporation system. Recently, Maier *et al.* (1993) purified the *hypB* gene product and showed that it bound guanine nucleotides and hydrolyzed GTP. These results bolster the hypothesis that UreG binds to and hydrolyzes GTP. The hypothetical UreG-catalyzed reaction may be concomitant with urease metallocenter assembly, or it may be involved in activation of another accessory protein.

UreD and UreF sequences exhibit no significant homology when compared to other sequences available in various data bases. Furthermore, both the *ureD* and *ureF* genes are expressed at very low levels that complicate purification and characterization of these gene products. The properties and role of UreF in metallocenter assembly remain unknown; however, by manipulation of the *ureD* ribosome binding site using site-directed mutagenesis techniques, this gene has been overexpressed and the gene product has been partially purified and characterized (I.-S. Park and R. P. Hausinger, unpublished results). For cells grown in the absence of nickel, UreD was found to be tightly associated with urease apoenzyme, consistent with a possible role for the complex in nickel incorporation. We propose that UreD induces a conformational change in urease apoenzyme that exposes the metal binding site to the nickel donor.

The speculative roles for the urease accessory proteins are combined in the model shown in Fig. 3-5.

3.5.4 Accessory Genes for Urease Activation in Eucaryotes

Several studies of fungal ureases are consistent with the presence of, and requirement for, urease genes that do not encode the urease subunits. For

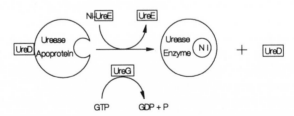

Figure 3-5. Speculative model indicating possible functions of the urease accessory proteins. UreE has been shown to be a nickel-binding protein; hence, it is reasonable to propose a role for this protein as a nickel donor to urease apoprotein. Because (i) the UreG protein sequence includes a potential ATP- or GTP-binding motif, (ii) *in vivo* reconstitution of urease apoprotein is energy-dependent, and (iii) the UreG sequence is related to that of HypB, a protein that is known to bind and hydrolyze GTP, the model proposes that UreG somehow couples the energy of GTP hydrolysis to nickel incorporation into urease or to activation of one of the accessory proteins. UreD is suggested to bind to and induce a conformational change in the urease apoprotein that is required for nickel incorporation. The role for UreF in the functional assembly of the urease metallocenter remains unknown.

example, four complementation groups were identified and shown to be required for *Aspergillus nidulans* urease activity (Mackay and Pateman, 1982). Similarly, four loci (encoding unknown functions) are required for functional urease activity in *Neurospora crassa* (Benson and Howe, 1978) and in *Schizosaccharomyces pombe* (Kinghorn and Fluri, 1984). In *A. nidulans, ureA* is the structural gene for the urea transport protein, *ureB* encodes the single-subunit urease enzyme, *ureC* encodes a product needed for enzyme activity but of unknown function, and *ureD* is suspected to be necessary for synthesis or incorporation of the nickel cofactor. A mutation in *ureD* can be overcome by growth in the presence of 0.1 mM nickel sulfate (Mackay and Pateman, 1980). In unpublished work, the *A. nidulans ureD* gene was cloned and sequenced; the protein encoded by this gene exhibits significant sequence similarity to the *K. aerogenes* UreG protein (Gareth Wyn Griffith, personal communication).

The most detailed genetic analyses reported for any eucaryotic urease are those of soybean, a plant that is associated with three distinct urea-degrading isozymes: the embryo-specific, ubiquitous, and background enzymes (Holland and Polacco, 1992; Holland *et al.*, 1987; Meyer-Bothling and Polacco, 1987; Meyer-Bothling *et al.*, 1987; Polacco *et al.*, 1989; Stebbins *et al.*, 1991; and accompanying references). The embryo-specific or seed urease is expressed at very high levels in the developing embryo and is encoded by the *Eu1* locus (Holland *et al.*, 1987; Meyer-Bothling and Polacco, 1987; Polacco and Havir, 1979). The ubiquitous isozyme is constitutively expressed at low levels in all soybean tissues from the *Eu4* locus (Polacco *et al.*, 1989). A portion of the ubiquitous urease gene has been cloned by Krueger *et al.* (1987), who used synthetic oligonucleotides based on the jack bean and soybean urease sequences to screen a soybean genomic library. The isolated subclone possesses only a partial urease sequence, but it matches 108 of 130 amino acids determined for the jack bean enzyme. Finally, a background urease was identified in double mutants defective in both the embryo-specific and ubiquitous ureases (Holland and Polacco, 1992). This background urease appears to be produced not by the plant, but rather by a phylloplane-associated bacterium, *Methylobacterium mesophilicum*. The isolated bacterium possesses urease with properties that are very similar to the urease activity in the plant double mutant (Holland and Polacco, 1992). All three soybean-associated ureases appear to require nickel ion, as shown by the near absence of activity when plants are grown in nickel-depleted medium.

Accessory genes that facilitate nickel incorporation into urease appear to be required for the plant enzymes. Meyer-Bothling *et al.* (1987) isolated pleiotropic mutants that are deficient in both plant-derived ureases. These mutations map to two distinct loci, *Eu2* and *Eu3*, neither of which is closely associated with the urease structural loci *Eu1* and *Eu4*. They proposed that these loci may encode genes that are required for a common urease maturation function such as nickel ion emplacement (Meyer-Bothling *et al.,* 1987). Curiously, mutants at either of the two plant accessory gene loci also eliminate urease activity in the phylloplane-associated bacterium (Holland and Polacco, 1992). Bacteria isolated from these plant mutants were shown to possess transient urease and hydrogenase deficiencies that could be corrected by addition of nickel ion to the cultures. Any relationship between the genes associated with the *Eu2* or *Eu3* loci and the bacterial *ureD, ureE, ureF,* and *ureG* genes is obscure at this time. In barley, however, the partial sequence of an unidentified gene reveals an open reading frame with significant sequence similarity to the *K. aerogenes* UreG protein (Brian Forde, personal communication). Thus, plant and bacterial urease accessory proteins are likely to be related.

3.6 Perspective

Urease, the first enzyme ever to be crystallized and the first enzyme shown to contain nickel ion, continues to be a source of perplexing and fascinating questions. Did the gene encoding the single-subunit plant enzyme undergo disruption to yield the multiple genes encoding the two or three bacterial subunits? Or did bacterial genes fuse to form the gene encoding the plant subunit? What is so special about the chemistry of nickel ion, compared to zinc or other metal ions, that it is required for all ureases? What is the detailed structure and geometry of the binuclear nickel site and what other protein groups are involved in catalysis? What are the roles for the multiple gene products that facilitate assembly of a functional metallocenter in urease? These and other questions are stimulating urease-related research in many areas. By combining three-dimensional structural characterization of urease crystals, chemical modification and site-directed mutagenesis of the urease protein and structural genes, and the application of spectroscopic and biophysical techniques to probe the metallocenter, a clear picture of the active-site structure and function should emerge. Chemical modeling of the metallocenter will assist in understanding the unique nickel-based chemistry. Finally, purification and characterization of the accessory proteins, along with development of an *in vitro* reconstitution system, will elucidate the requirements for assembly of the urease active site.

References

Alagna, L., Hasnain, S. S., Piggott, B., and Williams, D. J., 1984. The nickel environment in jack bean urease, *Biochem. J.* 220:591–595.

Andrews, R. K., Blakeley, R. L., and Zerner, B., 1984. Urea and urease, in *Advances in Inorganic Biochemistry,* Vol. 6 (G. L. Eichhorn and L. G. Marzilli, eds.), Elsevier Science Publishing, New York, pp. 245–283.

Andrews, R. K., Blakeley, R. L., and Zerner, B., 1988. Urease—a Ni(II) metalloenzyme, in *The Bioinorganic Chemistry of Nickel* (J. R. Lancaster, Jr., ed.), VCH Publishers, New York, pp. 141–165.

Argall, M. E., Smith, G. D., Stamford, N. P. J., and Youens, B. N., 1992. Purification and properties of urease from the cyanobacterium *Anabaena cylindrica, Biochem. Int.* 27:1027–1036.

Austin, J. W., Doig, P., Stewart, M., and Trust, T. J., 1992. Structural comparison of urease and a GroEL analog from *Helicobacter pylori, J. Bacteriol.* 174:7470–7473.

Bast, E., 1988. Nickel requirement for the formation of active urease in purple sulfur bacteria (Chromatiaceae), *Arch. Microbiol.* 150:6–10.

Benchemsi-Bekkari, N., and Pizelle, G., 1992. *In vivo* urease activity in *Robinia pseudoacacia, Plant Physiol. Biochim.* 30:187–192.

Benson, E. W., and Howe, H. B., Jr., 1978. Reversion and interallelic complementation at four urease loci in *Neurospora crassa, Mol. Gen. Genet.* 165:277–288.

Blakeley, R. L., and Zerner, B., 1984. Jack bean urease: The first nickel enzyme, *J. Mol. Catal.* 23:263–292.

Blakeley, R. L., Treston, A., Andrews, R. K., and Zerner, B., 1982. Nickel (II) promoted ethanolysis and hydrolysis of *N*-(2-pyridylmethyl)urea. A model for urease, *J. Am. Chem. Soc.* 104:612–614.

Blakeley, R. L., Dixon, N. E., and Zerner, B., 1983. Jack bean urease. VII. Light scattering and nickel (II) spectrum. Thiolate → nickel(II) charge transfer peaks in the spectrum of the β-mercaptoethanol–urease complex, *Biochim. Biophys. Acta* 744:219–229.

Blanchard, A., 1990. *Ureaplasma urealyticum* urease genes; use of a UGA tryptophan codon, *Mol. Microbiol.* 4:669–676.

Blattler, D. P., Contaxis, C. C., and Reithel, F. J., 1967. Dissociation of urease by glycol and glycerol, *Nature (London)* 216:274–275.

Booth, J. L., and Vishniac, H. S., 1987. Urease testing and yeast taxonomy, *Can. J. Microbiol.* 33:396–404.

Braude, A. I., and Siemienski, J., 1960. Role of bacterial urease in experimental pyelonephritis, *J. Bacteriol.* 80:171–179.

Breitenbach, J. M., and Hausinger, R. P., 1988. *Proteus mirabilis* urease: Partial purification and inhibition by boric acid and boronic acids, *Biochem. J.* 250:917–920.

Bremner, J. M., and Mulvaney, R. L., 1978. Urease activity in soils, in *Soil Enzymes* (R. G. Burns, ed.), Academic Press, New York, pp. 146–196.

Buchanan, R. M., Mashuta, M. S., Oberhausen, K. J., Richardson, J. F., Li, Q., and Hendrickson, D. N., 1989. Active site model of urease: Synthesis, structure, and magnetic properties of a binuclear Ni(II) complex containing a polyimidazole ligand, *J. Am. Chem. Soc.* 111:4497–4498.

Christians, S., and Kaltwasser, H., 1986. Nickel-content of urease from *Bacillus pasteurii, Arch. Microbiol.* 145:51–55.

Christians, S., Jose, J., Schäfer, U., and Kaltwasser, H., 1991. Purification and subunit determination of the nickel-dependent *Staphylococcus xylosus* urease, *FEMS Microbiol. Lett.* 80:271–276.

Clark, P. A., and Wilcox, D. E., 1989. Magnetic properties of the nickel enzymes urease, nickel-substituted carboxypeptidase A, and nickel-substituted carbonic anhydrase, *Inorg. Chem.* 28:1326–1333.

Clark, P. A., Wilcox, D. E., and Scott, R. A., 1990. X-ray absorption spectroscopic evidence for binding of the competitive inhibitor 2-mercaptoethanol to the nickel sites of jack bean urease. A new Ni–Ni interaction in the inhibited enzyme, *J. Am. Chem. Soc.* 29:579–581.

Clayton, C. L., Pallen, M. J., Kleanthous, H., Wren, B. W., and Tabaqchali, S., 1990. Nucleotide sequence of two genes from *Helicobacter pylori* encoding for urease subunits, *Nucleic Acids Res.* 18:362.

Collins, C. M., and Falkow, S., 1988. Genetic analysis of an *Escherichia coli* urease locus: Evidence of DNA rearrangement, *J. Bacteriol.* 170:1041–1045.

Collins, C. M., and Gutman, D. M., 1992. Insertional inactivation of an *Escherichia coli* urease gene by IS*3411, J. Bacteriol.* 174:883–888.

Contaxis, C. C., and Reithel, F. J., 1971. Studies on protein multimers. II. A study of the mechanism of urease dissociation in 1,2-propanediol: Comparative studies with ethylene glycol and glycerol, *J. Biol. Chem.* 246:677–685.

Creaser, E. H., and Porter, R. L., 1985. The purification of urease from *Aspergillus nidulans, Int. J. Biochem.* 17:1339–1341.

Curtis, N. J., Dixon, N. E., and Sargeson, A. M., 1983. Synthesis, linkage isomerism, and ligand reactivity of (urea)pentaamminerhodium(III) complexes, *J. Am. Chem. Soc.* 105:5347–5353.

Cussac, V., Ferrero, R. I., and Labigne, A., 1992. Expression of *Helicobacter pylori* urease genes in *Escherichia coli* grown under nitrogen-limiting conditions, *J. Bacteriol.* 174:2466–2473.

Dalton, D. A., Evans, H. J., and Hanus, F. J., 1985. Stimulation by nickel of soil microbial urease activity and urease and hydrogenase activities in soybeans grown in low-nickel soil, *Plant Soil* 88:245–258.

Dalton, D. A., Russell, S. A., and Evans, H. J., 1988. Nickel as a micronutrient for plants, *BioFactors* 1:11–16.

Day, E. P., Peterson, J., Sendova, M., Todd, M. J., and Hausinger, R. P., 1993. Saturation magnetization of ureases from *Klebsiella aerogenes* and jack bean: No evidence for exchange coupling between the two active site nickel ions in the native enzyme, *Inorg. Chem.* 32:634–638.

Dixon, N. E., Gazzola, C., Watters, J. J., Blakeley, R. L., and Zerner, B., 1975a. Inhibition of jack bean urease (EC 3.5.1.5) by acetohydroxamic acid and by phosphoramidate. An equivalent weight for urease, *J. Am. Chem. Soc.* 97:4130–4131.

Dixon, N. E., Gazzola, C., Blakeley, R. L., and Zerner, B., 1975b. Jack bean urease (EC 3.5.1.5.). A metalloenzyme. A simple biological role for nickel?, *J. Am. Chem. Soc.* 97:4131–4133.

Dixon, N. E., Blakeley, R. L., and Zerner, B., 1980a. Jack bean urease (EC 3.5.1.5). I. A simple dry ashing procedure for the microdetermination of trace metals in proteins. The nickel content of urease, *Can. J. Biochem.* 58:469–473.

Dixon, N. E., Gazzola, C., Asher, C. J., Lee, D. S. W., Blakeley, R. L., and Zerner, B., 1980b. Jack bean urease (EC 3.5.1.5). II. The relationship between nickel, enzymatic activity, and the "abnormal" ultraviolet spectrum. The nickel content of jack beans, *Can. J. Biochem.* 58:474–480.

Dixon, N. E., Blakeley, R. L., and Zerner, B., 1980c. Jack bean urease (EC 3.5.1.5). III. The involvement of active-site nickel ion in inhibition by β-mercaptoethanol, phosphoramidate, and fluoride, *Can. J. Biochem.* 58:481–488.

Dixon, N. E., Hinds, J. A., Fihelly, A. K., Gazzola, C., Winzor, D. J., Blakeley, R. L., and Zerner, B., 1980d. Jack bean urease (EC 3.5.1.5). IV. The molecular size and the mechanism of inhibition by hydroxamic acids. Spectrophotometric titration of enzymes with reversible inhibitors, *Can. J. Biochem.* 58:1323–1334.

Dixon, N. E., Riddles, P. W., Gazzola, C., Blakeley, R. L., and Zerner, B., 1980e. Jack bean urease (EC 3.5.1.5). V. On the mechanism of action of urease on urea, formamide, acetamide, *N*-methylurea, and related compounds, *Can. J. Biochem.* 58:1335–1344.

Dunn, B. E., Campbell, G. P., Perez-Perez, G., and Blaser, M. J., 1990. Purification and characterization of urease from *Helicobacter pylori, J. Biol. Chem.* 265:9464–9469.

Eaton, K. A., Brooks, C. L., Morgan, D. R., and Krakowka, S., 1991. Essential role of urease in pathogenesis of gastritis induced by *Helicobacter pylori* in gnotobiotic piglets, *Infect. Immun.* 59:2470–2475.

Eskew, D. L., Welch, R. M., and Cary, E. E., 1983. Nickel: An essential micronutrient for legumes and possibly all higher plants, *Science* 222:621–623.

Eskew, D. L., Welch, R. M., and Norvell, W. A., 1984. Nickel in higher plants. Further evidence for an essential role, *Plant Physiol.* 76:691–693.

Evans, D. J., Jr., Evans, D. G., Kirkpatrick, S. S., and Graham, D. Y., 1991. Characterization of the *Helicobacter pylori* urease and purification of its subunits, *Microb. Pathogenesis* 10:15–26.

Finnegan, M. G., Kowal, A., Werth, M. T., Clark, P. A., Wilcox, D. E., and Johnson, M. K., 1991. Variable-temperature magnetic circular dichroism spectroscopy as a probe of the electronic and magnetic properties of nickel in jack bean urease, *J. Am. Chem. Soc.* 113:4030–4032.

Fishbein, W. N., Nagarajan, K., and Scurzi, W., 1973. Urease catalysis and structure. IX. The half-unit and hemipolymers of jack bean urease, *J. Biol. Chem.* 248:7870–7877.

Fishbein, W. N., Engler, W. F., Griffin, J. L., Scurzi, W., and Bahr, G. F., 1977. Electron microscopy of negatively stained jack bean urease at three levels of quaternary structure, and comparison with hydrodynamic studies, *Eur. J. Biochem.* 73:185–190.

Gerlach, G.-F., Clegg, S., and Nichols, W. A., 1988. Characterization of the genes encoding urease activity of *Klebsiella pneumoniae, FEMS Microbiol. Lett.* 50:131–135.

Goodwin, C. S., Armstrong, J. A., and Marshall, B. J., 1986. *Campylobacter pyloridis,* gastritis, and peptic ulceration, *J. Clin. Pathol.* 38:353–365.

Gordon, W. R., Schwemmer, S. S., and Hillman, W. S., 1978. Nickel and the metabolism of urea by *Lemna paucicostata* Hegelm. 6746, *Planta* 140:265–268.

Granick, S., 1937. Urease distribution in plants, *Plant Physiol.* 12:471–486.

Griffith, D. P., Musher, D. M., and Hin, C., 1976. Urease: The primary cause of infection-induced urinary stones, *Invest. Urol.* 13:346–350.

Hasnain, S. S., and Piggott, B., 1983. An EXAFS study of jack bean urease, a nickel metalloenzyme, *Biochem. Biophys. Res. Commun.* 112:279–283.

Hausinger, R. P., 1986. Purification of a nickel-containing urease from the rumen anaerobe *Selenomonas ruminantium, J. Biol. Chem.* 261:7866–7870.

Hawtin, P. R., Delves, H. T., and Newell, D. G., 1991. The demonstration of nickel in the urease of *Helicobacter pylori* by atomic absorption spectroscopy, *FEMS Microbiol. Lett.* 77:51–54.

Hazell, S. L., and Lee, A., 1986. *Campylobacter pyloridis,* urease, hydrogen ion back diffusion, and gastric ulcers, *Lancet* ii:15–17.

Holland, M. A., and Polacco, J. A., 1992. Urease-null and hydrogenase-null phenotypes of a phylloplane bacterium reveal altered nickel metabolism in two soybean mutants, *Plant Physiol.* 98:942–948.

Holland, M. A., Griffin, J. D., Meyer-Bothling, L. E., and Polacco, J. C., 1987. Developmental genetics of the soybean urease isozymes, *Dev. Genet.* 8:375–387.

Hu, L.-T., and Mobley, H. L. T., 1990. Purification and N-terminal analysis of urease from *Helicobacter pylori, Infect. Immun.* 58:992–998.

Huntington, G. B., 1986. Uptake and transport of nonprotein nitrogen by the ruminant gut, *Fed. Proc.* 45:377–383.

Jabri, E., Lee, M. H., Hausinger, R. P., and Karplus, P. A., 1992. Preliminary crystallographic studies of urease from jack bean and from *Klebsiella aerogenes, J. Mol. Biol.* 227:934–937.

Jones, B. D., and Mobley, H. L. T., 1989. *Proteus mirabilis* urease: Nucleotide sequence determination and comparison with jack bean urease, *J. Bacteriol.* 171:6414–6422.

Jose, J., Christians, S., Rosenstein, R., Götz, F., and Kaltwasser, H., 1991. Cloning and expression of various staphylococcal genes encoding urease in *Staphylococcus carnosus, FEMS Microbiol. Lett.* 80:277–282.

Kakimoto, S., Sumino, Y., Akiyama, S.-I., and Nakao, Y., 1989. Purification and characterization of acid urease from *Lactobacillus reuteri, Agric. Biol. Chem.* 53:1119–1125.

Kakimoto, S., Sumino, Y., Kawahara, K., Yamazaki, E., and Nakatsui, I., 1990. Purification and characterization of acid urease from *Lactobacillus fermentum, Appl. Microbiol. Biotechnol.* 32:538–543.

Kinghorn, J. R., and Fluri, R., 1984. Genetic studies of purine breakdown in the fission yeast *Schizosaccharomyces pombe, Curr. Genet.* 8:99–105.

Krueger, R. W., Holland, M. A., Chisholm, D., and Polacco, J. C., 1987. Recovery of a soybean urease genomic clone by sequential library screening with two synthetic oligodeoxynucleotides, *Gene* 54:41–50.

Labigne, A., Cussac, V., and Courcoux, P., 1991. Shuttle cloning and nucleotide sequences of *Helicobacter pylori* genes responsible for urease activity, *J. Bacteriol.* 173:1920–1931.

Larson, A. D., and Kallio, R. E., 1954. Purification and properties of bacterial urease, *J. Bacteriol.* 68:67–73.

Lee, M. H., Mulrooney, S. B., and Hausinger, R. P., 1990. Purification, characterization, and *in vivo* reconstitution of *Klebsiella aerogenes* urease apoenzyme, *J. Bacteriol.* 172:4427–4431.

Lee, M. H., Mulrooney, S. B., Renner, M. J., Markowicz, Y., and Hausinger, R. P., 1992. *Klebsiella aerogenes* urease gene cluster: Sequence of *ureD* and demonstration that four accessory genes (*ureD, ureE, ureF,* and *ureG*) are involved in nickel metallocenter biosynthesis, *J. Bacteriol.* 174:4324–4330.

Lee, M. H., Pankratz, H. S., Wang, S., Scott, R. A., Finnegan, M. G., Johnson, M. K., Ippolito, J. A., Christianson, D. W., and Hausinger, R. A., 1993. Purification and characterization of *Klebsiella aerogenes* UreE protein: A nickel-binding protein that functions in urease metallocenter assembly, *Protein Sci.* 2:1042–1052.

Loyola-Vargas, V., Roman, M. E., Quiroz, J., Oropeza, C., Robert, M. L., and Scorer, K. N., 1988. Nitrogen metabolism in *Canavalia ensiformis* DC. I. Arginase and urease ontogeny, *J. Plant Physiol.* 132:284–288.

Lutz, S., Jacobi, A., Schlensog, V., Böhm, R., Sawers, G., and Böck, A., 1991. Molecular characterization of an operon (*hyp*) necessary for the activity of the three hydrogenase isoenzymes in *Escherichia coli, Mol. Microbiol.* 5:123–135.

Mackay, E. M., and Pateman, J. A., 1980. Nickel requirement of a urease-deficient mutant in *Aspergillus nidulans, J. Gen. Microbiol.* 116:249–251.

Mackay, E. M., and Pateman, J. A., 1982. The regulation of urease activity in *Aspergillus nidulans, Biochem. Genet.* 20:763–776.

Mackerras, A. H., and Smith, G. D., 1986. Urease activity of the cyanobacterium *Anabaena cylindrica, J. Gen. Microbiol.* 132:2749–2752.

Maier, T., Jacobi, A., Sauter, M., and Böck, A., 1993. The product of the *hypB* gene, which is required for nickel incorporation into hydrogenases, is a novel guanine nucleotide-binding protein, *J. Bacteriol.* 175:630–635.

Martin, P. R., and Hausinger, R. P., 1992. Site-directed mutagenesis of the active site cysteine in *Klebsiella aerogenes* urease, *J. Biol. Chem.* 267:20024–20027.

Maslak, P., Sczepanske, J. J., and Parvez, M., 1991. Complexation through nitrogen in copper and nickel complexes of substituted ureas, *J. Am. Chem. Soc.* 113:1062–1063.

McCoy, D. D., Cetin, A., and Hausinger, R. P., 1992. Characterization of urease from *Sporosarcina ureae, Arch. Microbiol.* 157:411–416.

McDonald, J. A., Vorhaben, J. E., and Campbell, J. W., 1980. Invertebrate urease: Purification and properties of the enzyme from a land snail, *Otala lactea, Comp. Biochem. Physiol.* 66B: 223–231.

Mégraud, F., Neman-Simha, V., and Brügmann, D., 1992. Further evidence of the toxic effect of ammonia produced by *Helicobacter pylori* urease on human epithelial cells, *Infect. Immun.* 60:1858–1863.

Meyer-Bothling, L. E., and Polacco, J. C., 1987. Mutational analysis of the embryo-specific urease locus of soybean, *Mol. Gen. Genet.* 209:439–444.

Meyer-Bothling, L. E., Polacco, J. C., and Cianzio, S. R., 1987. Pleiotropic soybean mutants defective in both urease isozymes, *Mol. Gen. Genet.* 209:432–438.

Mobley, H. L. T., and Hausinger, R. P., 1989. Microbial ureases: Significance, regulation, and molecular characterization, *Microbiol. Rev.* 53:85–108.

Mobley, H. L. T., and Warren, J. W., 1987. Urease-positive bacteriuria and obstruction of long-term urinary catheters, *J. Clin. Microbiol.* 25:2216–2217.

Mörsdorf, G., and Kaltwasser, H., 1990. Cloning of the genes encoding urease from *Proteus vulgaris* and sequencing of the structural genes, *FEMS Microbiol. Lett.* 66:67–74.

Mulrooney, S. B., and Hausinger, R. P., 1990. Sequence of the *Klebsiella aerogenes* urease genes and evidence for accessory proteins facilitating nickel incorporation, *J. Bacteriol.* 172:5837–5843.

Mulrooney, S. B., Lynch, M. J., Mobley, H. L. T., and Hausinger, R. P., 1988. Purification, characterization, and genetic organization of recombinant *Providencia stuartii* urease expressed in *Escherichia coli, J. Bacteriol.* 170:2202–2207.

Mulrooney, S. B., Pankratz, H. S., and Hausinger, R. P., 1989. Regulation of gene expression and cellular localization of cloned *Klebsiella aerogenes* (*K. pneumoniae*) urease, *J. Gen. Microbiol.* 135:1769–1776.

Nakano, H., Takenishi, S., and Watanabe, Y., 1984. Purification and properties of urease from *Brevibacterium ammoniagenes, Agric. Biol. Chem.* 48:1495–1502.

Norris, R., and Brocklehurst, K., 1976. A convenient method of preparation of high-activity urease from *Canavalia ensiformis* by covalent chromatography and an investigation of its thiol groups with 2,2'-dipyridyl disulfide as a thiol titrant and reactivity probe, *Biochem. J.* 159:245–257.

Oliveira, L., and Antia, N. J., 1984. Evidence of nickel ion requirement for autotrophic growth of a marine diatom with urea serving as nitrogen source, *Br. Phycol. J.* 19:125–134.

Park, I.-S., and Hausinger, R. P., 1993a. Diethylpyrocarbonate reactivity of *Klebsiella aerogenes* urease: Effect of pH and active site ligands on rate of enzyme inactivation, *J. Prot. Chem.* 12:51–56.

Park, I.-S., and Hausinger, R. P., 1993b. Site-directed mutagenesis of *Klebsiella aerogenes* urease: Identification of histidine residues that appear to function in nickel ligation, substrate binding, and catalysis, *Protein Sci.* 2:1034–1041.

Pechman, K. J., Lewis, B. J., and Woese, C. R., 1976. Phylogenetic status of *Sporosarcina urease, Int. J. Syst. Bacteriol.* 26:305–310.

Pérez-Pérez, G. I., Olivares, A. Z., Cover, T. L., and Blaser, M. J., 1992. Characteristics of *Helicobacter pylori* variants selected for urease deficiency, *Infect. Immun.* 60:3658–3663.

Pérezurria, E., Estrella, M., and Vicente, C., 1986. Function of nickel in the urease activity of the lichen *Evernia prunastri, Plant Sci.* 43:37–43.

Polacco, J. C., 1977. Is nickel a universal component of plant ureases?, *Plant Sci. Lett.* 10:249–255.

Polacco, J. C., and Havir, E. A., 1979. Comparisons of soybean urease isolated from seed and tissue culture, *J. Biol. Chem.* 254:1707–1715.

Polacco, J. C., Krueger, R. W., and Winkler, R. G., 1985. Structure and possible ureide degrading function of the ubiquitous urease of soybean, *Plant Physiol.* 79:794–800.

Polacco, J. C., Judd, A. K., Dybing, J. K., and Cianzio, S. R., 1989. A new mutant class of soybean lacks urease in leaves but not in leaf-derived callus or in roots, *Mol. Gen. Genet.* 217:257–262.

Precious, B. L., Thirkell, D., and Russell, W. C., 1987. Preliminary characterization of the urease and a 96 kDa surface-expressed polypeptide of *Ureaplasma urealyticum, J. Gen. Microbiol.* 133:2659–2670.

Price, N. M., and Morel, F. M. M., 1991. Colimitation of phytoplankton growth by nickel and nitrogen, *Limnol. Oceanogr.* 36:1071–1077.

Rando, D., Steglitz, U., Mörsdorf, G., and Kaltwasser, H., 1990. Nickel availability and urease expression in *Proteus mirabilis, Arch. Microbiol.* 154:428–432.

Rees, T. A., V., and Bekheet, I. A., 1982. The role of nickel in urea assimilation by algae, *Planta* 156:385–387.

Riddles, P. W., Whan, V., Blakeley, R. L., and Zerner, B., 1991. Cloning and sequencing of a jack bean urease-encoding cDNA, *Gene* 108:265–267.

Sabbaj, J., Sutter, V. L., and Finegold, S. M., 1970. Urease and deaminase activities of fecal bacteria in hepatic coma, *Antimicrob. Agents Chemother.* 1970:181–185.

Sakaguchi, K., Mitsui, K., Kobashi, K., and Hase, J., 1983. Photo-oxidation of jack bean urease in the presence of methylene blue, *J. Biochem.* 93:681–686.

Sakaguchi, K., Mitsui, K., Nakai, N., and Kobashi, K., 1984. Amino acid sequence around a cysteine residue in the active center of jack bean urease, *J. Biochem.* 96:73–79.

Salata, C. A., Youinou, M.-T., and Burrows, C. J., 1989. (Template)^2 synthesis of a dinucleating macrocyclic ligand and crystal structure of its dicopper(II) imidazolate complex, *J. Am. Chem. Soc.* 111:9278–9279.

Salata, C. A., Youinou, M.-T., and Burrows, C. J., 1991. Preparation and structural characterization of dicopper(II) and dinickel(II) imidazolate-bridged macrocyclic Schiff base complexes, *Inorg. Chem.* 30:3454–3461.

Samtoy, B., and DeBreukelaer, M. M., 1980. Ammonia encephalopathy secondary to urinary tract infection with *Proteus mirabilis, Pediatrics* 65:294–297.

Saraste, M., Sibbald, P. T., and Wittinghofer, A., 1990. The P-loop: A common motif in ATP- and GTP-binding proteins, *Trends Biochem. Sci.* 15:430–434.

Schneider, J., and Kaltwasser, H., 1984. Urease from *Arthrobacter oxydans,* a nickel-containing enzyme, *Arch. Microbiol.* 139:355–360.

Segal, E. D., Shon, J., and Tompkins, L. S., 1992. Characterization of *Helicobacter pylori* urease mutants, *Infect. Immun.* 60:1883–1889.

Singh, S., 1990. Regulation of urease activity in the cyanobacterium *Anabaena doliolum, FEMS Microbiol. Lett.* 67:79–84.

Singh, S., 1991. Role of nickel and N-starvation in the regulation of urea metabolism in the cyanobacterium *Anacystis nidulans, J. Gen. Appl. Microbiol.* 37:325–330.

Smoot, D. T., Mobley, H. L. T., Chippendale, G. R., Lewison, J. F., and Resau, J. H., 1990. *Helicobacter pylori* urease activity is toxic to human gastric epithelial cells, *Infect. Immun.* 58:1992–1994.

Spears, J. W., and Hatfield, E. E., 1978. Nickel for ruminants. I. Influence of dietary nickel on ruminal urease activity, *J. Anim. Sci.* 47:1345–1350.

Spears, J. W., Smith, C. J., and Hatfield, E. E., 1977. Rumen bacterial urease requirement for nickel, *J. Dairy Sci.* 7:1073–1076.

Stebbins, N., Holland, M. A., Cianzio, S. R., and Polacco, J. C., 1991. Genetic tests of the roles of the embryonic ureases of soybean, *Plant Physiol.* 97:1004–1010.

Sumner, J. B., 1926. The isolation and crystallization of the enzyme urease, *J. Biol. Chem.* 69: 435–441.

Sumner, J. B., and Somers, G. F., 1953. *Chemistry and Methods of Enzymes,* Academic Press, New York, p. 156.

Takishima, K., Mamiya, G., and Hata, M., 1983. Amino acid sequence of a peptide containing an essential cysteine residue of jack bean urease, in *Frontiers in Biochemical and Biophysical Studies of Proteins and Membranes* (T.-Y. Liu, S. Sakakibara, A. N. Schechter, K. Yagi, H. Yajima, and K. T. Yasunobu, eds.), Elsevier, New York, pp. 193–201.

Takishima, K., Suga, T., and Mamiya, G., 1988. The structure of jack bean urease. The complete amino acid sequence, limited proteolysis and reactive cysteine residues, *Eur. J. Biochem.* 175:151–165.

Thirkell, D., Myles, A. D., Precious, B. L., Frost, J. S., Woodall, J. C., Burdon, M. G., and Russell, W. C., 1989. The urease of *Ureaplasma urealyticum, J. Gen. Microbiol.* 135:315–323.

Thompson, J. F., 1980. Arginine synthesis, proline synthesis and related processes, in *The Biochemistry of Plants, A Comprehensive Treatise,* Vol. 5 (P. K. Stumpf and E. E. Conn, eds.), Academic Press, New York, pp. 375–402.

Todd, M. J., and Hausinger, R. P., 1987. Purification and characterization of the nickel-containing multicomponent urease from *Klebsiella aerogenes, J. Biol. Chem.* 262:5963–5967.

Todd, M. J., and Hausinger, R. P., 1989. Competitive inhibitors of *Klebsiella aerogenes* urease. Mechanisms of interaction with the nickel active site, *J. Biol. Chem.* 264:15835–15842.

Todd, M. J., and Hausinger, R. P., 1991a. Reactivity of the essential thiol of *Klebsiella aerogenes* urease. Effect of pH and ligands on thiol modification, *J. Biol. Chem.* 266:10260–10267.

Todd, M. J., and Hausinger, R. P., 1991b. Identification of the essential cysteine residue in *Klebsiella aerogenes* urease, *J. Biol. Chem.* 266:24327–24331.

Visek, W. J., 1972. Effects of urea hydrolysis on cell life-span and metabolism, *Fed. Proc.* 31: 1178–1191.

Vogels, G., and van der Drift, C., 1976. Degradation of purines and pyrimidines by microorganisms, *Bacteriol. Rev.* 40:403–468.

Walker, C. D., Graham, R. D., Madison, J. T., Cary, E. E., and Welch, R. M., 1985. Effects of Ni deficiency on some nitrogen metabolites in cowpeas (*Vigna unguiculata* L. Walp), *Plant Physiol.* 79:474–479.

Walz, S. E., Wray, S. K., Hull, S. I., and Hull, R. E., 1988. Multiple proteins encoded within the urease gene complex of *Proteus mirabilis, J. Bacteriol.* 170:1027–1033.

Wang, S., Lee, M. H., Hausinger, R. P., Clark, P. A., Wilcox, D. E., and Scott, R. A., 1993. Structure of the dinuclear active site of urease. X-ray absorption spectroscopic study of native and 2-mercaptoethanol-inhibited bacterial and plant enzymes, *Inorg. Chem.* (in press).

Waugh, R., and Boxer, D. H., 1986. Pleiotropic hydrogenase mutants of *Escherichia coli* K-12: Growth in the presence of nickel can restore hydrogenase activity, *Biochimie* 68:157–166.

Winkler, R. G., Polacco, J. C., Eskew, D. L., and Welch, R. M., 1983. Nickel is not required for apourease synthesis in soybean seeds, *Plant Physiol.* 72:262–263.

Winkler, R. G., Blevins, D. G., Polacco, J. C., and Randall, D. D., 1987. Ureide catabolism of soybeans. II. Pathway of catabolism in intact leaf tissue, *Plant Physiol.* 83:585–591.

Wu, L.-F., 1992. Putative nickel-binding sites of microbial proteins, *Res. Microbiol.* 143:347–351.

Yamazaki, E., Kurasawa, T., Kakimoto, S., Sumino, Y., and Nakatsui, I., 1990. Characteristics of acid urease from *Streptococcus mitior, Agric. Biol. Chem.* 54:2433–2435.

Zawada, J. W., and Sutcliffe, J. F., 1981. A possible role for urease as a storage protein in *Aspergillus tamarii, Ann. Bot.* 48:797–810.

Zerner, B., 1991. Recent advances in the chemistry of an old enzyme, *Bioorg. Chem.* 19:116–131.

Hydrogenase 4

4.1 Introduction

Many microorganisms possess hydrogenase activity that catalyzes the reversible activation of hydrogen according to the following reaction:

$$H_2 \leftrightarrow 2H^+ + 2e^-$$

In vivo, hydrogenases function primarily in one direction, either consuming hydrogen to provide electrons for reductive metabolism or evolving hydrogen gas as a means to remove excess electrons. "Uptake" hydrogenases are present in the aerobic "Knallgas" bacteria and the anaerobic methanogenic archaea and sulfate-reducing bacteria, among others. These microbes make ATP via electron transport phosphorylation by coupling hydrogen oxidation to reduction of O_2, CO_2, and sulfate, respectively [reviewed by Adams *et al.* (1981)]. In addition, a special role for an uptake hydrogenase is hypothesized for nitrogen-fixing bacteria. The enzyme is thought to recover some of the energy wasted in the ATP-dependent hydrogen production that accompanies nitrogenase activity. By contrast, hydrogenase-catalyzed hydrogen production is observed in many fermentative reactions, including those of strict anaerobes such as *Clostridium* spp. and the formate-hydrogenlyase reaction of facultative enteric bacteria (Adams *et al.,* 1981). Hydrogen production and consumption may occur in the same microorganism, depending on the growth conditions. Furthermore, Odom and Peck (1984) have proposed that these reactions may occur simultaneously in sulfate-reducing bacteria, where cytoplasmic hydrogen gas production may be coupled energetically to periplasmic hydrogen consumption in a process called hydrogen cycling to provide extra energy to the cell. In addition, evidence for hydrogen cycling involving cytoplasmic and membrane-associated hydrogenases has been obtained in acetate-degrading methanogens (Kemner, 1993). Clearly, hydrogenases play varied and essential roles in bacteria.

Based on their metal contents, three broad classes of hydrogenase enzymes have been demonstrated: those that contain only iron in the form of iron–sulfur clusters, those that contain both nickel and iron, and those without nickel or iron–sulfur clusters. The former class has been reviewed recently by Adams (1990) and will not be further described. Similarly, the third class, found uniquely in methanogenic archaea, where it functions as a hydrogen-, forming methylenetetrahydromethanopterin dehydrogenase (Zirngibl *et al.,* 1992), will not be discussed further. Rather, this chapter will focus on the NiFe-hydrogenases and a selenium-containing subclass termed the NiFeSe-hydrogenases. Selected aspects of nickel-containing hydrogenases from methanogenic, sulfate-reducing, photosynthetic, and hydrogen-oxidizing bacteria have been reviewed recently (Bastian *et al.,* 1988; Cammack *et al.,* 1988; Fauque *et al.,* 1988; Moura *et al.,* 1988; Przybyla *et al.,* 1992). Below, I describe the general enzyme properties of NiFe- and NiFeSe-hydrogenases, detail the structural, spectroscopic, and oxidation–reduction characteristics of the nickel active site in these enzymes, compare these properties of the enzyme nickel metallocenter with those of nickel model compounds, and summarize molecular biological evidence that relates to identification of the nickel ligands and to metallocenter biosynthesis.

4.2 Evidence for Nickel in Hydrogenases

The clearest evidence for the presence of nickel in a hydrogenase comes from direct analysis of the isolated enzyme. Nickel-containing hydrogenases have been purified and characterized from each of the microorganisms listed in Table 4-1. In addition, nickel-dependent chemolithotrophic growth or nickel-dependent hydrogenase activity has been observed in the microbes listed in Table 4-2. Clearly, nickel is intimately connected to hydrogenase activity in a wide range of microorganisms.

The first hint that hydrogenases may contain nickel was the observation that chemolithoautotrophic growth of two strains of *Alcaligenes* sp. (formerly *Hydrogenomonas*) requires nickel ion in the medium (Bartha and Ordal, 1965). This initial observation was followed by the isolation of a nickel-dependent *Oscillatoria* sp. (van Baalen and O'Donnell, 1978) and the studies of Tabillion *et al.* (1980), who demonstrated the nickel-dependent chemolithoautotrophic growth of five strains of *Alcaligenes eutrophus,* two strains of *Xanthobacter autotrophicus, Pseudomonas flava,* and *Arthrobacter* sp. strains 11X and 12X. Shortly thereafter, a direct nickel requirement for hydrogenase activity was noted in *A. eutrophus* (Friedrich *et al.,* 1981) and *Rhodobacter capsulatus* (formerly *Rhodopseudomonas capsulata*) (Takakuwa and Wall, 1981). These and the other studies listed in Table 4-2 only demonstrate a requirement for nickel in lithotrophic growth or hydrogenase activity; they do not establish whether nickel is a component of hydrogenase in these microorganisms.

About the same time that the early nickel dependence studies were reported, Lancaster (1980) reported a novel electron paramagnetic resonance (EPR) spectroscopic signal with g values of 2.30, 2.23, and 2.02 in cell extracts from *Methanobacterium bryantii*. He boldly hypothesized that this signal arose from octahedrally coordinated nickel(III) and subsequently corroborated the paramagnetic nickel assignment (Lancaster, 1982) by demonstrating the presence of clearly defined hyperfine structure in extracts prepared from cells grown in the presence of ^{61}Ni, an isotope which possesses a nuclear spin of $\frac{3}{2}$. The disparate studies demonstrating nickel-dependent hydrogenase activity in many microbes and identifying nickel as the source of an unusual methanogen EPR signal came together with the demonstration that hydrogenase from *Methanobacterium thermoautotrophicum* strain Marburg contains nickel (Graf and Thauer, 1981) and with the finding that hydrogenase-bound nickel gives rise to the novel EPR spectrum (Albracht *et al.*, 1982). A flurry of reports have since described the presence and properties of nickel in purified hydrogenases, as described below.

4.3 General Properties of Nickel-Containing Hydrogenases

Table 4-1 summarizes the properties of purified nickel-containing hydrogenases. These enzymes have not been characterized to the same extent; thus, their cellular locations and many of the values for subunit M_r and metal content should be considered tentative. As an example of the problem in distinguishing between soluble and membrane-bound enzyme, the *Methanobacterium formicicum* hydrogenase responsible for reducing coenzyme F_{420} can be isolated from cell-free extracts as if it is a soluble protein (Jin *et al.*, 1983); however, this enzyme was later shown to be associated with membranes both by sucrose density-gradient centrifugation (Baron *et al.*, 1987) and by immunocytochemical localization (Baron *et al.*, 1989). Similarly, complications arise in determining subunit sizes and numbers. Harker *et al.* (1984) and Seefeldt *et al.* (1987) have demonstrated the sensitivity to proteolysis for the smaller subunits from *Bradyrhizobium japonicum* and *Rhodobacter capsulatus* hydrogenases. Proteolysis may also occur in other microorganisms. In the case of *Chromatium vinosum*, early studies utilized a protein that apparently contained a single band of M_r 60,000 (Albracht *et al.*, 1983), whereas more recent work with a different strain yielded a two-subunit species (van der Zwaan *et al.*, 1990; Coremans *et al.*, 1992b). The calculation of metal stoichiometry is dependent on obtaining reliable values for both the metal and the protein concentration. Atomic absorption spectroscopy, plasma

Table 4-1. Compositions of Purified Nickel-Containing Hydrogenases

Microorganism	Enzyme identification[a]	Subunit M_r	Metal and cofactor content[b]	Reference(s)
Alcaligenes eutrophus	Soluble, NAD-reducing	63,000, 56,000, 30,000, 26,000	2 Ni, 16 Fe, 1 FMN/205,000	Friedrich et al. (1982); Schneider and Piechulla (1986)
Alcaligenes eutrophus	Membrane-bound	67,000, 31,000	0.6–0.7 Ni, 7–9 Fe/98,000	Friedrich et al. (1982); Schneider et al. (1983)
Alcaligenes latus	Membrane-bound	67,000, 34,000	0.54 Ni, 1.7 Fe/101,000	Pinkwart et al. (1983)
Azotobacter vinelandii	Membrane-bound	67,000, 31,000	0.68 Ni, 6.6 Fe/98,000	Seefeldt and Arp (1986)
Bradyrhizobium japonicum[c] (free-living)	Membrane-bound	65,000, 33,000	NR[d,e]	Harker et al. (1984); Stults et al. (1984, 1986a)
Bradyrhizobium japonicum[c] (soybean nodules)	Membrane-bound	64,000, 35,000	0.59 Ni, 6.5 Fe/99,000	Arp (1985)
Chloroflexus aurantiacus	Soluble	35,000	0.7 Ni/35,000	Serebryakova et al. (1990)
Chromatium vinosum	Soluble	60,000, 34,000	1.8 Ni, 4 Fe/60,000 (or 1 Ni, 9–12 Fe/94,000)	Albracht et al. (1983); van der Zwaan et al. (1990)
Desulfovibrio africanus	Soluble	65,000, 27,000	0.9 Ni, 12 Fe/92,000	Nivière et al. (1986)
Desulfovibrio baculatus (DSM 1743)	Periplasm	49,000, 26,000	0.69 Fe, 9.25 Fe, 0.66 Se/75,000	Teixeira et al. (1987)
Desulfovibrio baculatus (DSM 1743)	Cytoplasm	54,000, 27,000	0.54 Ni, 7.7 Fe, 0.56 Se/81,000	Teixeira et al. (1987)
Desulfovibrio baculatus (DSM 1743)	Membrane-bound	62,000, 27,000	0.9 Ni, 10.3 Fe, 0.86 Se/89,000	Teixeira et al. (1987)
Desulfovibrio baculatus[f] (Norway)	Membrane-bound	59,800, 27,100	6 Fe/58,000[g]	Lalla-Maharajh et al. (1983)
Desulfovibrio baculatus[f] (Norway)	Soluble	56,400, 28,600	0.45–0.8 Ni, 5–10 Fe, 0.45–0.71 Se/85,000	Rieder et al. (1984)
Desulfovibrio desulfuricans (ATCC 27774)	Soluble, enzyme 1	NR	0.6 Ni, 7.8 Fe/77,600	Krüger et al. (1982)
Desulfovibrio desulfuricans (ATCC 27774)	Soluble, enzyme 2	NR	0.6 Ni, 10.9 Fe/75,500	Krüger et al. (1982)
Desulfovibrio fructosovorans	Periplasm	60,000, 28,500	0.9 Ni, 11 Fe/88,500	Hatchikian et al. (1990)
Desulfovibrio gigas	Periplasm	62,000, 26,000	0.64–0.95 Ni, 12 Fe/89,500	Cammack et al. (1982); LeGall et al. (1982); Teixeira et al., 1983

Organism	Location/type	Subunit molecular weights	Metal content	Reference
Desulfovibrio multispirans	Cytoplasm	58,000, 24,500	0.9 Ni, 11 Fe/82,500	Czechowski et al. (1984)
Desulfovibrio salexigens	Periplasmic	62,000, 36,000	1.03 Ni, 12–15 Fe, 1.08 Se/98,000	Teixeira et al. (1986)
Desulfovibrio vulgaris (Hildenborough)	Membrane-bound, enzyme 3[h]	86,000, 45,000	0.3 Ni, 4 Fe, 0.3 Se/131,000	Lissolo et al. (1986)
Desulfovibrio vulgaris (Miyazaki)	Membrane-bound enzyme	NR	1.0 Ni, 13 Fe/89,000	Asso et al. (1992)
Escherichia coli	Membrane-bound, enzyme 1	64,000, 35,000	0.64 Ni, 12.2 Fe/200,000	Sawers and Boxer (1986)
Escherichia coli	Membrane-bound, enzyme 2	61,000, 30,000	3.1 Ni, 12.5 Fe/180,000	Ballantine and Boxer (1986)
Methanobacterium formicicum	Soluble, F_{420}-reducing	42,600, 34,000, 23,500	3 Ni, 20 Fe/170,000[i]	Jin et al. (1983); Nelson et al. (1984)
Methanobacterium formicicum	Soluble, non-F_{420}-reducing	48,000, 38,000	0.49 Ni, 9.4 Fe, 1.6 Cu, 0.8 Zn/70,000	Adams et al. (1986); Jin et al. (1983)
Methanobacterium thermoautotrophicum (Marburg)	Soluble, non-F_{420}-reducing	52,000, 38,000	0.8 Ni, variable Fe/60,000	Graf and Thauer (1981); Coremans et al. (1989)
Methanobacterium thermoautotrophicum (ΔH)	Soluble, F_{420}-reducing	47,000, 31,000, 26,000	0.6–0.7 Ni, 13–14 Fe, 0.8–0.9 FAD/115,000	Kojima et al. (1983); Fox et al. (1987)
Methanobacterium thermoautotrophicum (ΔH)	Soluble, non-F_{420}-reducing	52,000, 40,000	NR[g]	Kojima et al. (1983)
Methanococcus vannielii	Soluble, F_{420}-reducing	42,000, 35,000, 27,000	1 Ni, 9–10 Fe, 1.9 Se, 2 FAD/170,000	Yamazaki (1982); Stadtman (1990)
Methanococcus voltae	Soluble, F_{420}-reducing	55,000, 45,000, 37,000, 27,000	0.63 Ni, 4.5 Fe, 0.66 Se, 0.85 FAD/105,000	Muth et al. (1987)
Methanosarcina barkeri (DSM 800)	Soluble, F_{420}-reducing	60,000	0.6–0.8 Ni, 8–10 Fe, 1 FMN/60,000	Fauque et al. (1984)
Methanosarcina barkeri (DSM 800)	Membrane-bound, non-F_{420}-reducing	57,000, 35,000	0.5 Ni, 8 Fe/98,000	Kemner (1993)
Methanosarcina barkeri (DSM 804)	Membrane-bound, non-F_{420}-reducing	46,000, 39,000, 28,000, 25,000, 23,000, 21,000, 20,000, 16,000, 15,000	1.2 Ni, 16–18 Fe, 0.14 FAD/233,000	Heiden et al. (1993)

(continued)

Table 4-1. (Continued)

Microorganism	Enzyme identification[a]	Subunit M_r	Metal and cofactor content[b]	Reference(s)
Methanosarcina strain Göl	Membrane-bound, non-F_{420}-reducing	60,000, 40,000	0.8 Ni, 15.11 Fe/100,000	Deppenmeier *et al.* (1992)
Methanospirillum hungatei	F_{420}-reducing	50,700, 32,900, 30,700	0.7–0.8 Ni/81,400[e]	Sprott *et al.* (1987); Choquet and Sprott (1991)
Nocardia opaca 1b	Soluble, NAD-reducing	64,000, 56,000, 31,000, 27,000	3.8 Ni, 13.6 Fe, 1 FMN/178,000	Schneider *et al.* (1984a)
Paracoccus denitrificans	Membrane-bound	64,000, 34,000	0.6 Ni, 7.3 Fe/100,000	Knüttel *et al.* (1989)
Pyrococcus furiosus	Cytoplasm	46,000, 27,000, 24,000	0.98 Ni, 31 Fe/185,000	Bryant and Adams (1989)
Pyrodictium brockii	Membrane-bound	66,000, 45,000	8.7 Fe/118,000[e]	Pihl and Maier (1991)
Rhodobacter capsulatus[j]	Membrane-bound	67,000, 31,000	0.2 Ni, 3.6 Fe/65,000	Colbeau *et al.* (1983); Colbeau and Vignais (1983); Seefeldt *et al.* (1987)
Thermodesulfobacterium mobile	Soluble	55,000, 15,000	0.6–0.7 Ni, 7–8 Fe/70,000	Fauque *et al.* (1992)
Thiocapsa roseopersicina	Membrane-bound	64,000 and/or 34,000	1.4 Ni, 6 Fe/98,000	Gogotov (1986); Zorin (1986); Cammack *et al.* (1989a)
Wolinella succinogenes[k]	Membrane-bound	60,000, 30,000	1 Ni, 11–20 Fe/100,000	Unden *et al.* (1982)
Wolinella succinogenes	Membrane-bound	60,000, 30,000, 23,000	Ni, Fe, 0.8 cytochrome b/113,000	Dross *et al.* (1992)

[a] The identity of the enzymes refers to the cellular location and, in some cases, the electron acceptor used (e.g., NAD or the methanogen coenzyme F_{420}) or a name given to that enzyme species.
[b] Numbers of metals or cofactors per M_r of the minimal form of the enzyme.
[c] Formerly *Rhizobium japonicum.*
[d] NR, Not reported.
[e] Nickel was shown to be present by comigration of hydrogenase activity and ^{63}Ni on native gels or by co-chromatography of ^{63}Ni and enzyme.
[f] Formerly *Desulfovibrio desulfuricans* (Norway).
[g] Nickel shown to be present by EPR spectroscopy.
[h] Two additional nickel-containing membrane-bound hydrogenases are present as well as an iron-only periplasmic hydrogenase.
[i] FAD also present.
[j] Formerly *Rhodopseudomonas capsulata.*
[k] Formerly *Vibrio succinogenes.*

Table 4-2. Additional Microorganisms That Exhibit Nickel-Dependent
Hydrogenase Activity

Microorganism	Reference(s)
Anabaena spp. strains CA and 1F	Xiankong *et al.* (1984)
Anabaena cylindrica	Daday and Smith (1983); Daday *et al.* (1985)
Anabaena variabilis	Almon and Böger (1984)
Anacystis nidulans	Papen *et al.* (1986)
Arthrobacter sp. strains 11X and 12X	Tabillion *et al.* (1980)
Azospirillum brasilense	Pedrosa and Yates (1983)
Azospirillum lipoferum	Pedrosa and Yates (1983)
Azotobacter chroococcum	Partridge and Yates (1982)
Bradyrhizobium japonicum	Klucas *et al.* (1983)
Chlorella emersonii	Soeder and Engelmann (1984)
Chloroflexus aurantiacus	Drutschmann and Klemme (1985)
Derxia gummosa	Pedrosa and Yates (1983)
Frankia strain KB5	Sellstedt and Smith (1990)
Mastigocladus laminosus	Pederson *et al.* (1986)
Methylosinus trichosporium OB3b	Chen and Yoch (1987)
Nostoc muscorum	Rai and Raizada (1986)
Pseudomonas flava	Tabillion *et al.* (1980)
Pseudomonas saccharophila	Barraquio and Knowles (1989)
Rhodospirillum rubrum	Bonam *et al.* (1988); Koch *et al.* (1992)
Xanthobacter autotrophicus	Nakamura *et al.* (1985); Tabillion *et al.* (1980)

emission spectroscopy, and colorimetric methods each have characteristic limitations for measurement of metal ions, and colorimetric methods for protein determination are often suspect and depend, in part, on the protein composition. For example, when the iron content for two iron-only hydrogenases from *Clostridium pasteurianum* was reevaluated by using quantitative amino acid analysis, increases from 12 to 20 Fe/hydrogenase I and from 8 to 14 Fe/hydrogenase II were observed (Adams *et al.,* 1989). Despite these caveats, comparison of the hydrogenase properties listed in Table 4-1 allows one to make several generalizations about these nickel-containing enzymes, as described below.

Although many of the purified hydrogenases are membrane-bound, others have been isolated from the periplasm or cytoplasm (and from a less well defined soluble fraction). In the case of *Desulfovibrio baculatus* DSM 1743, hydrogenases with similar properties have been isolated from all three compartments (Teixeira *et al.,* 1987). It is unclear whether these three enzymes result from three distinct gene products or if some type of differential processing of the same gene product may take place. For *Escherichia coli,* genetic and biochemical evidence is clearly consistent with the presence of at least three distinct sets of gene products (Sawers *et al.*, 1985). The record for the most hydrogenase isoenzymes in the same cell appears to belong to *Desulfovibrio vulgaris* Hildenborough, which possesses at least three nickel-containing, membrane-bound enzymes as well as a soluble iron-only hydrogenase (Lissolo *et al.*, 1986).

All of the membrane-bound and many of the soluble hydrogenases possess a large and a small subunit which range in M_r from 49,000 to 86,000 and 15,000 to 45,000, respectively. Furthermore, many of these dimeric enzymes are immunologically related (Kovacs *et al.*, 1989). The finding that nickel, monitored as radioactive [63]Ni, is associated with an immunoprecipitin arc that contains the large subunit, but not the small subunit, of *E. coli* hydrogenase isoenzyme 1 is consistent with localization of nickel to the M_r 64,000 peptide (Sawers and Boxer, 1986). Additionally, the two subunits of isoenzyme 2 of *E. coli* hydrogenase could be dissociated, and the inactive, M_r 61,000 subunit was found to contain nickel, again determined by using [63]Ni (Ballantine and Boxer, 1986). In contrast, Szökefalvi-Nagy *et al.* (1990) have proposed, on the basis of particle-induced X-ray emission studies, that nickel binds primarily to the small subunit and most of the iron binds to the large subunit of *Thiocapsa roseopersicina* hydrogenase. These results require corroboration because they were carried out on isolated subunits that were boiled in sodium dodecyl sulfate (SDS) for 5 min prior to polyacrylamide gel electrophoresis. Molecular biological evidence, combined with biophysical analyses, is most consistent with localization of nickel to the large subunit, as described in Section 4.5. A general model for the two-subunit hydrogenase core which may be present in all hydrogenases is shown in Fig. 4-1.

Several of the hydrogenases listed in Table 4-1 possess more than two subunits; however, the same basic two-subunit core may be present in each. For example, the four-subunit enzyme from *Nocardia opaca* 1b was dissociated by preparative gel electrophoresis in the absence of nickel ions into two types of subunit dimers: one dimer contains subunits of M_r 64,000 and 31,000, and the other possesses subunits of M_r 56,000 and 27,000 (Schneider *et al.*, 1984b). [The *N. opaca* hydrogenase is unique in requiring nickel ion in the assay buffer (Aggag and Schlegel, 1974). This metal ion apparently binds to and stabilizes the native, NAD-reducing tetrameric structure (Schneider *et*

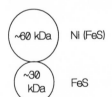

Figure 4-1. Model illustrating the two-subunit core found in nearly all hydrogenases. The large subunit is generally considered to be the nickel-binding region, whereas most of the iron is probably bound by the small subunit.

al., 1984a)]. The larger dimer possesses some iron and flavin mononucleotide (FMN), but has no hydrogenase activity. Notably, the smaller dimer retains hydrogenase activity, when assayed by using viologen dyes, and possesses 4 Fe and 2 Ni/83,000. Thus, the second dimer resembles several of the two-subunit hydrogenases. A model illustrating the subunit composition of the *N. opaca* hydrogenase is provided in Fig. 4-2.

Alcaligenes eutrophus hydrogenase possesses a similar four-subunit structure, and its subunits are related to those of *N. opaca* both immunologically and in amino-terminal sequences (Zaborosch *et al.*, 1989). Furthermore, when the enzyme is stored in the absence of nickel ions, a partial dissociation of the tetramer is observed (Johannssen *et al.*, 1991). Addition of nickel ions leads to subunit reassociation but does not restore hydrogenase activity. A mutant strain of *A. eutrophus* was shown to produce only the M_r 56,000 peptide of the four subunits, and this inactive protein was found to contain nickel and iron (Hornhardt *et al.*, 1986). Thus, the nickel location is consistent with the pattern seen in the two-subunit enzymes. Antibodies raised against the larger subunit from the *A. eutrophus, Bradyrhizobium japonicum,* or *Thiocapsa roseopersicina* dimeric, membrane-bound enzymes cross-react with the M_r 56,000 peptide of the four-subunit soluble *A. eutrophus* enzyme (Schneider and Piechulla, 1986; Kovács *et al.*, 1989). Although no immunological cross-reactivity was noted for the *A. eutrophus* peptides using antibodies directed against the small subunit of various two-subunit hydrogenases, sequence comparison of the *A. eutrophus* hydrogenase operon demonstrates that homology does exist (Tran-Betcke *et al.*, 1990). Furthermore, as will be discussed in Section 4.5, sequence data also are consistent

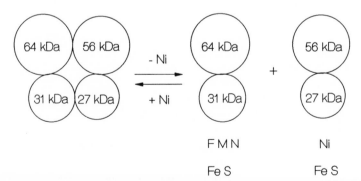

Figure 4-2. Two roles for nickel in *Nocardia opaca* hydrogenase. In the absence of stabilizing nickel, the NAD-reducing, tetrameric enzyme dissociates to form two dimers. One dimer, analogous to the two-subunit core shown in Fig. 4-1, contains nickel and is capable of reducing viologen dyes. The other dimer contains flavin mononucleotide (FMN) and is likely to be the site of interaction for NAD.

with the three-subunit enzymes in Table 4.1 retaining the basic two-subunit core.

Most of the enzymes that contain more than two subunits also were found to possess a flavin, which, depending on the enzyme, may be either flavin mononucleotide (FMN) or flavin adenine dinucleotide (FAD). One obvious hypothesis is that a separate, flavin-containing subunit is present in these enzymes. The flavins probably serve as an interface between the single electron transfer centers within the enzyme and the requisite two-electron transfer to the substrates for these enzymes, namely, NAD (in *A. eutrophus* or *N. opaca*) or coenzyme F_{420} (in methanogens). The *Methanosarcina barkeri* (DSM 800) F_{420}-reducing hydrogenase could possibly be an exception to this generalization because it was suggested to consist of a single subunit (Fauque *et al.*, 1984); however, the subunit number and sizes are confounded by the presence of bands at $\sim$50–70 kDa, $\sim$120–140 kDa, and $\sim$200 kDa after denaturing gel electrophoresisis. Each of the other nickel-containing F_{420}-reducing hydrogenases listed in Table 4-1, as well as F_{420}-reducing hydrogenases not assayed for nickel after purification from *Methanosarcina barkeri* (strain Fusaro) (Fiebig and Friedrich, 1989) and *Methanococcus jannaschii* (Shah and Clark, 1990), possess three or more subunits and probably contain flavins. As an extreme case of subunit complexity, the membrane-bound hydrogenase from *Methanosarcina barkeri* (DSM 804) has been shown to exist in a complex of nine polypeptides (Heiden *et al.*, 1993). In addition to exhibiting hydrogenase activity as assayed with various chemical dyes, this enzyme complex is capable of catalyzing reduction of the disulfide CoM-S-S-HTP (see Chapter 6) using hydrogen gas. The authors speculated that seven components of the complex include a three-subunit hydrogenase (containing Ni, FeS centers, and FAD), a three-subunit heterodisulfide reductase, and a cytochrome *b* that may function in electron transfer between these two enzymes.

As an alternative to a flavin-containing subunit, a three-subunit form of *Wolinella succinogenes* hydrogenase was shown to possess cytochrome *b* (Dross *et al.*, 1992). This cofactor may be important for interaction with the quinone electron acceptor of this protein. DNA sequence analysis (Section 4.5) has revealed that similar proteins are encoded in membrane-associated hydrogenase gene clusters of *Alcaligenes eutrophus, Azotobacter chroococcum, Azotobacter vinelandii, Desulfovibrio vulgaris, Escherichia coli, Rhizobium leguminosarum, Rhodobacter capsulatus,* and *Rubrivivax gelatinosus* among others. Hence, quinones may serve as electron acceptors in these species as well. Furthermore, the membrane association of these proteins may arise, at least in part, through protein–protein interaction with the autologous membrane-bound cytochromes.

As described above, the values for metal content in the listed hydrogenases must be viewed with suspicion. Nevertheless, all of these enzymes do contain nickel and iron, and certain hydrogenases contain selenium. In several cases, the evidence for nickel involves comigration of enzyme activity and ^{63}Ni during native gel electrophoresis, co-chromatography of enzyme and ^{63}Ni radioisotope, or EPR spectroscopic analysis, rather than direct metal analysis. Nonheme iron is present in all of these proteins as iron–sulfur clusters. Although iron content was not reported for *Methanospirillum hungatei* hydrogenase, the enzyme is brown in color (Sprott *et al.*, 1987), consistent with the presence of this type of metallocenter. The number and types of iron–sulfur clusters may vary among the proteins. In the best characterized example, the enzyme from *Desulfovibrio gigas,* the presence of two [4Fe-4S] and one [3Fe-4S] clusters has been established (Teixeira *et al.*, 1983). Selenium is found in each of the three *Desulfovibrio baculatus* (DSM 1743) isoenzymes (Teixeira *et al.*, 1987), the soluble hydrogenases from *Desulfovibrio baculatus* strain Norway (formerly *Desulfovibrio desulfuricans*) (Rieder *et al.*, 1984), *Methanococcus vannielii* (Yamazaki, 1982), and *Methanococcus voltae* (Muth *et al.*, 1987), the periplasmic enzyme from *Desulfovibrio salexigens* (Teixeira *et al.*, 1986), and two of the membrane-bound isoenzymes from *Desulfovibrio vulgaris* Hildenborough (Lissolo *et al.*, 1986). In each case, the selenium is probably present as selenocysteine, a modified amino acid which is encoded by a UGA codon (see Section 4.5 on hydrogenase structure). In addition, selenium increases hydrogenase expression in *Bradyrhizobium japonicum* and is incorporated into the enzyme (Boursier *et al.*, 1988). In contrast to that in the other hydrogenases, however, the selenium is not incorporated as selenocysteine, but rather it is present in a labile, undefined form (Hsu *et al.*, 1990).

The presence of nickel in this wide range of hydrogenases raises questions concerning its role in hydrogen activation. Functional analysis of the nickel center has involved a variety of spectroscopic methods, as detailed in the following section.

4.4 Characterization of the Hydrogenase Nickel Center

A variety of spectroscopic and physical methods have been used to probe the nickel center in hydrogenases. Selected key findings from these studies will be summarized in this section. It is important to note, however, that for some hydrogenases the results obtained by using these techniques are dependent not only on the conditions under which the sample was studied, but also on the history of the protein. Thus, it is necessary to briefly discuss isolation and activation of these enzymes.

4.4.1 Inactive and Active Forms of Hydrogenase

Purified hydrogenases are inhibited or irreversibly inactivated by oxygen (Seefeldt and Arp, 1989), yet many of these enzymes are routinely purified aerobically in an inactive form that is subsequently activated by reductive procedures (e.g., Fox *et al.*, 1987; Hatchikian *et al.*, 1978; Schneider and Schlegel, 1976; Stults *et al.*, 1986a; van Heerikhuizen *et al.*, 1981). The activation process has been best studied for *Desulfovibrio gigas* hydrogenase, where it was shown early on that at least two steps are involved: removal of oxygen, followed by reduction of the enzyme (Berlier *et al.*, 1982). More recently, two forms of oxidized enzyme were shown to exist. These forms exhibit great differences in their rates of reductive activation and have been termed "unready" and "ready" (Fernandez *et al.*, 1985). In work done primarily with the unready state, the rate or extent of activation was shown to depend both on pH and reduction potential of the medium, with a midpoint reduction potential of -360 to -310 mV, versus the normal hydrogen electrode, at pH 7.0 (Fernandez *et al.*, 1984; Lissolo *et al.*, 1984; Mege and Bourdillon, 1985). In the following discussion, evidence is summarized that shows that the two inactive states and various forms of the active enzyme have very different spectroscopic properties. A working model for the redox states of *D. gigas* hydrogenase is illustrated in Fig. 4-3. This model will serve as a useful reference with regard to the EPR spectroscopic results in the next section. Note, however, that other hydrogenases may have different numbers of metallocenters with corresponding differences in the complexity of redox states (Cammack *et al.*, 1986). Furthermore, many of the published midpoint potential values obtained during redox titrations must be considered tentative because redox mediators have been found to affect the oxidation–reduction behavior of the hydrogenase metallocenters (Coremans *et al.*, 1992a).

4.4.2 EPR Spectroscopy of Hydrogenases

The seminal studies of Lancaster clearly established the utility of EPR spectroscopy as a probe of nickel in biological systems. He attributed a novel EPR signal ($g = 2.30$, 2.23, and 2.02) arising from aerobic extracts of *Methanobacterium bryantii* cells to Ni(III) (Lancaster, 1980) and later used samples prepared in the presence of ^{61}Ni to verify this assignment on the basis of hyperfine broadening of the signal (Lancaster, 1982). Immediately afterward, Albracht *et al.* (1982) applied this spectroscopic method to partially purified samples of *Methanobacterium thermoautotrophicum* strain Marburg hydrogenase, and LeGall *et al.* (1982) and Cammack *et al.* (1982) used this technique to study purified *Desulfovibrio gigas* hydrogenase. These and many other

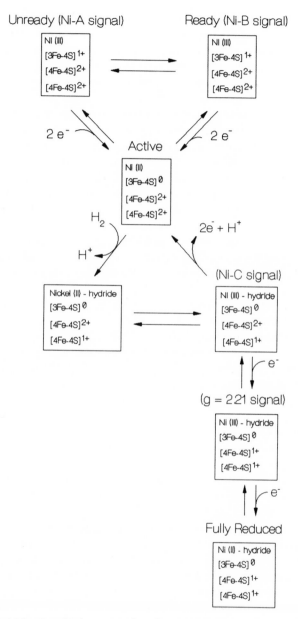

Figure 4-3. Model for the oxidation states of the four metallocenters in *Desulfovibrio gigas* hydrogenase. The nickel(III)-hydride and nickel(II)-hydride species have alternatively been considered as nickel(I)-proton and nickel(O)-proton species by some investigators. See text for a detailed discussion. Only the EPR signals that arise from Ni are indicated.

hydrogenases in a wide variety of states have now been probed by EPR spectroscopy. Indeed, examination by EPR spectroscopy is used routinely in the characterization of nickel-containing enzymes (see references in Table 4-1). Note, however, that not all nickel-containing hydrogenases give rise to an EPR signal from nickel at the temperatures routinely used to examine metallocenters (Bryant and Adams, 1989).

I will not attempt to provide an exhaustive summary of the intricate EPR spectroscopic differences observed among hydrogenases (cf. Cammack et al., 1988; Coremans et al., 1992a,b; Fauque et al., 1988; Moura et al., 1988), although some of these differences are likely to be important. Rather, key findings that provide insight into the structure and function of the nickel center in one enzyme—namely, the *D. gigas* enzyme—will be highlighted. I have selected this enzyme because it is the most intensively studied and best understood of the hydrogenases. To some extent, the characteristics of the iron–sulfur centers also must be considered in elucidating the properties of the nickel center. Thus, the following discussion will necessarily include spectroscopic characterization of both the nickel and the three distinct iron–sulfur centers in the protein.

Using *D. gigas* hydrogenase, Cammack et al. (1982) and LeGall et al. (1982) demonstrated that a Ni(III) signal (g = 2.31, 2.23, and 2.02; hereafter referred to as the Ni-A signal) and a $[3Fe-4S]^{1+}$ signal (g = 2.01) are observed for the oxidized enzyme. This form of the enzyme also contains two $[4Fe-4S]^{2+}$ centers (Teixeira et al., 1983). Redox titration of the enzyme yielded pH-dependent midpoint potentials for the Ni-A signal of -145 mV at pH 7.2 and -220 mV at pH 8.5 (Cammack et al., 1982; Teixeira et al., 1983). The Ni-A signal arises from the "unready" state of the enzyme described above (Fig. 4-3).

In addition to the Ni-A signal, the isolated enzyme and the reduced enzyme samples exhibit several other signals, some of which are attributed to nickel on the basis of the observation of hyperfine interactions when the cells were grown in the presence of ^{61}Ni (Moura et al., 1982). One of these signals (g = 2.33, 2.16, and 2.02; hereafter referred to as Ni-B) is present at low intensity in sample as isolated, or this signal is generated by placing reduced samples under argon (i.e., conditions in which the reduced sample is oxidized anaerobically). This signal is also thought to arise from a nickel(III) species, but one which differs from that giving rise to the Ni-A signal either by dissociation of oxygen (Teixeira et al., 1985) or by a change in the coordination state of nickel (Fernandez et al., 1986; Cammack et al., 1987) to yield the "ready" state. The [3Fe-4S] center and the two [4Fe-4S] centers remain oxidized in the state giving rise to the Ni-B signal (Teixeira et al., 1985).

Reduction of the oxidized enzyme by hydrogen gas initially results in loss of the nickel(III) and $[3Fe-4S]^{1+}$ spectral species and formation of a

$g = 12$ species (Cammack *et al.*, 1982; LeGall *et al.*, 1982; Teixeira *et al.*, 1985; Fernandez *et al.*, 1986; Teixeira *et al.*, 1989). Disappearance of the Ni-A or Ni-B signals was thought to occur either through the reduction of nickel(III) to nickel(II) (Cammack *et al.*, 1982) or by the reduction of an [4Fe-4S] cluster, which then couples to the nickel(III) center, yielding an EPR-silent state (Teixeira *et al.*, 1985). Recent studies focusing on the states of the iron–sulfur centers (Huynh *et al.*, 1987; Teixeira *et al.*, 1989) have ruled out the latter hypothesis and proven that the nickel is indeed reduced in this EPR-silent state. The [3Fe-4S] center clearly was shown to be reduced to the zero oxidation state possessing a spin $S = 2$, giving rise to the $g = 12$ signal (Huynh *et al.*, 1987), and the [4Fe-4S] centers remain oxidized in this enzyme state. This form of the enzyme is referred to as the "active" state (Fig. 4-3).

When the reduction potential drops below -270 mV (at pH 8.5) the "active" state is transformed into a species which possesses an EPR signal arising from nickel ($g = 2.19, 2.14$, and 2.02) in what is termed the Ni-C state (Teixeira *et al.*, 1985). The Ni-C state is thought to be formed by an internal electron transfer after hydride binds to nickel(II). Cammack *et al.* (1987) proposed that the nickel(II)-hydride species may be converted to a nickel(I)-proton species with concomitant reduction of a [4Fe-4S] center. Alternatively, Teixeira *et al.* (1989) proposed that the required electron redistribution results in reduction of a [4Fe-4S] center and formation of a nickel(III)-hydride species. The overall oxidation states of these two proposed nickel species are equivalent and cannot be distinguished by EPR spectroscopy; thus, for simplicity, Fig. 4-3 illustrates only the latter case. The evidence does seem to be clear, however, that nickel in the Ni-C state is associated with some species of hydrogen. Following methods developed in the pioneering studies of the *Chromatium vinosum* enzyme (van der Zwaan *et al.*, 1985), illumination of the *D. gigas* Ni-C state was shown to convert it to another species ($g = 2.28, 2.12$, and 2.03) (Cammack *et al.*, 1987). In *C. vinosum* this photolysis exhibits a kinetic isotope effect when carried out in ^2H_2O versus H_2O, consistent with the splitting of a bond between nickel and a hydrogen species (van der Zwaan *et al.*, 1985). Further evidence that nickel is directly involved in hydrogen binding is available from inhibitor studies. Carbon monoxide was shown to competitively inhibit hydrogen binding to the *D. gigas* enzyme, and the CO ligand induces a new EPR signal (Cammack *et al.*, 1987). By using ^{13}CO in more extensive EPR spectroscopic studies with the *C. vinosum* enzyme, van der Zwaan *et al.* (1990) have shown that carbon monoxide binds directly to the nickel, probably as an axial ligand.

Further reduction of the Ni-C state of the *D. gigas* enzyme leads to the disappearance of the Ni-C signal and appearance of a broad $g = 2.21$ signal, followed by the appearance of a fully reduced sample exhibiting no nickel EPR signals (Teixeira *et al.*, 1989). The complex $g = 2.21$ signal in the *D.*

gigas hydrogenase has been suggested to arise from splitting of the Ni-C signal by magnetic dipole interaction with a spin $\frac{1}{2}$ system, such as a nearby reduced [4Fe-4S] center (Cammack *et al.*, 1987; Teixeira *et al.*, 1989), based in part on work with the *C. vinosum* enzyme (Albracht *et al.*, 1984; van der Zwaan *et al.*, 1987). This [4Fe-4S] center must be distinct from that involved in the electron redistribution mentioned above. The oxidation state of the nickel giving rise to the $g = 2.21$ signal would be identical to that in Ni-C. Reduction of this still not fully characterized species would leave the enzyme with a reduced nickel species [nickel(II)-hydride or, less likely, nickel(0)-proton] and fully reduced iron–sulfur centers. There is no evidence that the state exhibiting the $g = 2.21$ signal or the fully reduced state is important in hydrogenase catalysis.

EPR spectroscopy has provided important information concerning the ligands to nickel. With regard to nickel geometry, each of the nickel signals discussed above is consistent with a distorted octahedral, square-pyramidal, or trigonal bipyramidal ligand field. No hyperfine interaction arises from nitrogen ligands, indicating that nitrogen atoms probably do not bind to the nickel in an axial position. By contrast, EPR spectra of *Wolinella succinogenes* hydrogenase enriched in ^{33}S demonstrate hyperfine interaction from this $S = \frac{3}{2}$ nuclear spin, consistent with the presence of at least one sulfur ligand to nickel (Albracht *et al.*, 1986). Analogous studies carried out with ^{77}Se-enriched *Desulfovibrio baculatus* hydrogenase, a selenium-containing enzyme, also demonstrate broadening of the nickel EPR signal, consistent with nickel ligation by selenium in that enzyme (He *et al.*, 1989). Results obtained by EPR spectroscopic analysis of hydrogenase are complemented by the use of alternative spectroscopic methods. The following section summarizes several of these other spectroscopic studies.

4.4.3 Other Spectroscopic Studies of Hydrogenases

X-ray absorption spectroscopy (XAS), a technique that allows one to estimate the number, size, and distance of the ligands to a metal, has been used to characterize the nickel site in four hydrogenases. The nickel site in oxidized F_{420}-reducing *Methanobacterium thermoautotrophicum* enzyme was found to possess approximately three sulfur ligands at 2.25 Å, and no evidence for closer nitrogen or oxygen ligands was observed (Lindahl *et al.*, 1984). The sulfur ligands were proposed to be equatorial, with one or two axial ligands loosely held to nickel. No evidence for nickel–iron interaction was detected. Similarly, the *Desulfovibrio gigas* hydrogenase was examined by XAS both in the oxidized and the reduced forms (Scott *et al.*, 1984). The only significant difference between the two spectra is a 2-eV shift in the spectral edge, consistent with reduction of nickel(III) to nickel(II). In both reduction states, four sulfur atoms are present at 2.20 Å, and no other scatterers could be positively identified. In contrast, no edge shift is observed when the Ni-A, Ni-B, and Ni-C or two EPR-silent states of *Thiocapsa roseopersicina* hydrogenase are compared (Whitehead *et al.,* 1991; Bagyinka *et al.,* 1993). Indeed, the latter investigators suggested that the inability to detect a change in electron density at the nickel site indicates that nickel does not have a redox role in the enzyme. Rather, they proposed that the redox changes must occur at the ligands or at other centers in the protein. Data from the oxidized *T. roseopersicina* samples were originally thought to be consistent with six-coordinate geometry whereas the data for the Ni-C state were thought to be more consistent with trigonal bipyramidal coordination (Colpas *et al.,* 1991). The Ni-C form of the *T. roseopersicina* hydrogenase nickel center was proposed to be bound by 1–3 sulfur atoms at 2.22 Å and 2–4 nitrogen or oxygen atoms at 2.05 Å (Maroney *et al.,* 1990), whereas the Ni-A and Ni-B states are best fit by including an additional Ni—S bond of 2.40 and 2.50 Å, respectively (Whitehead *et al.,* 1991). More recently, Bagyinka *et al.* (1993) concluded that XAS data from all five oxidation states accessible to the enzyme can be fit to a five- or six-coordinate nickel site that possesses 3 ± 1 N/O donors at 2.00 ± 0.06 Å and 2 ± 1 S donors at 2.23 ± 0.03 Å. In addition to the direct ligands to the nickel, Maroney *et al.* (1991) proposed that scattering from the second and third coordination spheres is consistent with the presence of Ni–Fe distances of 4.3 and 6.0 Å. Finally, the active center in the selenium-containing *Desulfovibrio baculatus* hydrogenase (predominantly in the "active" state as typically isolated) offered the unique opportunity to perform both nickel and selenium XAS (Eidsness *et al.,* 1989). The nickel site was shown to be coordinated by 1–2 sulfur atoms at 2.17 Å, 3–4 oxygen or nitrogen atoms at 2.06 Å, and 1 selenium atom at 2.44 Å, whereas the selenium environment includes 1 carbon atom at 2.0 Å and a nickel or iron atom at 2.4 Å. These results are consistent with direct ligation of selenium to nickel in the selenium-containing hydrogenases.

Electron spin echo spectroscopy, a technique that can detect weak interactions between an electron paramagnet and a nucleus containing a small nuclear quadrupole, has been applied to four hydrogenases. Tan *et al.* (1984) demonstrated that the oxidized F_{420}-reducing hydrogenase from *M. thermoautotrophicum* possesses a nitrogen nucleus (^{14}N, nuclear spin of 1) located near but not directly bound to the nickel, whereas no evidence for an analogous nitrogen was observed in the non-F_{420}-reducing hydrogenase from the same microbe. They speculated that the nitrogen may be associated with FAD, present in the former enzyme but lacking in the latter. This suggestion was not supported by subsequent studies carried out with the oxidized *T. roseopersicina* hydrogenase (Cammack *et al.,* 1989b) or the Ni-A and Ni-C states of *D. gigas* hydrogenase (Chapman *et al.,* 1988) where nitrogen coupling was observed despite the absence of flavin in these enzymes. Rather, the observed nitrogen coupling was speculated to arise from a nearby imidazole, perhaps the distal nitrogen of a bound histidine residue. In addition to the spectroscopy carried out using samples dissolved in water, the Ni-A and Ni-C states of the *D. gigas* hydrogenase were examined in ^2H_2O (Chapman *et al.*, 1988). Whereas the paramagnetic nickel in the Ni-A state of the enzyme is not accessible to solvent, deuterium is able to interact with the Ni-C state.

Another method which has been used to examine the exchangeable protons interacting with the *D. gigas* hydrogenase nickel site is electron-nuclear double resonance (ENDOR) spectroscopy (Fan *et al.*, 1991). ENDOR measurements detected two distinct types of exchangeable protons bound to the Ni-C state. One of these protons appears to interact directly with the nickel, whereas the other was suggested to represent a nickel-bound water or hydroxide ion. The detailed type of interaction between nickel and the directly bound proton has not been determined; however, an axial hydride species is thought to be excluded. Three potential species are considered possible: an in-plane hydride, an $X-H$ proton interacting directly with nickel (X = O, S, or N), or hydrogen gas. No nickel-accessible protons were detected in the Ni-A species, consistent with the electron spin echo spectroscopic analyses. In contrast, when the Ni-C form is reoxidized to obtain at least partial Ni-B species, one exchangeable proton is observed. The small coupling in this proton is consistent with nickel-bound water or hydroxide ion (Fan *et al.*, 1991).

Low-temperature magnetic circular dichroism (MCD) spectroscopy has also been used to characterize the hydrogenase nickel center. Temperature-dependent optical transitions that do not arise from iron–sulfur centers have been assigned to nickel in the oxidized enzymes from both *M. thermoautotrophicum* (Johnson *et al.*, 1985) and *D. gigas* (Johnson *et al.*, 1986). Magnetization data demonstrated that the MCD transitions originate from the same ground state as the Ni-A EPR signal. No nickel-dependent MCD transitions were observed in a sample of extensively reduced *D. gigas* hydrogenase. The authors concluded that the reduced nickel is low-spin ($S = 0$) nickel(II).

A final physical method that has been used to probe the hydrogenase nickel center is multifield saturation magnetization measurements. Wang *et al.* (1992) used this approach to examine the nickel(II) site in the active state of *D. baculatus* hydrogenase lacking the [3Fe-4S] cluster found in the *D. gigas* enzyme. They found the nickel center to be diamagnetic, consistent with five-coordinate geometry but inconsistent with octahedral geometry.

Hypothetical structures for the hydrogenase nickel center in the Ni-A/ Ni-B, the active, and the Ni-C states, derived from the results of EPR, XAS, electron spin echo, and ENDOR spectroscopic studies and from saturation magnetization measurements, are illustrated in Fig. 4-4. It is hoped that these models will stimulate further structural studies.

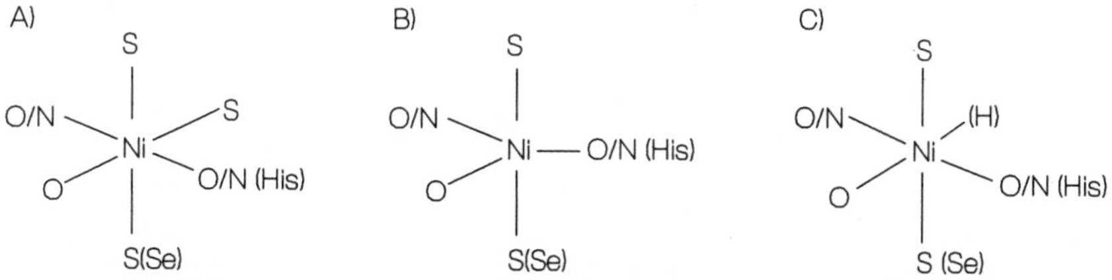

Figure 4-4. Hypothetical structures for the nickel center in hydrogenase. Results from electron paramagnetic resonance, electron spin echo, electron-nuclear double-resonance, and X-ray absorption spectroscopies as well as saturation magnetization measurements were combined in an effort to provide a working model for the hydrogenase nickel center. (A) The Ni-A and Ni-B states of the enzyme are best represented by distorted octahedral geometry in which ~3 cysteinyl sulfur and ~3 oxygen or nitrogen atoms act as nickel ligands. Although nitrogen ligands are not likely to be axially coordinated, equatorial ligation is allowed. Compelling evidence for histidinyl ligands is available for selected hydrogenases. In enzyme isolated from some microbes, one cysteine is replaced by a selenocysteine. (B) The "active" state of the enzyme is represented as a five-coordinate species in which one of the sulfur ligands in the Ni-A or Ni-B states is removed. (C) The Ni-C state of hydrogenase is most consistent with ligation by 2 cysteine thiols and 3 oxygen or nitrogen atoms. In addition, some form of hydrogen, e.g., a hydride, is likely to be bound, but not at an axial position.

An essential component in interpreting results from the spectroscopic studies described above is comparison with results from similar analyses of model compounds. In the following section, I will describe briefly some of the properties of selected model compounds that have been designed to mimic the hydrogenase nickel center.

4.4.4 Comparison to Nickel Model Compounds

Synthetic models of the hydrogenase nickel site have been used to improve our understanding of the chemistry, structure, and function of the enzyme center. This section will highlight some of the recent findings from model studies that shed light on the hydrogenase nickel site.

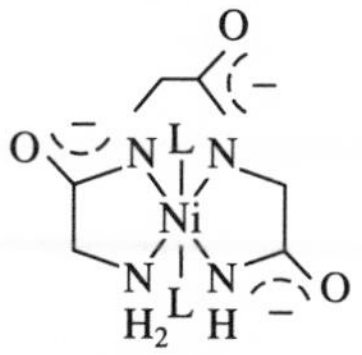

Scheme 4-1

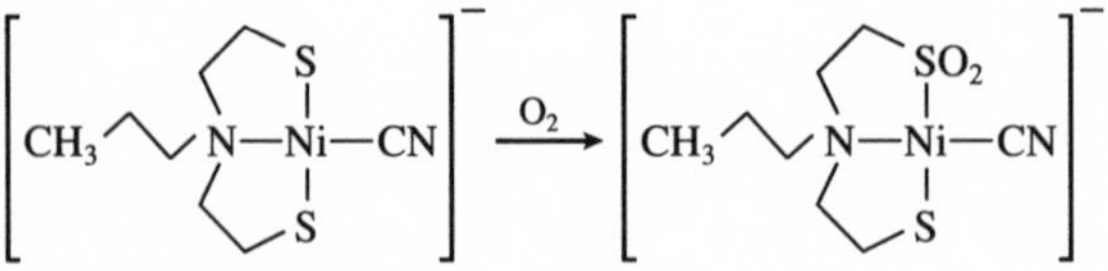

Scheme 4-2

Initial assignment of the EPR signal in oxidized, hydrogenase-containing samples to nickel(III) in a distorted octahedral geometry relied, in part, on similarity to the EPR spectra of nickel(III) model compounds such as nickel(III)-oligopeptides (Scheme 4-1; Lappin *et al.*, 1978).

As it became clear that thiolate ligands are a major contributor to the nickel coordination sphere in hydrogenase (Lindahl *et al.*, 1984; Scott *et al.*, 1984; Albracht *et al.*, 1986), the properties of nickel thiolate compounds came under intense scrutiny. However, early attempts to oxidize nickel(II) thiolates to the nickel(III) thiolate species were unsuccessful due to irreversible oxidation of the ligands, forming disulfides. Studying the chemistry of this reaction, Kumar *et al.* (1989a) suggested that an intermediate in disulfide formation is a species in which the nickel(II) (spin $S = 1$) is antiferromagnetically coupled to a thiyl radical ($S = \frac{1}{2}$) to yield a spin $\frac{1}{2}$ system ($g = 2.17, 2.11,$ and 2.07). From this work, the authors suggested that nickel(III) may not exist with simple alkyl thiol ligands and, furthermore, that the formal nickel(III) states in hydrogenases may actually involve nickel(II) interacting with an oxidized ligand. Although nickel(III)-thiolate model compounds have now been synthesized (discussed below), the existence of a nickel(II)-thiyl radical species in certain states of the hydrogenase has not been excluded. A separate form of thiol ligand oxidation also has been studied by Kumar *et al.* (1989b). Oxygen was demonstrated to react with a nickel(II)-thiolate compound to yield a sulfinato complex (Scheme 4-2). The authors speculated that this ligand oxidation reaction may be related to the unready state of hydrogenase.

The first stable nickel(III)-thiolate complex (Scheme 4-3) was shown by Krüger and Holm (1987) to possess an EPR spectrum with $g = 2.29, 2.11,$ and 2.04. These g values are clearly reminiscent of the hydrogenase signal. Although the nickel(III)/nickel(II) potential of this complex (-35 mV) was

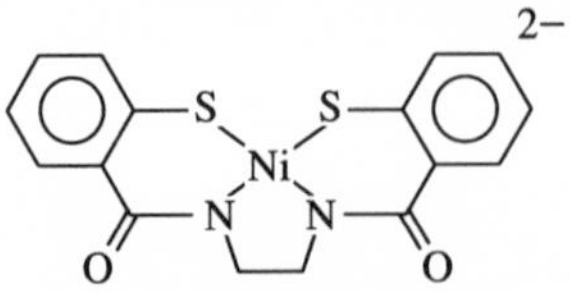

Scheme 4-3

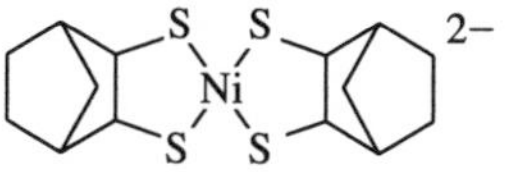

Scheme 4-4

lower than that of any other compound known at that time, it is very high compared to the values observed in hydrogenases. In contrast, Fox *et al.* (1990) demonstrated an exceptionally stable nickel(III) species (Scheme 4-4) with a midpoint potential of -760 mV.

Krüger and Holm (1990) and Krüger *et al.* (1991) have carried out a systematic analysis of the effects of ligand structure on redox potential of model compounds containing at least partial thiolate ligands. They established a requirement for the presence of polarizable, electron-rich donor groups in order to stabilize the low-potential nickel(III) state.

Analysis of XAS edge data for the nickel center in hydrogenase is highly dependent on comparison to analogous data obtained for model compounds. Colpas *et al.* (1991) collected data on six-coordinate (octahedral), five-coordinate (trigonal bipyramidal and square-pyramidal), and four-coordinate (planar and tetrahedral) nickel(II) and nickel(III) complexes containing various ligands. These analyses allowed them to conclude that the data obtained for the Ni-C form of *Thiocapsa roseopersicina* hydrogenase is most consistent with a distorted trigonal bipyramidal geometry (hydrogen cannot be detected by XAS measurements) containing a mixed-donor ligand environment (Colpas *et al.*, 1991). Of the six trigonal bipyramidal compounds examined, two possess single thiolate ligands and the others have no thiolate ligands. To more closely mimic the putative hydrogenase nickel center, pentacoordinated nickel(II) complexes containing two thiolate ligands (e.g., Scheme 4-5) were recently synthesized (Baidya *et al.*, 1991, 1992b). These samples are able to be reduced by dithionite to yield species with EPR spectra (e.g., for one particular complex, $g_{\parallel} = 2.247$ and $g_{\perp} = 2.123$), consistent with

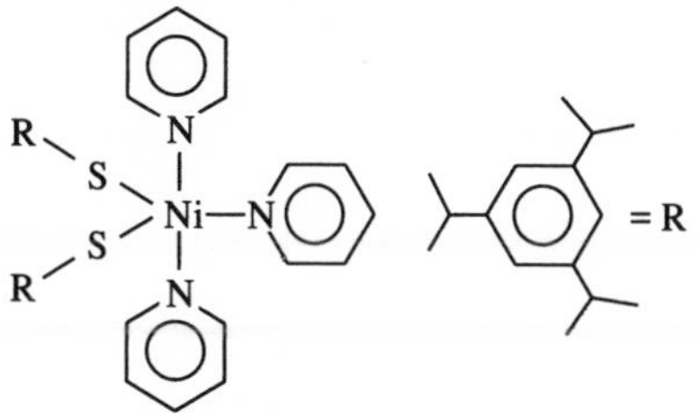

Scheme 4-5

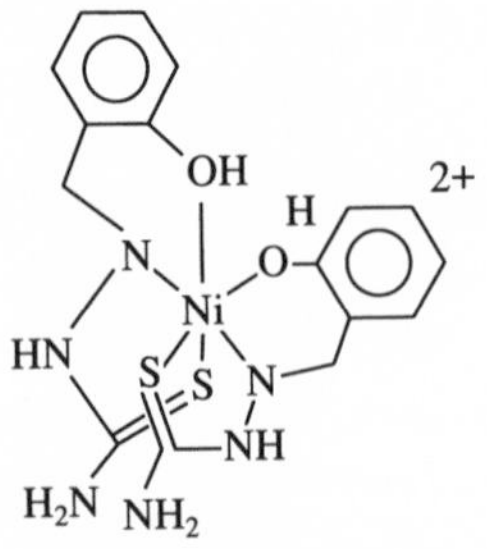

Scheme 4-6

formation of nickel(I). Furthermore, upon treatment with NaBH$_4$, the nickel(II) species generated EPR-active complexes (e.g., g = 2.238, 2.191, and 2.045) that were suggested to arise from nickel(I)-hydride species, although nickel(I)-proton complexes appear to be equally plausible. Furthermore, the reduced complexes reversibly bind CO, causing a shift in the EPR spectra (e.g., g = 2.247, 2.128, and 2.025). The results from these model studies raise the possibility that the Ni-C state in hydrogenases could arise from a nickel(I)-hydride species as an alternative to the nickel(III)-hydride or nickel(I)-proton species described earlier. The same research group also characterized a mononuclear nickel complex that can assume the Ni(I), Ni(II), and Ni(III) states (Baidya *et al.*, 1993) as well as complexes with NiN$_x$Se$_y$ coordination that may model the NiFeSe-hydrogenase active center (Baidya *et al.*, 1992a). Continued analysis of the structure and reactivity of nickel model complexes will facilitate understanding the chemistry of the hydrogenase metallocenter.

Functional modeling of the nickel center in hydrogenase apparently has been achieved (Zimmer *et al.*, 1991) by using the nickel(II) compound shown in Scheme 4-6. Although this crystalline compound is octahedral, the phenol ligands dissociate in solution. Deuterium nuclear magnetic resonance spectroscopy was used to demonstrate 7.5 turnovers of D$_2$/H exchange by this compound in dimethyl sulfoxide/ethanol solution. Emphasizing again the possible role of nickel(I) in hydrogen activation, electrochemical reduction of this model compound results in an air-stable nickel(I) species.

This section is an appropriate place to mention the exceptional chemistry observed in nickel-substituted rubredoxin, although it is not a synthetic model compound. Kowal *et al.* (1988) replaced the native iron with nickel in the *Desulfovibrio gigas*, *D. vulgaris*, and *Clostridium pasteurianum* proteins and characterized their properties by optical and variable-temperature circular dichroism spectroscopies. The metal is bound tetrahedrally by four cysteines, a geometry not thought to be present in hydrogenase; yet Saint-Martin *et al.*

(1988) showed that the nickel-substituted *D. gigas* protein is capable of catalyzing hydrogen production and deuterium–proton exchange reactions. Furthermore, carbon monoxide was shown to inhibit these reactions. Moreover, Mus-Veteau *et al.* (1991) demonstrated that the *D. vulgaris* protein can be oxidized by potassium ferricyanide to provide an EPR signal that resembles those observed in hydrogenases. The significance of this chemistry to the hydrogenase active site remains to be established, but these studies may be useful in better understanding the requirements for hydrogen activation.

In this section, the ligand geometries and chemistries have been described for various nickel model compounds with reference to the hydrogenase nickel center. The next section will describe attempts to identify amino acid residues in the hydrogenase protein that could serve as ligands to the nickel and iron–sulfur centers.

4.5 Hydrogenase Structure: Characterization of Nickel Ligands

Although crystals of *Desulfovibrio gigas* hydrogenase were obtained and found to diffract to 3 Å (Nivière *et al.*, 1987), no high-resolution three-dimensional structure is available for any nickel-containing hydrogenase. Low-resolution structures have been obtained, however, for the enzymes from *Methanobacterium thermoautotrophicum* (Wackett *et al.*, 1987) and *Thiocapsa roseopersicina* (Sherman *et al.*, 1991) by using electron microscopy. The protein structures were shown to be large, ring-shaped complexes, but no information could be obtained concerning the metal centers. In the absence of a three-dimensional structure, this section will focus on the known hydrogenase primary sequences (reviewed by Wu and Mandrand, 1993) in an attempt to deduce structural information, such as the likely metallocenter ligands. Because iron–sulfur centers and the nickel center are thought to possess at least partial cysteine coordination, I will focus on the conserved cysteine residues.

Comparisons of the DNA sequences encoding 20 nickel-containing hydrogenases clearly demonstrate the presence of several highly conserved cysteine-containing regions, as illustrated in Fig. 4-5. In contrast, no sequence similarity was observed between these enzymes and the sequence for an iron-only hydrogenase (Voordouw and Brenner, 1985) or the putative sequence of one subunit from the *Anabaena cylindrica* hydrogenase (Ewart *et al.*, 1990), of unknown metal content. All of the hydrogenase genes encoding two-subunit nickel-containing enzymes appear to be present in operons where the two structural genes are adjacent, usually with the gene for the small subunit preceding that for the large subunit. In addition, other hydrogenase-related genes often are found closely associated with the structural genes (Colbeau *et al.*, 1993; Dernedde *et al.*, 1993; Hidalgo *et al.*, 1990, 1992; Kortlüke *et al.*,

Small subunit

```
              ▼   ▼              ▼           ▼              ▼
A.e.1  (38)..CGCWGCTL..(64)..GACAV..(61)..PGCPP..(28)    [209]
A.e.2  (57)..LECTCCSE..(90)..GSCAS..(29)..PGCPP..(36)..DKCYR..
A.c.   (48)..LECTCCSE..(87)..GSCAS..(29)..PGCPP..(36)..DKSYR..
A.v.   (59)..LECTCCSE..(90)..GSCAS..(29)..PGCPP..(36)..DKCYR..
B.j.   (60)..LECTCCSE..(90)..GACAS..(29)..PGCPP..(36)..DKCYR..
D.b.   (57)..QGCTGCSV..(90)..GTCSA..(33)..PGCPP..(42)..ENCPY..
D.f.   (64)..AECTGCTE..(85)..IICIR..(31)..PGCPP..(35)..DNCPR..
D.g.   (64)..AECTGCSE..(87)..GTCAT..(31)..AGCPP..(35)..DNCPR..
D.v.   (64)..AECTGCSE..(89)..GTCAT..(31)..AGCPP..(36)..EQCPR..
E.c.1  (59)..LECTCCTE..(90)..GTCAS..(29)..PGCPP..(36)..DKCYR..
M.t.1  (12)..GGCSGCHL..(66)..GTCAV..(59)..PGCPP..(26)..EVCPR..
M.f.                                              (?)..EECER..
R.l.   (59)..LECTCCSE..(90)..GACAS..(29)..PGCPP..(36)..DKCYR..
R.c.   (14)..LECTCCSE..(90)..GACAS..(29)..PGCPP..(36)..DKCYR..
R.g.   (56)..LECTCCSE..(89)..GACAS..(29)..PGCPP..(36)..DKCYR..
W.s.   (50)..AECTVVAE..(89)..GSCSS..(67)..............DLCER..
```

Small subunit (cont.)

```
              ▼           ▼               ▼   ▼
A.e.2  (20)..GFCLYKMGCKGPTTYNACST..(14)..HGCIGCSE..(64)  [360]
A.c.   (20)..GYCLYKVGCKGPTSYNACST..(14)..HGCIGCSE..(59)  [344]
A.v.   (20)..GYCLKYVGCKGPTSYNACST..(14)..HGCIGCSE..(59)  [358]
B.j.   (20)..GYCLYKMGCKGPTTYNACST..(14)..HGCIGCSE..(63)  [363]
D.b.   (15)..PGCKAELGCKGPSTYADCAK..(13)..AVCIGCVE..(14)  [315]
D.f.   (20)..GFCLYELGCKGPVTYNNCPL..(13)..HPCLGCSE..(15)  [314]
D.g.   (20)..GYCLYELGCKGPDTYNNCPK..(13)..HPCIACSE..(13)  [314]
D.v.   (20)..GWCLYELGCKGPVTMNNCPK..(13)..HPCIGCSE..(13)  [317]
E.c.1  (20)..GYCLYKMGCKGPTTYNACSS..(14)..HGCLGCAE..(73)  [372]
                PQGL
M.t.1  (23)..DLCLI\/ICMGPATVSICGA...(6)..IPCRGCYG..(63)  [308]
                AQGL
M.f.   (23)..DLCLI\/VCMGPATTSICGA...(6)..IPCQGCYG..(63)[127+N-term.]
R.l.   (20)..GYCLYKMGCKGPTTYNACST..(14)..HGCIGCSE..(61)  [360]
R.c.   (20)..GYCLYKMGCKGPTTYNACST..(14)..HGCIGCSE..(59)  [313+leader]
R.g.   (20)..GFCLYKVGCKGPTTYNACST..(14)..HGARR-SE..(66)  [360]
W.s.   (20)..GYCLYKVGCKGPYTFNNCSK..(14)..HGCIGCSE..(44)  [330]
```

Figure 4-5. Comparison of hydrogenase sequences. Partial sequences located near conserved cysteine residues are compared for the small and large subunits from several hydrogenases. The two-letter species abbreviations stand for sequences derived from the following hydrogenases: A.e.1, soluble enzyme from *Alcaligenes eutrophus* (Tran-Betcke *et al.,* 1990); A.e.2, membrane-bound enzyme from *A. eutrophus* (Kortlüke *et al.,* 1992); A.c., *Azotobacter chroococcum* (Ford *et al.,* 1990); A.v., *Azotobacter vinelandii* (Menon *et al.,* 1990a); B.j., *Bradyrhizobium japonicum* (Sayavedra-Soto *et al.,* 1988); D.b., *Desulfovibrio baculatus* (DSM 1743) (Menon *et al.,* 1987; Voordouw *et al.,* 1989); D.f., *Desulfovibrio fructosovorans* (Rousset *et al.,* 1990); D.g., *Desulfovibrio gigas* (Li *et al.,* 1987; Voordouw *et al.,* 1989); D.v., *Desulfovibrio vulgaris* Miyazaki (Deckers *et al.,* 1990); E.c.1, isoenzyme 1 of *Escherichia coli* (Menon *et al.,* 1990b); E.c.2, isoenzyme 2 of *E. coli* (Przybyla *et al.,* 1992); E.c.3, isoenzyme 3 of *E. coli* (Böhm *et al.,* 1990); M.t.1, non-F$_{420}$-reducing enzyme from *Methanobacterium thermoautotrophicum* (Reeve *et al.,* 1989); M.t.2, F$_{420}$-reducing hydrogenase from *M. thermoautotrophicum* (Alex *et al.,* 1990); M.v., F$_{420}$-reducing enzyme from *Methanococcus voltae* (Halboth and Klein, 1992); M.f., partial sequence from *Methanothermus fervidus* (Steigerwald *et al.,* 1990); R.l., *Rhizobium leguminosarum* (Hidalgo *et al.,* 1990); R.c., *Rhodobacter capsulatus* (Leclerc *et al.,* 1988); R.g., *Rubrivivax gelatinosus* (formerly *Rhodocyclus gelatinosus*) (Uffen *et al.,* 1990); W.c., *Wolinella succinogenes* (Dross *et al.,* 1992). The numbers in parentheses indicate the number of unseen amino acids either at the termini or between the sequences shown. The total number of amino acids for each peptide is shown in square brackets. Any nonconserved cysteine (or histidine in the large subunit) residue is underlined or, in the case of selenocysteine, shown as a bold **U**. NR, Not reported.

Large subunit

```
                ▼   ▼              ▼   ▼
A.e.1   ( 58)..RICGICFV..(388)..DPCLSCATH..(24)  [487]
A.e.2   ( 72)..RICGVCTG..(388)..DPCLACSTH..(15)  [617]
A.c.    ( 71)..RICGVCTG..(498)..DPCLACSTH..(15)  [601]
A.v.    ( 70)..RICGVCTG..(500)..DPCLACSTH..(15)  [602]
B.j.    ( 72)..RICGVCTG..(492)..DPCLACSTH..(15)  [596]
D.b.    ( 68)..RICGVCPT..(414)..DPULGCAVH..(15)  [514]
D.f.    ( 69)..RACGVCTY..(463)..DPCIACGVH..(15)  [564]
D.g.    ( 62)..RACGVCTY..(457)..HDCIACGVH..(15)  [551]
D.v.    ( 78)..RTCGVCTY..(457)..DPCIACGVH..(15)  [567]
E.c.1   ( 74)..RICGVCTG..(491)..DPCLACSTH..(15)  [597]
E.c.2   ( NR)..RICGVCTT..( NR)...DPCMACAVH..(NR)  [NR]
E.c.3   (238)..RVCGICGF..(282)..DPCYSCTDR..(32)  [569]
M.t.1   ( 58)..RICGICDV..(373)..DPCLSCATH..(24)  [472]
M.t.2   ( 60)..RICGVCPI..(309)..DPCLSCATH..(19)  [405]
M.v.    ( 60)..RICGICQA..(317)..DIUASCATH..(17)  [410]
M.f.    ( 58)..RICGICQV..(373)..DPCLSCATH..(24)  [472]
R.l.    ( 72)..RICGVCTG..(492)..DPCLACSTH..(15)  [596]
R.c.    ( 71)..RICGVCTG..(493)..DPCLACSTH..(35)  [616]
R.g.    ( 71)..RICGVCTG..(514)..DPCLACSTH..(15)  [617]
W.s.    ( 59)..RICGVCTY..(474)..DPCIACAVH..(23)  [575]
```

Figure 4-5. (*Continued*)

1992; Menon *et al.,* 1990a,b, 1991, 1992; Reeve *et al.,* 1989; Rey *et al.,* 1992; Rousset *et al.,* 1990). In *Wolinella succinogenes,* one of these accessory genes was shown to encode cytochrome *b* (Dross *et al.,* 1992). This role also has been postulated for the *hupC* gene of *Rhizobium leguminosarum* (Hidalgo *et al.,* 1992), and sequences similar to the *W. succinogenes* cytochrome are present in numerous other hydrogenase gene clusters [e.g., *Alcaligenes eutrophus* (Kortlüke *et al.,* 1992), *Azotobacter chroococcum* (Ford *et al.,* 1990), *Azotobacter vinelandii* (Menon *et al.,* 1990a), *Bradyrhizobium japonicum* (Sayavedra-Soto *et al.,* 1988), *Desulfovibrio vulgaris* (Deckers *et al.,* 1990), *Escherichia coli* hydrogenase 1 (Menon *et al.,* 1990b), *Rhodobacter capsulatus* (Leclerc *et al.,* 1988), and *Rhodocyclus gelatinosus* (Uffen *et al.,* 1990)]. In the case of *A. vinelandii,* deletion of the center of this gene led to a decrease in the rate of hydrogen oxidation with oxygen as the electron acceptor, consistent with a role for this gene product in electron transport (Sayavedra-Soto and Arp, 1992). Several of the small subunits possess leader sequences that are proteolytically removed in the mature proteins. The four *A. eutrophus* subunit genes are also present in an operon, and two of these encode proteins that are similar in sequence to the two-subunit hydrogenases (Tran-Betcke *et al.,* 1990). Of the 14 cysteine residues that are highly conserved in Fig. 4-5, 10 are present in the small subunit and 4 in the large subunit. However, these cysteines are not conserved in all hydrogenases. For example, the *A. eutrophus* small subunit is much smaller than others; hence, it contains only 4 of the 10 conserved cysteine residues. Furthermore, *E. coli* isoenzyme 3 is encoded in an operon containing eight genes, only one of which (that encoding the large subunit) exhibits sequence similarity to other hydrogenase sequences (Böhm *et al.*, 1990). Two other genes in this operon encode peptides rich in cysteine residues that occur in ferredoxin-like sequence motifs appropriate for binding iron–sulfur clusters, and one of these proteins may function as the small subunit in that protein. Similarly, the operon of the F_{420}-reducing hydrogenase of *Methanobacterium thermoautotrophicum* was shown to encode a peptide homologous to the other large subunits and a cysteine-rich peptide (likely to bind iron–sulfur centers) that is probably analogous to the usual small subunit (Alex *et al.*, 1990). Amino-terminal sequence analysis of the isolated small subunit demonstrated that the peptide containing the ferredoxin-like sequence serves as the small subunit in this enzyme. A similar situation exists in the case of the *Methanococcus voltae* hydrogenase; however, in that case the microbe possesses four highly related gene clusters (two that could potentially encode NiFe-hydrogenases and two for NiFeSe-hydrogenases), of which only one is shown in Fig. 4-5 (Halboth and Klein, 1992). Finally, it is interesting to note that the *A. chroococcum, R. gelatinosus,* and *W. succino-genes* sequences are missing one, two, and two of the conserved cysteine residues, respectively.

With regard to nickel ligation, the *Desulfovibrio baculatus* and *Methanococcus voltae* hydrogenase sequences present special insights because the penultimate conserved cysteine codon is replaced by UGA. This codon, which normally functions as a termination signal, encodes a selenocysteine residue in selected proteins (Stadtman, 1990), and these enzymes have been shown to possess selenocysteine (Teixeira *et al.*, 1987; Muth *et al.*, 1987). Moreover, the selenium in the *D. baculatus* enzyme directly coordinates nickel, as shown by XAS studies (Eidsness *et al.*, 1989) and by EPR spectroscopy with [77]Se-substituted enzyme (He *et al.*, 1989). These results raise the reasonable possibility that the cysteine residue at this position in the other hydrogenases may serve as a nickel ligand. In addition to this residue, other conserved cysteines may function as nickel ligands, especially the cysteine located two residues away. Ligation of nickel solely by the large subunit is consistent with evidence presented earlier (Section 4.3). Furthermore, three of the conserved large-subunit cysteines have been individually changed to serine residues by site-directed mutagenesis with complete abolishment of activity for isoenzyme 1 of *E. coli* (Przybyla *et al.*, 1992). These mutagenesis studies do not demonstrate a role for the large-subunit cysteines in nickel ligation, but the data are consistent with the conserved thiol residues possessing essential functions. Evidence for the presence of a histidinyl residue near, and perhaps coordinating, the active-site nickel has been presented (Cammack *et al.*, 1989b; Chapman *et al.*, 1988). The highly conserved histidine two residues beyond the last conserved cysteine in the large subunit may be important in this regard, as suggested by Rousset *et al.* (1990). As shown in Fig. 4-5, this histidine residue is conserved in all enzymes except for isoenzyme 3 of *E. coli*. However, in site-directed mutagenesis studies, conversion of this histidine to a leucine residue in *E. coli* isoenzyme 1 only decreased the activity rather than eliminating hydrogenase function (Przybyla *et al.*, 1992). Other highly conserved large-subunit residues that, when mutated, lead to the loss of hydrogenase activity include the arginine near the amino terminus (see Fig. 4-5) and the aspartic acid and proline residues near the carboxyl terminus (Przybyla *et al.*, 1992). None of these mutated proteins has been purified, and, thus, metal quantitation to assess the amount of nickel has not been possible.

The ligands to the iron–sulfur centers are thought to be primarily or solely cysteine residues, and the small subunit is likely to play the major role in ligation of iron based on its number of conserved cysteines. Although the pattern of cysteines does not follow a precise ferredoxin-like grouping, there is clustering of cysteine residues. Sayavedra-Soto and Arp (1993) reported that substitution of any of four cysteine residues (positions 62, 65, 294, and 297) in the small subunit of the *A. vinelandii* hydrogenase by serine leads to the partial loss of activity; however, they did not purify the mutant proteins and demonstrate that their changes affect the FeS content. Spectroscopic studies of *Desulfovibrio gigas* hydrogenase have been interpreted as indicating the presence of three iron–sulfur clusters: two [4Fe-4S] and one [3Fe-4S] center (Teixeira *et al.*, 1983), requiring a total of 11 cysteine residues. The number of cysteine residues participating in binding nickel is still unclear. If three or fewer are utilized, the 14 conserved cysteines are adequate for binding all metals. Several other hydrogenases are thought to possess fewer iron–sulfur centers and would require fewer cysteine ligands. Indeed, the greatly shortened sequence of the small subunit from *A. eutrophus* hydrogenase would preclude the two-subunit core from binding three iron–sulfur clusters. Tran-Betcke *et al.* (1990) have suggested that only two iron–sulfur centers may be bound to this protein—a [4Fe-4S] and [3Fe-4S] center, requiring only seven ligands. In summary, the sequence comparisons highlight the location of several putative metal-binding ligands. These residues will continue to be appropriate targets for site-directed mutagenesis analysis; however, the true picture of which residues bind each metallocenter must await a crystal structure.

This section has focused primarily on the structural genes for hydrogenase. As described below, however, additional genes are required for proper assembly or activation of the hydrogenase enzyme, and at least some of these accessory genes function in nickel metallocenter biosynthesis.

4.6 Biosynthesis of the Hydrogenase Nickel Center

The biosynthesis of active hydrogenase is currently best understood from a genetics standpoint, with the most detailed analyses available in *Escherichia coli.* Unfortunately, because its hydrogenase genes have been studied by so many investigators, the same names have been used for different loci and genes. To minimize confusion I will adopt the nomenclature suggested by Lutz *et al.* (1991): *E. coli* hydrogenase isoenzyme 1 is encoded by the *hya* operon, which contains six open reading frames (Menon *et al.*, 1990b), isoenzyme 2 is encoded by the poorly characterized *hyb* locus, and isoenzyme 3 is encoded in the *hyc* operon of eight genes (Böhm *et al.*, 1990). Furthermore, *E. coli* possesses the *hyp* operon (so named because mutations in this operon affect all three *hy*drogenases in a *p*leiotropic manner) and the *nik* operon (involved in nickel transport), which do not encode any hydrogenase structural gene (Lutz *et al.*, 1991; Wu *et al.*, 1991). A map of the *E. coli* chromosome highlighting these regions is shown in Fig. 4-6. In this section, the roles, focusing on nickel metabolism, for hydrogenase-related accessory genes in the *hyp*, *nik*, *hya*, *hyb*, and *hyc* operons in *E. coli* will be described, and then the evidence for analogous genes in other microorganisms will be briefly reviewed.

4.6.1 Hydrogenase Accessory Genes in *E. coli*

The *hyp* operon contains six genes (*hypABCDE* and *fhlA*) that have been partially characterized by deletion analysis (Lutz *et al.*, 1991; Jacobi *et al.*, 1992). The *fhlA* gene, encoding a regulatory protein that controls expression of the formate dehydrogenase gene that functions in the formate-hydrogenase lyase complex, will not be further discussed. The *hypA* gene has a positive regulatory function for both the *hyp* operon and the *hyc* operon, while acting as a repressor of the *hyb* operon. The *hypC* gene is essential for generation of hydrogenase 3 activity and enhances the activity levels of hydrogenase 1. The three remaining genes (*hypB*, *hypD*, and *hypE*) are essential for the synthesis

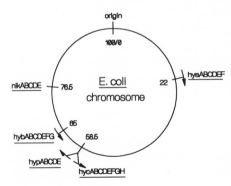

Figure 4-6. *Escherichia coli* chromosome map indicating the positions for gene clusters related to hydrogenase nickel metabolism. The three nickel-containing hydrogenase isoenzymes are encoded by genes within *hya*, *hyb*, and *hyc* operons. The *hyp* gene cluster encodes at least one protein that is related to nickel metabolism and required for obtaining each of the active hydrogenases. The *nik* gene cluster (previously termed *hydC*) encodes a high-affinity nickel transport system.

of all three *E. coli* hydrogenases (Jacobi *et al.,* 1992). Of greatest interest regarding nickel metabolism is the *hypB* gene. Mutations in this gene (previously referred to as *hydB* or *hydE*) were shown to be complemented by high concentrations of nickel (0.5–0.6 m*M*) in the medium (Waugh and Boxer, 1986; Chaudhuri and Krasna, 1987), and it was initially suggested that this gene may encode a protein that is required for nickel transport or for facilitating incorporation of nickel into the hydrogenase. The ability of an *hypB* mutant containing a plasmid encoding the urease operon to synthesize an active urease enzyme demonstrates that nickel transport is not affected in the mutant (Jacobi *et al.,* 1992). Thus, the HypB protein appears to function directly or indirectly in nickel delivery to the hydrogenase apoproteins. The *hypB* gene product has recently been purified and characterized (Maier *et al.,* 1993). The protein was shown to bind guanine nucleotides and was able to hydrolyze GTP. The authors suggested that HypB-dependent GTP hydrolysis may be required for nickel incorporation into hydrogenase. Although the precise roles for the five *hyp* genes remain unknown, the presence of each is required for generating the three hydrogenase activities and for successful processing of the large subunits from the three hydrogenases as visualized by changes in migration behavior on denaturing gel electrophoresis; that is, deletions of any *hyp* gene led to more slowly migrating hydrogenase peptides, and the effects were reversed by providing appropriate plasmid subclones (Lutz *et al.,* 1991; Jacobi *et al.,* 1992). Additional insight into the roles of the *E. coli hyp* genes is available from analysis of homologous genes in other organisms, as discussed in Section 4.6.2.

A less well characterized locus, previously termed *hydC,* possesses several properties that are analogous to those of *hypB.* Namely, *hydC* mutants are defective in all three *E. coli* hydrogenases, and this mutation can be overcome by 0.5 m*M* nickel (Wu and Mandrand-Berthelot, 1986). This locus, located at 77 min in the *E. coli* chromosome, is distinct from the *hyp* or *hyc* operons at 58–59 min (Lutz *et al.,* 1991), from the *hya* operon at 21 min (Böhm *et al.,* 1990), and from the *hyb* operon at 65 min. Stoker *et al.* (1989) proposed that this locus is a *cis*-acting regulatory gene based on its inability to be complemented in *trans;* however, Wu *et al.* (1989) suggested that the former group may have studied a second gene near *hydC* which regulates its expression. The finding of insignificant cellular nickel levels in *hydC* mutants and the ability to suppress this mutation by growth in magnesium-limited medium was taken as evidence that *hydC* encodes a specific transport system for nickel (Wu *et al.,* 1989). This proposal has been confirmed by cloning and sequence analysis (Wu *et al.,* 1991). Five genes are present and exhibit significant homology to other transport operons. The authors proposed that the name of *hydC* be revised to the *nikABCDE* genes. These genes only provide nickel to the enzyme and are not true hydrogenase genes.

In contrast to the above cases, no direct role in nickel metabolism has been identified for any of the ancillary genes in the *hya, hyb,* and *hyc* operons. The nonstructural hydrogenase genes associated with *E. coli* isoenzyme 1 include *hyaC, hyaD, hyaE,* and *hyaF* (Menon *et al.,* 1990b). Deletion analysis has shown that *hyaA* through *hyaE* are required to obtain functional hydrogenase isoenzyme 1, but, except for the hydrogenase structural genes, their detailed roles are unknown. In low-nickel medium, the inclusion of *hyaF* increases levels of this hydrogenase isozyme to wild-type levels, whereas enhancement of activity is not observed in nickel-supplemented growth medium. Thus, *hyaF* may have a role in facilitating nickel incorporation into this enzyme (Menon *et al.,* 1991). Furthermore, examination of the size and location of hydrogenase subunits in the deletion mutants allowed Menon *et al.* (1991) to conclude that the *hyaDEF* genes are involved in some type of hydrogenase subunit processing events, similar to that described for the *hyp* genes, that leads both to an increased mobility of the large subunit for hydrogenase 1 on denaturing gels and membrane insertion. The structural subunit of isoenzyme 3 is encoded by the *hycE* gene (Böhm *et al.,* 1990). This operon also encodes two proteins (*hycB* and *hycF* genes) that possess ferredoxin-like sequences, a protein (*hycG*) with homology to a chloroplast electron transport protein, and two very hydrophobic proteins (*hycC* and *hycD*) that resemble subunits of NADH-ubiquinol oxidoreductases. Sauter *et al.* (1992) carried out a mutational analysis of the *hyc* operon and showed that *hycH* is required for processing of the hydrogenase into the mature form. Further studies are needed to determine whether *hycH* or any of the *hya, hyb,* or *hyc* hydrogenase-related genes participate in nickel metabolism.

4.6.2 Hydrogenase Accessory Genes in Other Microorganisms

Several other microbial hydrogenases are encoded in operons that include more than the hydrogenase structural genes. The following paragraphs summarize the evidence that some of these genes have roles in hydrogenase activation.

In *Alcaligenes eutrophus,* the soluble and membrane-bound hydrogenases are encoded by the *hox* (hydrogen *ox*idation) gene cluster [reviewed by Friedrich (1990)]. This ~100-kilobase pair (kbp) region of DNA includes the *hoxS* locus encoding the four-subunit soluble hydrogenase (*hoxFUYK;* Tran-Betcke *et al.,* 1990) near one end, the *hoxP* locus encoding the two-subunit membrane-bound hydrogenase and at least eight other genes (*hoxZMLOQRTV;* Kortlüke *et al.,* 1992) at the other end, and a middle region that includes regulatory (*hoxAX*) and pleiotropic (*hoxBCDE*) genes (Eberz and Friedrich, 1991; Kortlüke and Friedrich, 1992; Dernedde *et al.,* 1993) and a nickel transport (*hoxN*) gene (Eberz *et al.*, 1989; Eitinger and Friedrich, 1991). Functions for several of the membrane-bound hydrogenase accessory genes have been suggested, and sequence relationships to hydrogenase genes in other microorganisms have been identified (Kortlüke *et al.*, 1992; Dernedde *et al.*, 1993). For example, *hoxZ* is likely to encode a cytochrome based on its predicted sequence similarity to homologous genes including that of *Wolinella succinogenes* (Dross *et al.*, 1992). Related in sequence to *E. coli hyaD* (Menon *et al.*, 1991), *hoxM* mediates attachment of hydrogenase to the membrane as shown by immunogold localization studies (Kortlüke *et al.*, 1992). Although its function is unknown, *hoxL* is related to *hypC*. The *hoxO* and *hoxQ* genes are homologous to *hyaE* and *hyaF*, found to be involved in maturation of *E. coli* hydrogenase 1 (Menon *et al.*, 1991). A rubredoxin is apparently encoded by the *hoxR* gene. Finally, *hoxT* and *hoxV* of the *hoxP* locus have unknown roles. The pleiotropic gene region has been renamed to *hyp* because sequence analysis revealed the presence of genes homologous to *E. coli hypA*, *hypB*, *hypC*, *hypD*, and *hypE* (Dernedde *et al.*, 1993). These investigators further demonstrated that *hypB* and *hypD* were needed for incorporating nickel into the soluble and membrane-bound hydrogenases. In separate studies, the *hypD* and *hypE* genes were shown to be essential for maturation of both subunits of the membrane-bound hydrogenase as visualized by Western blots after denaturing gel electrophoresis (Kortlüke and Friedrich, 1992). The final point to be made about hydrogenase biosynthesis in *A. eutrophus* concerns deletion mutants that were used to define the *hoxN* locus (Eberz *et al.*, 1989). These mutants were shown to require higher concentrations of nickel (0.5 μM) for hydrogenase activity than the wild-type species. Furthermore, the nickel requirement is exacerbated by the presence of high concentrations of magnesium. In addition, the rate of ^{63}Ni uptake by the mutant cells is diminished compared to the wild-type level. Finally, the activity levels of a different nickel-containing enzyme, urease, are also diminished in the *hoxN* mutants. The authors concluded that the *hoxN* locus is involved in a high-affinity nickel transport system of *A. eutrophus* (Eberz *et al.*, 1989). More recently, the *hoxN* gene was subcloned, and its sequence was determined (Eitinger and Friedrich, 1991). The sequence of this nickel transport gene, shown to encode a hydrophobic peptide containing five probable transmembrane regions, exhibits no homology to any other genes including various hydrogenase-related genes and the *E. coli nik* genes.

Homologs of *hypB*, one of the three *hyp* genes that are required for activation of all three *E. coli* hydrogenases, are present in hydrogenase gene clusters of several organisms: ORF4 in *Rhodobacter capsulatus* (Xu and Wall, 1991), *hypB* (formerly *hoxB*) in *A. eutrophus* (Kortlüke and Friedrich, 1992; Jacobi *et al.*, 1992; Dernedde *et al.*, 1993), *hoxL* in *Azotobacter vinelandii*

(Chen and Mortenson, 1992), *hypB* in *Azotobacter chroococcum* (Tibelius *et al.*, 1993) and *hypB* of *Rhizobium leguminosarum* (Rey *et al.*, 1992; Rey *et al.*, 1993), among others. A role for the products of these genes in donating nickel to the hydrogenase apoproteins of these microbes is reasonable but remains to be demonstrated. More surprisingly, these hydrogenase accessory genes share ~25% sequence identity at the protein level to *ureG*, a gene required for nickel incorporation into urease (Lee *et al.*, 1992; Wu, 1992; see Chapter 3). This result is consistent with the presence of an ancestral nickel incorporation gene which diverged to give rise to *hypB* and *ureG* in the hydrogenase and urease systems.

As described above, the *hypD* and *hypE* genes of *E. coli* (Lutz *et al.*, 1991) and *A. eutrophus* (Kortlüke and Friedrich, 1992) appear to be important in processing of hydrogenase subunits to yield peptides of different electrophoretic mobility. These genes have homologs in several other microorganisms including *Azotobacter chroococcum* (Du *et al.*, 1992), *A. vinelandii* (Chen and Mortenson, 1992), *Rhizobium leguminosarum* (Rey *et al.*, 1992; Rey *et al.*, 1993) and *Rhodobacter capsulatus* (Colbeau *et al.*, 1993). In the case of *A. vinelandii,* the processing step has been at least partially characterized. Gollin *et al.* (1992) purified the large subunit of the *A. vinelandii* hydrogenase and subjected it to electrospray mass spectrometry to show a M_r of 64,942 ± 20; that is, 1663 less than that calculated from the gene sequence. The amino-terminal sequence was found to be intact except for removal of the methionine terminus; hence, the investigators concluded that the subunit must undergo carboxyl-terminal processing. Interestingly, the size of the peptide removed (15 amino acids) would place the cleavage site immediately after the conserved histidine shown in the sequence of the large subunit (Fig. 4-5) or three residues after the last conserved cysteine, a likely nickel ligand. Whether nickel plays a role in the processing event, perhaps by an oxidative cleavage process (see Chapter 9), is unknown. A genetic region associated with hydrogenase processing also has been identified in *Bradyrhizobium japonicum* (Fu and Maier, 1993), and a well-defined cleavage site immediately following the conserved histidine residue has been demonstrated in one of the hydrogenases from *Methanococcus voltae* (Sorgenfrei *et al.*, 1993). In the latter enzyme, the gene for the large subunit is split into two separate open reading frames. Cleavage of the peptide encoded by the second region results in a 25-residue peptide that contains a portion of the nickel-binding site.

In *B. japonicum*, Fu and Maier (1992) have demonstrated the ability to incorporate nickel into hydrogenase apoenzyme *in vivo,* but not *in vitro.* They suggested that cell disruption-labile factors are required for metal ion incorporation. Furthermore, by construction of appropriate mutants, they identified a 0.6-kbp chromosomal region that appears to encode a factor involved in this process. Whether this locus involves the cell disruption-labile component is unclear. Furthermore, the sequence relationship of this region to hydrogenase-related genes in other organisms is unknown.

This section has summarized evidence for the presence of accessory genes that are involved in nickel transport, incorporation of nickel into hydrogenase apoenzyme, or participation in other unknown roles to yield active enzyme. As described below, the involvement of nickel in hydrogenase genetics is further complicated for some microorganisms that possess a mechanism to sense nickel concentration and transcriptionally regulate hydrogenase.

4.7 Nickel-Dependent Transcriptional Regulation of Hydrogenase

Using *Bradyrhizobium japonicum,* Stults *et al.* (1986b) demonstrated by immunological methods that hydrogenase protein does not accumulate in cells that are grown in the absence of nickel. Increasing nickel levels correlate with increasing cross-reactive material and with increasing hydrogenase activity. These results were interpreted to suggest that nickel is involved in the regulation of hydrogenase biosynthesis. Furthermore, when nickel is added simultaneously with rifampicin, an inhibitor of transcription, no hydrogenase activity is observed. This finding led to the suggestion that hydrogenase-specific mRNA is not present under these conditions; that is, regulation was proposed to be at the transcriptional level (Stults *et al.,* 1986b). More recent work from the same lab (Kim and Maier, 1990) supports this hypothesis by demonstrating higher levels of hydrogenase-specific mRNA in cells grown in the presence of $5 \mu M$ nickel than in those grown in the absence of nickel. Using a gene fusion with the β-galactosidase structural gene, they identified the upstream region of DNA that interacts with the postulated nickel-dependent regulatory protein. Amazingly, the 50-bp *cis*-acting region involved in nickel-dependent regulation is also required for regulation by hydrogen and oxygen (Kim *et al.,* 1991). A nickel metabolism-related locus located within the *B. japonicum* hydrogenase gene cluster has been identified (Fu and Maier, 1991) and was suggested to be involved in incorporating nickel into the *trans*-acting regulatory factor as well as into hydrogenase (discussed in the Section 4.6.2). These experiments provide the first clear evidence for nickel-dependent transcriptional regulation of any protein. Further studies are needed to identify and characterize the regulatory protein involved in this process.

Although other microorganisms may also possess mechanisms for nickel-dependent transcriptional regulation of hydrogenase, the evidence is much less clear. For example, Doyle and Arp (1988) showed that expression of cross-reactive material corresponding to the *Alcaligenes latus* hydrogenase protein appears to require nickel. They concluded that their results are consistent with either a role for nickel in regulation or rapid degradation of nickel-free enzyme in the cell. In contrast to the above studies, an absence of nickel-dependent regulation has been shown for several microbial hydrogenases. For example, Friedrich *et al.* (1984) and Chen and Yoch (1987) have shown that hydrogenase apoenzyme does accumulate in *Alcaligenes eutrophus* or *Methylosinus trichosporium* when grown in the absence of nickel. Furthermore, by using β-galactosidase fusions analogous to those described above, Zinoni *et al.* (1984) have presented evidence that nickel does not control hydrogenase transcription in the case of *E. coli* isoenzyme 3. The evolutionary significance of nickel-dependent transcriptional regulation of hydrogenase is currently unclear, and an understanding of it will require an improved knowledge of the scope of microbes which possess this regulatory mechanism.

4.8 Perspective

This chapter has summarized the evidence that nickel is an essential component of many hydrogenases, described the properties of the nickel center, discussed genetic studies relating to hydrogenase biosynthesis, and indicated that nickel concentration can regulate hydrogenase expression in selected microorganisms. Although the properties of the nickel center have been extensively analyzed, the precise structure and role of this metal ion remain elusive. The number and identification of the nickel ligands is still in question for the various forms of the enzyme. The number of enzyme intermediates that occur during catalysis and the corresponding oxidation–reduction states of nickel remain controversial. The number and roles of accessory proteins that are required for nickel metallocenter biosynthesis are not clear. Finally, the physical process responsible for nickel-dependent transcriptional regulation of certain hydrogenase genes is unknown. Clearly, these and other exciting questions regarding nickel metabolism related to this central bacterial enzyme require further exploration.

References

Adams, M. W. W., 1990. The structure and mechanism of iron-hydrogenases, *Biochim. Biophys. Acta* 1020:115–145.

Adams, M. W. W., Mortenson, L. E., and Chen, J.-S., 1981. Hydrogenase, *Biochim. Biophys. Acta* 594:105–176.

Adams, M. W. W., Jin, S.-L. C., Chen, J.-S., and Mortenson, L. E., 1986. The redox properties and activation of the F_{420}-non-reactive hydrogenase of *Methanobacterium formicicum, Biochim. Biophys. Acta* 869:37–47.

Adams, M. W. W., Eccleston, E. C., and Howard, J. B., 1989. Iron–sulfur clusters of hydrogenase I and hydrogenase II of *Clostridium pasteurii, Proc. Natl. Acad. Sci. USA* 86:4932–4936.

Aggag, M., and Schlegel, H. G., 1974. Studies on a gram-positive hydrogen bacterium, *Nocardia opaca* 1b. III. Purification, stability and some properties of the soluble hydrogen dehydrogenase, *Arch. Microbiol.* 100:25–39.

Albracht, S. P. J., Graf, E.-G., and Thauer, R. K., 1982. The EPR properties of nickel in hydrogenase from *Methanobacterium thermoautotrophicum, FEBS Lett.* 140:311–313.

Albracht, S. P. J., Kalkman, M. L., and Slater, E. C., 1983. Magnetic interaction of nickel (III) and the iron–sulfur cluster in hydrogenase from *Clostridium vinosum, Biochim. Biophys. Acta* 724:309–316.

Albracht, S. P. J., van der Zwaan, J. W., and Fontijn, R. D., 1984. EPR spectrum at 4, 9, and 35 GHz of hydrogenase from *Chromatium vinosum.* Direct evidence for spin–spin interaction between Ni(III) and the iron–sulfur cluster, *Biochim. Biophys. Acta* 766:245–258.

Albracht, S. P. J., Kröger, A., van der Zwaan, J. W., Unden, G., Böcher, R., Mell, H., and Fontijn, R. D., 1986. Direct evidence for sulfur as a ligand to nickel in hydrogenase: An EPR study of the enzyme from *Wolinella succinogenes* enriched in ^{33}S, *Biochim. Biophys. Acta* 874: 116–127.

Alex, L. A., Reeve, J. N., Orme-Johnson, W. H., and Walsh, C. T., 1990. Cloning, sequence determination, and expression of the genes encoding the subunits of the nickel-containing 8-hydroxy-5-deazaflavin reducing hydrogenase from *Methanobacterium thermoautotrophicum, Biochemistry* 29:7237–7244.

Almon, H., and Böger, P., 1984. Nickel dependent uptake hydrogenase activity in the blue-green alga *Anabaena variabilis, Z. Naturforsch., C* 39:90–92.

Arp, D. J., 1985. *Rhizobium japonicum* hydrogenase: Purification to homogeneity from soybean nodules, and molecular characterization, *Arch. Biochem. Biophys.* 237:504–512.

Asso, M., Guigliarelli, B., Yagi, T., and Bertrand, P., 1992. EPR and redox properties of *Desulfovibrio vulgaris* Miyazaki hydrogenase: Comparison with the Ni-Fe enzyme from *Desulfovibrio gigas, Biochim. Biophys. Acta* 1122:50–56.

Bagyinka, C., Whitehead, J. P., and Maroney, M. J., 1993. An X-ray absorption spectroscopic study of nickel redox chemistry in hydrogenase, *J. Am. Chem. Soc.* 115:3576–3585.

Baidya, N., Olmstead, M., and Mascharak, P. K., 1991. Pentacoordinated nickel(II) complexes with thiolato ligation: Synthetic strategy, structures, and properties, *Inorg. Chem.* 30:929–937.

Baidya, N., Noll, B. C., Olmstead, M. M., and Mascharak, P. K., 1992a. Nickel(II) complexes with the [NiN$_x$Se$_y$] chromophore in different coordination geometries: Search for a model of the active site of [NiFeSe] hydrogenases, *Inorg. Chem.* 31:2999–3000.

Baidya, N., Olmstead, M. M., Whitehead, J. P., Bagyinka, C., Maroney, M. J., and Mascharak, P. K., 1992b. X-ray absorption spectra of nickel complexes with N$_3$S$_2$ chromophores and spectroscopic studies on H$^-$ and CO binding at these nickel centers: Relevance to the reactivity of the nickel site(s) in [NiFe] hydrogenases, *Inorg. Chem.* 31:3612–3619.

Baidya, N., Olmstead, M. M., and Mascharak, P. K., 1993. A mononuclear nickel(II) complex with [NiN$_3$S$_2$] chromophore that readily affords the Ni(I) and Ni(III) analogues: Probe into the redox behavior of the nickel site in [FeNi] hydrogenases, *J. Am. Chem. Soc.* 114:9666–9668.

Ballantine, S. P., and Boxer, D. H., 1986. Isolation and characterization of a soluble active fragment of hydrogenase isoenzyme 2 from the membranes of anaerobically grown *Escherichia coli, Eur. J. Biochem.* 156:277–284.

Baron, S. F., Brown, D. P., and Ferry, J. G., 1987. Locations of the hydrogenases of *Methanobacterium formicicum* after subcellular fractionation of cell extract, *J. Bacteriol.* 169:3823–3825.

Baron, S. F., Williams, D. S., May, H. D., Patel, P. S., Aldrich, H. C., and Ferry, J. G., 1989. Immunogold localization of coenzyme F_{420}-reducing formate dehydrogenase and coenzyme F_{420}-reducing hydrogenase in *Methanobacterium formicicum, Arch. Microbiol.* 151:307–313.

Barraquio, W. L., and Knowles, R., 1989. Beneficial effects of nickel on *Pseudomonas saccharophila* under nitrogen-limited chemolithotrophic conditions, *Appl. Environ. Microbiol.* 55:3197–3201.

Bartha, R., and Ordal, E. J., 1965. Nickel-dependent chemolithotrophic growth of two *Hydrogenomonas* strains, *J. Bacteriol.* 89:1015–1019.

Bastian, N. R., Wink, D. A., Wackett, L. P., Livingston, D. J., Jordan, L. M., Fox, J. Orme-Johnson, W. H., and Walsh, C. A., 1988. Hydrogenases of *Methanobacterium thermoautotrophicum* strain ΔH, in *The Bioinorganic Chemistry of Nickel* (J. R. Lancaster, Jr., ed.), VCH Publishers, New York, pp. 227–247.

Berlier, Y. M., Fauque, G., Lespinat, P. A., and LeGall, J., 1982. Activation, reduction and role of proton–deuterium exchange reaction of the periplasmic hydrogenase from *Desulfovibrio gigas* in relation with the role of cytochrome c_3, *FEBS Lett.* 140:185–188.

Böhm, R., Sauter, M., and Böck, A., 1990. Nucleotide sequence and expression of an operon in *Escherichia coli* coding for formate hydrogenlyase components, *Mol. Microbiol.* 4:231–243.

Bonam, D., McKenna, M. C., Stephens, P. J., and Ludden, P. W., 1988. Nickel-deficient carbon monoxide dehydrogenase from *Rhodospirillum rubrum: In vivo* and *in vitro* activation by exogenous nickel, *Proc. Natl. Acad. Sci. USA* 85:31–35.

Boursier, P., Hanus, F. J., Becker, M. M., Russell, S. A., and Evans, H. J., 1988. Selenium increases hydrogenase expression in autotrophically cultured *Bradyrhizobium japonicum* and is a constituent of the purified enzyme, *J. Bacteriol.* 170:5594–5600.

Bryant, F. O., and Adams, M. W. W., 1989. Characterization of hydrogenase from the hyperthermophilic archaebacterium, *Pyrococcus furiosus, J. Biol. Chem.* 264:5070–5079.

Cammack, R., Patil, D., Aguirre, R., and Hatchikian, E. C., 1982. Redox properties of the ESR-detectable nickel in hydrogenase from *Desulfovibrio gigas, FEBS Lett.* 142:289–292.

Cammack, R., Fernandez, V. M., and Schneider, K., 1986. Activation and active-sites of nickel-containing hydrogenases, *Biochimie* 68:85–91.

Cammack, R., Patil, D. S., Hatchikian, E. C., and Fernandez, V. M., 1987. Nickel and iron-sulphur centres in *Desulfovibrio gigas* hydrogenase: ESR spectra, redox properties and interactions, *Biochim. Biophys. Acta* 912:98–109.

Cammack, R., Fernandez, V. M., and Schneider, K., 1988. Nickel in hydrogenases from sulfate-reducing, photosynthetic, and hydrogen-oxidizing bacteria, in *The Bioinorganic Chemistry of Nickel* (J. R. Lancaster, Jr., ed.), VCH Publishers, New York, pp. 167–190.

Cammack, R., Bagyinka, C., and Kovacs, K. L., 1989a. Spectroscopic characterization of the nickel and iron–sulphur clusters of hydrogenase from the purple photosynthetic bacterium *Thiocapsa roseopersicina*. 1. Electron spin resonance spectroscopy, *Eur. J. Biochem.* 182:357–362.

Cammack, R., Kovacs, K. L., McCracken, J., and Peisach, J., 1989b. Spectroscopic characterization of the nickel and iron–sulphur clusters of hydrogenase from the purple photosynthetic bacterium *Thiocapsa roseopersicina*. 2. Electron spin-echo spectroscopy, *Eur. J. Biochem.* 182:363–366.

Chapman, A., Cammack, R., Hatchikian, E. C., McCracken, J., and Peisach, J., 1988. A pulsed EPR study of redox-dependent hyperfine interactions for the nickel centre of *Desulfovibrio gigas* hydrogenase, *FEBS Lett.* 242:134–138.

Chaudhuri, A., and Krasna, A. I., 1987. Isolation of genes required for hydrogenase synthesis in *Escherichia coli, J. Gen. Microbiol.* 133:3289–3298.

Chen, J. C., and Mortenson, L. E., 1992. Identification of six open reading frames from a region of the *Azotobacter vinelandii* genome likely involved in dihydrogen metabolism, *Biochim. Biophys. Acta* 1131:199–202.

Chen, Y.-P., and Yoch, D. C., 1987. Regulation of two nickel-requiring (inducible and constitutive) hydrogenases and their coupling to nitrogenase in *Methylosinus trichosporium* OB3b, *J. Bacteriol.* 169:4778–4783.

Choquet, C. G., and Sprott, G. D., 1991. Metal chelate affinity chromatography for the purification of the F_{420}-reducing (Ni,Fe) hydrogenase of *Methanospirillum hungatei*, *J. Microbiol. Methods* 13:161–169.

Colbeau, A., and Vignais, P. M., 1983. The membrane-bound hydrogenase of *Rhodopseudomonas capsulatus* is inducible and contains nickel, *Biochim. Biophys. Acta* 748:128–138.

Colbeau, A., Chabert, J., and Vignais, P. M., 1983. Purification, molecular properties and localization in the membrane of the hydrogenase of *Rhodopseudomonas capsulata, Biochim. Biophys. Acta* 748:116–127.

Colbeau, A., Richaud, P., Toussaint, B., Caballero, F. J., Elster, C., Delphin, C., Smith, R. L., Chabert, J., and Vignais, P. M., 1993. Organization of the genes necessary for hydrogenase expression in *Rhodobacter capsulatus.* Sequence analysis and identification of two *hyp* regulatory mutants, *Mol. Microbiol.* 8:15–29.

Colpas, G. J., Maroney, M. J., Bagyinka, C., Kumar, M., Willis, W. S., Suib, S. L., Baidya, N., and Mascharak, P. K., 1991. X-ray spectroscopic studies of nickel complexes, with application to the structure of nickel sites in hydrogenases, *Inorg. Chem.* 30:920–928.

Coremans, J. M. C. C., van der Zwaan, J. W., and Albracht, S. P. J., 1989. Redox behaviour of nickel in hydrogenase from *Methanobacterium thermoautotrophicum* (strain Marburg). Correlation between the nickel valence state and enzyme activity, *Biochim. Biophys. Acta* 997:256–267.

Coremans, J. M. C. C., van Garderen, C. J., and Albracht, S. P. J., 1992a. On the redox equilibrium between H_2 and hydrogenase, *Biochim. Biophys. Acta* 1119:148–156.

Coremans, J. M. C. C., van der Zwaan, J. W., and Albracht, S. P. J., 1992b. Distinct redox behaviour of prosthetic groups in ready and unready hydrogenase from *Chromatium vinosum, Biochim. Biophys. Acta* 1119:157–168.

Czechowski, M. H., He, S. H., Nacro, M. DerVartanian, D. V., Peck, H. D., Jr., and LeGall, J., 1984. A cytoplasmic nickel-iron hydrogenase with high specific activity from *Desulfovibrio multispirans* sp. n., a new species of sulfate reducing bacterium, *Biochem. Biophys. Res. Commun.* 125:1025–1032.

Daday, A., and Smith, G. D., 1983. The effect of nickel on the hydrogen metabolism of the cyanobacterium *Anabaena cylindrica, FEMS Microbiol. Lett.* 20:327–330.

Daday, A., MacKerras, A. H., and Smith, G. D., 1985. The effect of nickel on hydrogen metabolism and nitrogen fixation in the cyanobacterium *Anabaena cylindrica, J. Gen. Microbiol.* 131:231–238.

Deckers, H. M., Wilson, F. R., and Voordouw, G., 1990. Cloning and sequencing of a [NiFe] hydrogenase operon from *Desulfovibrio vulgaris* Miyazaki F, *J. Gen. Microbiol.* 136:2021–2028.

Deppenmeier, U., Blaut, M., Schmidt, B., and Gottschalk, G., 1992. Purification and properties of a F_{420}-nonreactive, membrane-bound hydrogenase from *Methanosarcina* strain Göl, *Arch. Microbiol.* 157:505–511.

Dernedde, J., Eitinger, M., and Friedrich, B., 1993. Analysis of a pleiotropic gene region involved in formation of catalytically active hydrogenases in *Alcaligenes eutrophus* H16, *Arch. Microbiol.* 159:545–553.

Doyle, C. M., and Arp, D. J., 1988. Nickel affects expression of the nickel-containing hydrogenase of *Alcaligenes latus, J. Bacteriol.* 170:3891–3896.

Dross, F., Geisler, V., Lenger, R., Theis, F., Kraft, T., Fahrenholz, F., Kojro, E., Duchene, A., Tripier, D., Juvenal, K., and Kröger, A., 1992. The quinone-reactive Ni/Fe-hydrogenase of *Wolinella succinogenes, Eur. J. Biochem.* 206:93–102.

Drutschmann, M., and Klemme, J.-H., 1985. Sulfide-repressed, membrane-bound hydrogenase in the thermophilic facultative phototroph, *Chloroflexus aurantiacus, FEMS Microbiol. Lett.* 28:231–235.

Du, L., Stejskal, F., and Tibelius, K. H., 1992. Characterization of two genes (*hupD* and *hupE*) required for hydrogenase activity in *Azotobacter chroococcum, FEMS Microbiol. Lett.* 96: 93–102.

Eberz, G., and Friedrich, B., 1991. Three *trans*-acting regulatory functions control hydrogenase synthesis in *Alcaligenes eutrophus, J. Bacteriol.* 173:1845–1854.

Eberz, G., Eitinger, T., and Friedrich, B., 1989. Genetic determinants of a nickel-specific transport system are part of the plasmid-encoded hydrogenase gene cluster in *Alcaligenes eutrophus, J. Bacteriol.* 171:1340–1345.

Eidsness, M. K., Scott, R. A., Prickril, B. C., DerVartanian, D. V., LeGall, J., Moura, I., Moura, J. J. G., and Peck, H. D., Jr., 1989. Evidence for selenocysteine coordination to the active site nickel in the [NiFeSe]hydrogenases from *Desulfovibrio baculatus, Proc. Natl. Acad. Sci. USA* 86:147–151.

Eitinger, T., and Friedrich, B., 1991. Cloning, nucleotide sequence, and heterologous expression of a high-affinity nickel transport gene from *Alcaligenes eutrophus, J. Biol. Chem.* 266:3222–3227.

Ewart, G. D., Reed, K. C., and Smith, G. D., 1990. Soluble hydrogenase of *Anabaena cylindrica.* Cloning and sequencing of a potential gene encoding the tritium exchange subunit, *Eur. J. Biochem.* 187:215–223.

Fan, C., Teixeira, M., Moura, J., Moura, I., Huynh, B.-H., Le Gall, J., Peck, H. D., Jr., and Hoffman, B. M., 1991. Detection and characterization of exchangeable protons bound to the hydrogen-activation nickel site of *Desulfovibrio gigas* hydrogenase: A ^{1}H and ^{2}H Q-band ENDOR study, *J. Am. Chem. Soc.* 113:20–24.

Fauque, G., Teixeira, M., Moura, I., Lespinat, P. A., Xavier, A. V., DerVartanian, D. V., Peck, H. D., Jr., LeGall, J., and Moura, J. G., 1984. Purification, characterization and redox properties of hydrogenase from *Methanosarcina barkeri* (DSM 800), *Eur. J. Biochem.* 142:21–28.

Fauque, G., Peck, H. D., Jr., Moura, J. J. G., Huynh, B. H., Berlier, Y., DerVartanian, D. V., Teixeira, M., Przybyla, A. E., Lespinat, P. A., Moura, I., and LeGall, J., 1988. Three classes of hydrogenases from sulfate-reducing bacteria of the genus *Desulfovibrio, FEMS Microbiol. Rev.* 54:299–344.

Fauque, G., Czechowski, M., Berlier, Y. M., Lespinat, P. A., LeGall, J., and Moura, J. J. G., 1992. Partial purification and characterization of the first hydrogenase isolated from a thermophilic sulfate-reducing bacterium, *Biochem. Biophys. Res. Commun.* 184:1256–1260.

Fernandez, V. M., Aguirre, R., and Hatchikian, E. C., 1984. Reductive activation and redox properties of hydrogenase from *Desulfovibrio gigas, Biochim. Biophys. Acta* 790:1–7.

Fernandez, V. M., Hatchikian, E. C., and Cammack, R., 1985. Properties and reactivation of two different deactivated forms of *Desulfovibrio gigas* hydrogenase, *Biochim. Biophys. Acta* 832:69–79.

Fernandez, V. M., Hatchikian, E. C., Patil, D. S., and Cammack, R., 1986. ESR-detectable nickel and iron–sulfur centres in relation to the reversible activation of *Desulfovibrio gigas* hydrogenase, *Biochim. Biophys. Acta* 883:145–154.

Fiebig, K., and Friedrich, B., 1989. Purification of the F_{420}-reducing hydrogenase from *Methanosarcina barkeri* (strain Fusaro), *Eur. J. Biochem.* 184:79–88.

Ford, C. M., Garg, N., Garg, R. P., Tibelius, K. H., Yates, M. G., Arp, D. J., and Seefeldt, L. C., 1990. The identification, characterization, sequencing and mutagenesis of the genes (*hupSL*) encoding the small and large subunits of the H₂-uptake hydrogenase of *Azotobacter chroococcum, Mol. Microbiol.* 4:999–1008.

Fox, J. A., Livingston, D. J., Orme-Johnson, W. H., and Walsh, C. T., 1987. 8-Hydroxy-5-deazaflavin-reducing hydrogenase from *Methanobacterium thermoautotrophicum:* 1. Purification and characterization, *Biochemistry* 26:4219–4227.

Fox, S., Wang, Y., Silver, A., and Miller, M., 1990. Viability of the [Ni^{III}(SR)$_4$]$^-$ unit in classical coordination compounds and in the nickel–sulfur center of hydrogenases, *J. Am. Chem. Soc.* 112:3218–3220.

Friedrich, B., 1990. The plasmid-encoded hydrogenase gene cluster in *Alcaligenes eutrophus, FEMS Microbiol. Rev.* 87:425–430.

Friedrich, B., Heine, E., Finck, A., and Friedrich, C. G., 1981. Nickel requirement for active hydrogenase formation in *Alcaligenes eutrophus, J. Bacteriol.* 145:1144–1149.

Friedrich, C. G., Schneider, K., and Friedrich, B., 1982. Nickel in the catalytically active hydrogenase of *Alcaligenes eutrophus, J. Bacteriol.* 152:42–48.

Friedrich, C. G., Suetin, S., and Lohmeyer, M., 1984. Nickel and iron incorporation into soluble hydrogenase of *Alcaligenes eutrophus, Arch. Microbiol.* 140:206–211.

Fu, C., and Maier, R. J., 1991. Identification of a locus within the hydrogenase gene cluster involved in intracellular nickel metabolism in *Bradyrhizobium japonicum, Appl. Environ. Microbiol.* 57:3502–3510.

Fu, C., and Maier, R., 1992. Nickel-dependent reconstitution of hydrogenase apoprotein in *Bradyrhizobium japonicum* Hup^c mutants and direct evidence for a nickel metabolism locus involved in nickel incorporation into the enzyme, *Arch. Microbiol.* 157:493–498.

Fu, C., and Maier, R. J., 1993. A genetic region downstream of the hydrogenase structural genes of *Bradyrhizobium japonicum* that is required for nydrogenase processing, *J. Bacteriol.* 175:295–298.

Gogotov, I. N., 1986. Hydrogenases of phototrophic microorganisms, *Biochimie* 68:181–187.

Gollin, D. J., Mortenson, L. E., and Robson, R. L., 1992. Carboxyl-terminal processing may be essential for production of active NiFe hydrogenase in *Azotobacter vinelandii, FEBS Lett.* 309:371–375.

Graf, E.-G., and Thauer, R. K., 1981. Hydrogenase from *Methanobacterium thermoautotrophicum,* a nickel-containing enzyme, *FEBS Lett.* 136:165–169.

Halboth, S., and Klein, A., 1992. *Methanococcus voltae* harbors four gene clusters potentially encoding two [NiFe] and two [NiFeSe] hydrogenases, each of the cofactor F$_{420}$-reducing or F$_{420}$-non-reducing types, *Mol. Gen. Genet.* 233:217–224.

Harker, A. R., Xu, L.-S., Hanus, F. J., and Evans, H. J., 1984. Some properties of the nickel-containing hydrogenase of chemolithotrophically grown *Rhizobium japonicum, J. Bacteriol.* 159:850–856.

Hatchikian, E. C., Bruschi, M., and LeGall, J., 1978. Characterization of the periplasmic hydrogenase from *Desulfovibrio gigas, Biochem. Biophys. Res. Commun.* 82:451–461.

Hatchikian, C. E., Traore, A. S., Fernandez, V. M., and Cammack, R., 1990. Characterization of the nickel–iron periplasmic hydrogenase from *Desulfovibrio fructosovorans, Eur. J. Biochem.* 187:635–643.

He, S. H., Teixeira, M., LeGall, J., Patil, D. S., Moura, I., Moura, J. J. G., DerVartanian, D. V., Huynh, B. H., and Peck, H. D., Jr., 1989. EPR studies with ⁷⁷Se-enriched (NiFeSe) hydrogenase of *Desulfovibrio baculatus.* Evidence for a selenium ligand to the active site nickel, *J. Biol. Chem.* 264:2678–2682.

Heiden, S., Hedderich, R., Setzke, E., and Thauer, R. K., 1993. Purification of a cytochrome *b* containing H₂:heterodisulfide oxidoreductase complex from membranes of *Methanosarcina barkeri, Eur. J. Biochem.* 213:529–535.

Hidalgo, E., Leyva, A., and Ruiz-Argüeso, T., 1990. Nucleotide sequence of the hydrogenase structural genes from *Rhizobium leguminosarum, Plant Mol. Biol.* 15:367–370.

Hidalgo, E., Palacios, J. M., Murillo, J., and Ruiz-Argüeso, T., 1992. Nucleotide sequence and characterization of four additional genes of the hydrogenase structural operon from *Rhizobium leguminosarum* bv. viciae, *J. Bacteriol.* 174:4130–4139.

Hornhardt, S., Schneider, K., and Schlegel, H. G., 1986. Characterization of a native subunit of the NAD-linked hydrogenase isolated from a mutant *Alcaligenes eutrophus* H16, *Biochimie* 68:15–24.

Hsu, J.-C., Beilstein, M. A., Whanger, P., and Evans, H. J., 1990. Investigation of the form of selenium in the hydrogenase from chemolithotrophically cultured *Bradyrhizobium japonicum, Arch. Microbiol.* 154:215–220.

Huynh, B. H., Patil, D. S., Moura, I., Teixeira, M., Moura, J. J. G., DerVartanian, D. V., Czechowski, M. H., Prickril, B. C., Peck, H. D., Jr., and LeGall, J., 1987. On the active sites of the [NiFe] hydrogenase from *Desulfovibrio gigas.* Mössbauer and redox titration studies, *J. Biol. Chem.* 262:795–800.

Jacobi, A., Rossman, R., and Böck, A., 1992. The *hyp* operon gene products are required for maturation of catalytically active hydrogenase isoenzymes in *Escherichia coli, Arch. Microbiol.* 158:444–451.

Jin, S.-L. C., Blanchard, D. K., and Chen, J.-S., 1983. Two hydrogenases with distinct electron carrier specificity and subunit composition in *Methanobacterium formicicum, Biochim. Biophys. Acta* 748:8–20.

Johannssen, W., Gerberding, H., Rohde, M., Zaborosch, C., and Mayer, F., 1991. Structural aspects of the soluble NAD-dependent hydrogenase isolated from *Alcaligenes eutrophus* H16 and from *Nocardia opaca* 1b, *Arch. Microbiol.* 155:303–308.

Johnson, M. K., Zambrano, I. C., Czechowski, M. H., Peck, H. D., Jr., DerVartanian, D. V., and LeGall, J., 1985. Low temperature magnetic circular dichroism spectroscopy as a probe for the optical transitions of paramagnetic nickel in hydrogenase, *Biochem. Biophys. Res. Commun.* 128:220–225.

Johnson, M. K., Zambrano, I. C., Czechowski, M. H., Peck, H. D., Jr., DerVartanian, D. V., and LeGall, J., 1986. Magnetic circular dichroism and electron paramagnetic resonance studies of nickel-containing hydrogenases, in *Frontiers in Bioinorganic Chemistry* (A. V. Xavier, ed.), VCH Publishers, New York, pp. 36–44.

Kemner, J. M., 1993. Characterization of Electron Transfer Activities Associated with Acetate Dependent Methanogenesis by *Methanosarcina barkeri* MS, Ph.D. thesis, Michigan State University.

Kim, H., and Maier, R. J., 1990. Transcriptional regulation of hydrogenase synthesis by nickel in *Bradyrhizobium japonicum, J. Biol. Chem.* 265:18729–18732.

Kim, H., Yu, C., and Maier, R. J., 1991. Common *cis*-acting region responsible for transcriptional regulation of *Bradyrhizobium japonicum* hydrogenase by nickel, oxygen, and hydrogen, *J. Bacteriol.* 173:3993–3999.

Klucas, R. V., Hanus, F. J., Russell, S. A., and Evans, H. J., 1983. Nickel: A micronutrient element for hydrogen-dependent growth of *Rhizobium japonicum* and for expression of urease activity in soybean leaves, *Proc. Natl. Acad. Sci. USA* 80:2253–2257.

Knüttel, K., Schneider, K., Schlegel, H. G., and Müller, A., 1989. The membrane-bound hydrogenase from *Paracoccus denitrificans.* Purification and molecular characterization, *Eur. J. Biochem.* 179:101–108.

Koch, H.-G., Kern, M., and Klemme, J.-H., 1992. Reinvestigation of regulation of biosynthesis and subunit composition of nickel-dependent Hup-hydrogenase of *Rhodospirillum rubrum*, *FEMS Microbiol. Lett.* 91:193–198.

Kojima, N., Fox, J. A., Hausinger, R. P., Daniels, L., Orme-Johnson, W. H., and Walsh, C., 1983. Paramagnetic centers in the nickel-containing, deazaflavin-reducing hydrogenase from *Methanobacterium thermoautotrophicum, Proc. Natl. Acad. Sci. USA* 80:378–382.

Kortlüke, C., and Friedrich, B., 1992. Maturation of membrane-bound hydrogenase of *Alcaligenes eutrophus* H16, *J. Bacteriol.* 174:6290–6293.

Kortlüke, C., Horstmann, K., Schwartz, E., Rohde, M., Binsack, R., and Friedrich, B., 1992. A gene complex coding for the membrane-bound hydrogenase of *Alcaligenes eutrophus* H16, *J. Bacteriol.* 174:6277–6289.

Kovacs, K. L., Seefeldt, L. C., Tigyi, G., Doyle, C. M., Mortenson, L. E., and Arp, D. J., 1989. Immunological relationships among hydrogenases, *J. Bacteriol.* 171:430–435.

Kowal, A. T., Zambrano, I. C., Moura, I., Moura, J. J. G., LeGall, J., and Johnson, M. K., 1988. Electronic and magnetic properties of nickel-substituted rubredoxin: A variable-temperature magnetic circular dichroism study, *Inorg. Chem.* 27:1162–1166.

Krüger, H.-J., and Holm, R. H., 1987. Stabilization of nickel(III) in a classical N_2S_2 coordination environment containing anionic sulfur, *Inorg. Chem.* 26:3645–3647.

Krüger, H.-J., and Holm, R. H., 1990. Stabilization of trivalent nickel in tetragonal NiS_4N_2 and NiN_6 environments: Synthesis, structures, redox potentials, and observations related to [NiFe]-hydrogenases, *J. Am. Chem. Soc.* 112:2955–2963.

Krüger, H.-J., Huynh, B. H., Ljungdahl, P. O., Xavier, A. V., DerVartanian, D. V., Moura, I., Peck, H. D., Jr., Teixeira, M., Moura, J. J. G., and LeGall, J., 1982. Evidence for nickel and a three-iron center in the hydrogenase of *Desulfovibrio desulfuricans, J. Biol. Chem.* 257: 14620–14623.

Krüger, H.-J., Peng, G., and Holm, R. H., 1991. Low-potential nickel(III,II) complexes: New systems based on tetradentate amidate-thiolate ligands and the influence of ligand structure on potentials in relation to the nickel site in [NiFe]-hydrogenases, *Inorg. Chem.* 30:734–742.

Kumar, M., Day, R. O., Colpas, G. J., and Maroney, M. J., 1989a. Ligand oxidation in a nickel thiolate complex, *J. Am. Chem. Soc.* 111:5974–5976.

Kumar, M., Colpas, G. J., Day, R. O., and Maroney, M. J., 1989b. Ligand oxidation in a nickel thiolate complex: A model for the deactivation of hydrogenase by O_2, *J. Am. Chem. Soc.* 111:8323–8325.

Lalla-Maharajh, W. V., Hall, D. O., Cammack, R., Rao, K. K., and LeGall, J., 1983. Purification and properties of the membrane-bound hydrogenase from *Desulfovibrio desulfuricans, Biochem. J.* 209:445–454.

Lancaster, J. R., Jr., 1980. Soluble and membrane-bound paramagnetic centers in *Methanobacterium bryantii, FEBS Lett.* 115:285–288.

Lancaster, J. R., Jr., 1982. New biological paramagnetic center: Octahedrally coordinated nickel(III) in the methanogenic bacteria, *Science* 216:1324–1325.

Lappin, A. G., Murray, C. K., and Margerum, D. W., 1978. Electron paramagnetic resonance studies of nickel(III)-oligopeptide complexes, *Inorg. Chem.* 17:1630–1634.

Leclerc, M., Colbeau, A., Cauvin, B., and Vignais, P., 1988. Cloning and sequencing of the genes encoding the large and the small subunits of the H_2 uptake hydrogenase (*hup*) of *Rhodobacter capsulatus, Mol. Gen. Genet.* 214:97–107.

Lee, M. H., Mulrooney, S. B., Renner, M. J., Markowicz, Y., and Hausinger, R. P., 1992. *Klebsiella aerogenes* urease gene cluster: Sequence of *ureD* and demonstration that four accessory genes (*ureD, ureE, ureF,* and *ureG*) are involved in nickel metallocenter biosynthesis, *J. Bacteriol.* 174:4324–4330.

LeGall, J., Ljungdahl, P. O., Moura, I., Peck, H. D., Jr., Xavier, A. V., Moura, J. J. G., Teixeira, M., Huynh, B. H., and DerVartanian, D. V., 1982. The presence of redox-sensitive nickel in the periplasmic hydrogenase from *Desulfovibrio gigas, Biochem. Biophys. Res. Commun.* 106:610–616.

Li, C., Peck, H. D., Jr., LeGall, J., and Przybyla, A. E., 1987. Cloning, characterization, and sequencing of the genes encoding the large and small subunits of the periplasmic [NiFe]hydrogenase of *Desulfovibrio gigas, DNA* 6:539–551.

Lindahl, P. A., Kojima, N., Hausinger, R. P., Fox, J. A., Teo, B. K., Walsh, C. T., and Orme-Johnson, W. H., 1984. Nickel and iron EXAFS of F_{420}-reducing hydrogenase from *Methanobacterium thermoautotrophicum, J. Am. Chem. Soc.* 106:3062–3064.

Lissolo, T., Pulvin, S., and Thomas, D., 1984. Reactivation of the hydrogenase from *Desulfovibrio gigas* by hydrogen. Influence of redox potential, *J. Biol. Chem.* 259:11725–11729.

Lissolo, T., Choi, E. S., LeGall, J., and Peck, H. D., Jr., 1986. The presence of multiple intrinsic membrane nickel-containing hydrogenases in *Desulfovibrio vulgaris* (Hildenborough), *Biochem. Biophys. Res. Commun.* 139:701–708.

Lutz, S., Jacobi, A., Schlensog, V., Böhm, R., Sawers, G., and Böck, A., 1991. Molecular characterization of an operon (*hyp*) necessary for the activity of the three hydrogenase isoenzymes in *Escherichia coli, Mol. Microbiol.* 5:123–135.

Maier, T., Jacobi, A., Sauter, M., and Böck, A., 1993. The product of the *hypB* gene, which is required for nickel incorporation into hydrogenases, is a novel guanine nucleotide-binding protein, *J. Bacteriol.* 175:630–635.

Maroney, M. J., Colpas, G. J., and Bagyinka, C., 1990. X-ray absorption spectroscopic structural investigation of the Ni site in reduced *Thiocapsa roseopersicina* hydrogenase, *J. Am. Chem. Soc.* 112:7076–7068.

Maroney, M. J., Colpas, G. J., Bagyinka, C., Baidya, N., and Mascharak, P. K., 1991. EXAFS investigations of the Ni site in *Thiocapsa roseopersicina* hydrogenase: Evidence for a novel Ni,Fe,S cluster, *J. Am. Chem. Soc.* 113:3962–3972.

Mege, R.-M., and Bourdillon, C., 1985. Nickel controls the reversible anaerobic activation/inactivation of the *Desulfovibrio gigas* hydrogenase by the redox potential, *J. Biol. Chem.* 260:14701–14706.

Menon, N. K., Peck, H. D., Jr., LeGall, J., and Przybyla, A. E., 1987. Cloning and sequencing of the genes encoding the large and small subunits of the periplasmic (NiFeSe) hydrogenase of *Desulfovibrio baculatus, J. Bacteriol.* 169:5401–5407. [Erratum 170:4429.]

Menon, A. L., Stults, L. W., Robson, R. L., and Mortenson, L. E., 1990a. Cloning, sequencing and characterization of the [NiFe]hydrogenase-encoding structural genes (*hoxK* and *hoxG*) from *Azotobacter vinelandii*, Gene 96:67–74.

Menon, N. K., Robbins, J., Peck, H. D., Jr., Chatelus, C. Y., Choi, E.-S., and Przybyla, A. E., 1990b. Cloning and sequencing of a putative *Escherichia coli* [NiFe] hydrogenase-a operon containing six open reading frames, *J. Bacteriol.* 172:1969–1977.

Menon, N. K., Robbins, J., Wendt, J. C., Shanmugan, K. T., and Przybyla, A. E., 1991. Mutational analysis and characterization of the *Escherichia coli hya* operon which encodes [NiFe] hydrogenase 1, *J. Bacteriol.* 173:4851–4861.

Menon, A. L., Mortenson, L. E., and Robson, R. L., 1992. Nucleotide sequences and genetic analysis of hydrogen oxidation (*hox*) genes in *Azotobacter vinelandii, J. Bacteriol.* 174:4549–4557.

Moura, J. J. G., Moura, I., Huynh, B. H., Krüger, H.-J., Teixeira, M., DuVarney, R. C., Der Vartanian, D. V., Xavier, A. V., Peck, H. D., Jr., and LeGall, J., 1982. Unambiguous identification of the nickel EPR signal in [61]Ni-enriched *Desulfovibrio gigas* hydrogenase, *Biochem. Biophys. Res. Commun.* 108:1388–1393.

Moura, J. J. G., Teixeira, M., Moura, I., and LeGall, J., 1988. (Ni,Fe) hydrogenases from sulfate-reducing bacteria: Nickel catalytic and regulatory roles, in *The Bioinorganic Chemistry of Nickel* (J. R. Lancaster, Jr., ed.), VCH Publishers, New York, pp. 191–226.

Mus-Veteau, I., Diaz, D., Gracia-Mora, J., Guigliarelli, B., Chottard, G., and Bruschi, M., 1991. Spectroscopic studies of the nickel-substituted *Desulfovibrio vulgaris* Hildenborough rubredoxin: Implication for the nickel site in hydrogenases, *Biochim. Biophys. Acta* 1060:159–165.

Muth, E., Mörschel, E., and Klein, A., 1987. Purification and characterization of an 8-hydroxy-5-deazaflavin-reducing hydrogenase from the archaebacterium *Methanococcus voltae, Eur. J. Biochem.* 169:571–577.

Nakamura, Y., Someya, J.-I., and Suzuki, T., 1985. Nickel requirement of oxygen-resistant hydrogen bacterium, *Xanthobacter autotrophicus* strain Y38, *Agric. Biol. Chem.* 49:1711–1718.

Nelson, M. J. K., Brown, D. P., and Ferry, J. G., 1984. FAD requirement for the reduction of coenzyme F_{420} by hydrogenase from *Methanobacterium formicicum, Biochem. Biophys. Res. Commun.* 120:775–781.

Nivière, V., Forget, N., Gayda, J. P., and Hatchikian, E. C., 1986. Characterization of the soluble hydrogenase from *Desulfovibrio africanus, Biochem. Biophys. Res. Commun.* 139:658–665.

Nivière, V., Hatchikian, E., Cambillau, C., and Frey, M., 1987. Crystallization, preliminary X-ray study and crystal activity of the hydrogenase from *Desulfovibrio gigas, J. Mol. Biol.* 195:969–971.

Odom, J. M., and Peck, H. D., Jr., 1984. Hydrogenase, electron-transfer proteins, and energy coupling in the sulfate-reducing bacteria *Desulfovibrio, Annu. Rev. Microbiol.* 38:551–592.

Papen, H., Kentemich, T., Schmülling, T., and Bothe, H., 1986. Hydrogenase activities in cyanobacteria, *Biochimie* 68:121–132.

Partridge, C. D. P., and Yates, M. G., 1982. Effect of chelating agents on hydrogenase in *Azotobacter chroococcum*. Evidence that nickel is required for hydrogenase synthesis, *Biochem. J.* 204:339–344.

Pederson, D. M., Daday, A., and Smith, G. D., 1986. The use of nickel to probe the role of hydrogen metabolism in cyanobacterial nitrogen fixation, *Biochimie* 68:113–120.

Pedrosa, F. O., and Yates, M. G., 1983. Effect of chelating agents and nickel ions on hydrogenase activity in *Azospirillum brasilense, A. lipoferum* and *Derxia gummosa, FEMS Microbiol. Lett.* 17:101–106.

Pihl, T. D., and Maier, R. J., 1991. Purification and characterization of the hydrogen uptake hydrogenase from the hyperthermophilic archaebacterium *Pyrodictium brockii, J. Bacteriol.* 173:1839–1844.

Pinkwart, M., Schneider, K., and Schlegel, H. G., 1983. Purification and properties of the membrane-bound hydrogenase from N_2-fixing *Alcaligenes latus, Biochim. Biophys. Acta* 745:267–278.

Przybyla, A. E., Robbins, J., Menon, N., and Peck, H. D., Jr., 1992. Structure/function relationships among the nickel-containing hydrogenases, *FEMS Microbiol. Rev.* 88:109–136.

Rai, L. C., and Raizada, M., 1986. Nickel induced stimulation of growth, heterocyst differentiation, $^{14}CO_2$ uptake and nitrogenase activity in *Nostoc muscorum, New Phytol.* 104:111–114.

Reeve, J. N., Beckler, G. S., Cram, D. S., Hamilton, P. T., Brown, J. W., Krzycki, J. A., Kolodziej, A. F., Alex, L., Orme-Johnson, W. H., and Walsh, C. T., 1989. A hydrogenase-linked gene in *Methanobacterium thermoautotrophicum* strain ΔH encodes a polyferredoxin, *Proc. Natl. Acad. Sci. USA* 86:3031–3035.

Rey, L., Hidalgo, E., Palacios, J., and Ruiz-Argüeso, T., 1992. Nucleotide sequence and organization of an H_2-uptake gene cluster from *Rhizobium leguminosarum* bv. viciae containing a rubredoxin-like gene and four additional open reading frames, *J. Mol. Biol.* 228:998–1002.

Rey, L., Murillo, J., Hernando, Y., Hildalgo, E., Cabrera, E., Imperial, J., and Ruiz-Argüeso, T., 1993. Molecular analysis of a microaerobically induced operon required for hydrogenase synthesis in *Rhizobium leguminosasum* biovar viciae, *Mol. Microbiol.* 8:471–481.

Rieder, R., Cammack, R., and Hall, D. O., 1984. Purification and properties of the soluble hydrogenase from *Desulfovibrio desulfuricans* (strain Norway 4), *Eur. J. Biochem.* 145:637–643.

Rousset, M., Dermoun, Z., Hatchikian, C. E., and Bélaich, J.-P., 1990. Cloning and sequencing of the locus encoding the large and small subunit genes of the periplasmic [NiFe]hydrogenase from *Desulfovibrio fructosovorans, Gene* 94:95–101.

Saint-Martin, P., Lespinat, P. A., Fauque, G., Berlier, Y., LeGall, J., Moura, I., Teixeira, M., Xavier, A. V., and Moura, J. J. G., 1988. Hydrogen production and deuterium–proton exchange reactions catalyzed by *Desulfovibrio* nickel(II)-substituted rubredoxins, *Proc. Natl. Acad. Sci. USA* 85:9378–9380.

Sauter, M., Böhm, R., and Böck, A., 1992. Mutational analysis of the operon (*hyc*) determining hydrogenase 3 formation in *Escherichia coli, Mol. Microbiol.* 6:1523–1532.

Sawers, R. G., and Boxer, D. H., 1986. Purification and properties of membrane-bound hydrogenase isoenzyme 1 from anaerobically grown *Escherichia coli* K12, *Eur. J. Biochem.* 156:265–275.

Sawers, R. G., Ballantine, S. P., and Boxer, D. H., 1985. Differential expression of hydrogenase isoenzymes in *Escherichia coli* K-12: Evidence for a third isozyme, *J. Bacteriol.* 164:1324–1331.

Sayavedra-Soto, L. A., and Arp, D. J., 1992. The *hoxZ* gene of *Azotobacter vinelandii* hydrogenase operon is required for activation of hydrogenase, *J. Bacteriol.* 174:5295–5301.

Sayavedra-Soto, L. A., and Arp, D. J., 1993. In *Azotobacter vinelandii* hydrogenase, substitution of serine for the cysteine residues at positions 62, 65, 289, and 292 in the small (HoxK) subunit affects H_2 oxidation, *J. Bacteriol.* 175:3414–3421. [Erratum: 175:5744]

Sayavedra-Soto, L. A., Powell, G. K., Evans, H. J., and Morris, R. O., 1988. Nucleotide sequence of the genetic loci encoding subunits of *Bradyrhizobium japonicum* uptake hydrogenase, *Proc. Natl. Acad. Sci. USA* 85:8395–8399.

Schneider, K., and Piechulla, B., 1986. Isolation and immunological characterization of the four non-identical subunits of the soluble NAD-linked dehydrogenase from *Alcaligenes eutrophus, Biochimie* 68:5–13.

Schneider, K., and Schlegel, H. G., 1976. Purification and properties of soluble hydrogenase from *Alcaligenes eutrophus* H16, *Biochim. Biophys. Acta* 452:66–80.

Schneider, K., Patil, D. S., and Cammack, R., 1983. ESR properties of membrane-bound hydrogenases from aerobic hydrogen bacteria, *Biochim. Biophys. Acta* 748:353–361.

Schneider, K., Schlegel, H. G., and Jochim, K., 1984a. Effect of nickel on activity and subunit composition of purified hydrogenase of *Nocardia opaca* 1b, *Eur. J. Biochem.* 138:533–541.

Schneider, K., Cammack, R., and Schlegel, H. G., 1984b. Content and localization of FMN, Fe-S clusters and nickel in the NAD-linked hydrogenase of *Nocardia opaca* 1b, *Eur. J. Biochem.* 142:75–84.

Scott, R. A., Wallin, S. A., Czechowski, M., DerVartanian, D. V., LeGall, J., Peck, H. D., Jr., and Moura, I., 1984. X-ray absorption spectroscopy of nickel in the hydrogenase from *Desulfovibrio gigas, J. Am. Chem. Soc.* 106:6864–6865.

Seefeldt, L. C., and Arp, D. J., 1986. Purification to homogeneity of *Azotobacter vinelandii* hydrogenase: A nickel and iron containing $\alpha\beta$ dimer, *Biochimie* 68:25–34.

Seefeldt, L. C., and Arp, D. J., 1989. Oxygen effects on the nickel- and iron-containing hydrogenase from *Azotobacter vinelandii, Biochemistry* 28:1588–1596.

Seefeldt, L. C., McCollum, L. C., Doyle, C. M., and Arp, D. J., 1987. Immunological and molecular evidence for a membrane-bound, dimeric hydrogenase in *Rhodopseudomonas capsulata, Biochim. Biophys. Acta* 914:299–303.

Sellstedt, A., and Smith, G. D., 1990. Nickel is essential for active hydrogenase in free-living *Frankia* isolated from *Casuarina, FEMS Microbiol. Lett.* 70:137–140.

Serebryakova, L. T., Zorin, N. A., and Gogotov, I. N., 1990. Purification and properties of the hydrogenase of the green nonsulfur bacterium *Chloroflexus aurantiacus, Biokhimiya* 55: 372–380.

Shah, N. J., and Clark, D. S., 1990. Partial purification and characterization of two hydrogenases from the extreme thermophile *Methanococcus jannaschii, Appl. Environ. Microbiol.* 56:858–863.

Sherman, M. B., Orlova, E. V., Smirnova, E. A., Hovmöller, S., and Zorin, N. A., 1991. Three-dimensional structure of the nickel-containing hydrogenase from *Thiocapsa roseopersicina, J. Bacteriol.* 173:2576–2580.

Soeder, C. J., and Engelmann, G., 1984. Nickel requirement in *Chlorella emersonii, Arch. Microbiol.* 137:85–87.

Sorgenfrei, O., Linder, D., Karas, M., and Klein, A., 1993. A novel very small subunit of a selenium containing [NiFe] hydrogenase of *Methanococcus voltae* is posttranslationally processed by cleavage at a defined position, *Eur. J. Biochem.* 213:1355–1358.

Sprott, G. D., Shaw, K. M., and Beveridge, T. J., 1987. Properties of the particulate enzyme F_{420}-reducing hydrogenase isolated from *Methanospirillum hungatei, Can. J. Microbiol.* 33:896–904.

Stadtman, T. C., 1990. Selenium biochemistry, *Annu. Rev. Biochem.* 59:111–127.

Steigerwald, V. J., Beckler, G. S., and Reeve, J. N., 1990. Conservation of hydrogenase and polyferredoxin structures in the hyperthermophile archaebacterium *Methanothermus fervidus, J. Bacteriol.* 172:4715–4718.

Stoker, K., Oltmann, L. F., and Stouthamer, A. H., 1989. Randomly induced *Escherichia coli* K-12 Tn5 insertion mutants defective in hydrogenase activity, *J. Bacteriol.* 171:831–836.

Stults, L. W., O'Hara, E. B., and Maier, R. J., 1984. Nickel is a component of hydrogenase in *Rhizobium japonicum, J. Bacteriol.* 159:153–158.

Stults, L. W., Moshiri, F., and Maier, R. J., 1986a. Aerobic purification of hydrogenase from *Rhizobium japonicum* by affinity chromatography, *J. Bacteriol.* 166:795–800.

Stults, L. W., Sray, W. A., and Maier, R. J., 1986b. Regulation of hydrogenase biosynthesis by nickel in *Bradyrhizobium japonicum, Arch. Microbiol.* 146:280–283.

Szökefalvi-Nagy, Z., Bagyinka, C., Demeter, I., Kovács, K. L., and Quynh, L. H., 1990. Location and quantitation of metal ions in enzymes combining polyacrylamide gel electrophoresis and particle-induced X-ray emission, *Biol. Trace Elem. Res.* 93–101.

Tabillion, R., Weber, F., and Kaltwasser, H., 1980. Nickel requirement for chemolithotrophic growth in hydrogen-oxidizing bacteria, *Arch. Microbiol.* 124:131–136.

Takakuwa, S., and Wall, J. D., 1981. Enhancement of hydrogenase activity in *Rhodopseudomonas capsulata* by nickel, *FEMS Microbiol. Lett.* 12:359–363.

Tan, S. L., Fox, J. A., Kojima, N., Walsh, C. T., and Orme-Johnson, W. H., 1984. Nickel coordination in deazaflavin and viologen-reducing hydrogenases from *Methanobacterium thermoautotrophicum:* Investigation by electron spin echo spectroscopy, *J. Am. Chem. Soc.* 106: 3064–3066.

Teixeira, M., Moura, I., Xavier, A. V., DerVartanian, D. V., LeGall, J., Peck, H. D., Jr., Huynh, B. H., and Moura, J. J. G., 1983. *Desulfovibrio gigas* hydrogenase: Redox properties of the nickel and iron–sulfur centers, *Eur. J. Biochem.* 130:481–484.

Teixeira, M., Moura, I., Xavier, A. V., Huynh, B. H., DerVartanian, D. V., Peck, H. D., Jr., LeGall, J., and Moura, J. J. G., 1985. Electron paramagnetic resonance studies on the mechanism of activation and the catalytic cycle of the nickel-containing hydrogenase from *Desulfovibrio gigas, J. Biol. Chem.* 260:8942–8950.

Teixeira, M., Moura, I., Fauque, G., Czechowski, M., Berlier, Y., Lespinat, P. A., LeGall, J., Xavier, A. V., and Moura, J. J. G., 1986. Redox properties and activity studies on a nickel-containing hydrogenase isolated from a halophilic sulfate reducer *Desulfovibrio salexigens*, *Biochimie* 68:75–84.

Teixeira, M., Fauque, G., Moura, I., Lespinat, P. A., Berlier, Y., Prickril, B., Peck, H. D., Jr., Xavier, A. V., LeGall, J., and Moura, J. J. G., 1987. Nickel–[iron–sulfur]–selenium-containing hydrogenases from *Desulfovibrio baculatus* (DSM 1743). Redox centers and catalytic properties, *Eur. J. Biochem.* 167:47–58.

Teixeira, M., Moura, I., Xavier, A. V., Moura, J. J. G., LeGall, J., DerVartanian, D. V., Peck, H. D., Jr., and Huynh, B. H., 1989. Redox intermediates of *Desulfovibrio gigas* [NiFe] hydrogenase generated under hydrogen. Mössbauer and EPR characterization of the metal centers, *J. Biol. Chem.* 264:16435–16450.

Tibelius, K. H., Du, L., Tito, D., and Stejskal, F., 1993. The *Azotobacter chroococcum* hydrogenase gene cluster: Sequences and genetic analysis of four accessory genes, *hupA*, *hupB*, *hupY* and *hupC*, *Gene* 127:53–61.

Tran-Betcke, A., Warnecke, U., Böcker, C., Zaborosch, C., and Friedrich, B., 1990. Cloning and nucleotide sequences of the genes for the subunits of NAD-reducing hydrogenase of *Alcaligenes eutrophus* H16, *J. Bacteriol.* 172:2920–2929.

Uffen, R. L., Colbeau, A., Richaud, P., and Vignais, P. M., 1990. Cloning and sequencing the genes encoding uptake-hydrogenase subunits of *Rhodocyclus gelatinosus*, *Mol. Gen. Genet.* 221:49–58.

Unden, G., Böcher, R., Knecht, J., and Kröger, A., 1982. Hydrogenase from *Vibrio succinogenes*, a nickel protein, *FEBS Lett.* 145:230–234.

van Baalen, C., and O'Donnell, R., 1978. Isolation of a nickel-dependent blue-green alga, *J. Gen. Microbiol.* 105:351–353.

van der Zwaan, J. W., Albracht, S. P. J., Fontijn, R. D., and Slater, E. C., 1985. Monovalent nickel in hydrogenase from *Chromatium vinosum*, *FEBS Lett.* 179:271–277.

van der Zwaan, J. W., Albracht, S. P. J., Fontijn, R. D., and Mul, P., 1987. On the anomalous temperature behaviour of the EPR signal of monovalent nickel in hydrogenase, *Eur. J. Biochem.* 169:377–384.

van der Zwaan, J. W., Coremans, J. M. C. C., Bouwens, E. C. M., and Albracht, S. P. J., 1990. Effect of $^{17}O_2$ and ^{13}CO on EPR spectra of nickel in hydrogenase from *Chromatium vinosum*, *Biochim. Biophys. Acta* 1041:101–110.

van Heerikhuizen, H., Albracht, S. P. J., Slater, E. C., and Rheenen, P. S., 1981. Purification and some properties of the soluble hydrogenase from *Chromatium vinosum*, *Biochim. Biophys. Acta* 657:26–39.

Voordouw, G., and Brenner, S., 1985. Nucleotide sequence of the gene encoding the hydrogenase from *Desulfovibrio vulgaris* (Hildenborough), *Eur. J. Biochem.* 148:515–520.

Voordouw, G., Menon, N. K., LeGall, J., Choi, E.-S., Peck, H. D., Jr., and Przybyla, A. E., 1989. Analysis and comparison of nucleotide sequences encoding the genes for [NiFe] and [NiFeSe] hydrogenases from *Desulfovibrio gigas* and *Desulfovibrio baculatus*, *J. Bacteriol.* 171:2894–2899.

Wackett, L. P., Hartwieg, E. A., King, J. A., Orme-Johnson, W. H., and Walsh, C. T., 1987. Electron microscopy of nickel-containing methanogenic enzymes: Methyl reductase and F420-reducing hydrogenase, *J. Bacteriol.* 169:718–727.

Wang, C.-P., Franco, R., Moura, J. J. G., Moura, I., and Day, E. P., 1992. The nickel site in active *Desulfovibrio baculatus* [NiFeSe] hydrogenase is diamagnetic. Multifield saturation magnetization measurement of the spin state of Ni(II), *J. Biol. Chem.* 267:7378–7380.

Waugh, R., and Boxer, D. H., 1986. Pleiotropic hydrogenase mutants of *Escherichia coli* K12: Growth in the presence of nickel can restore hydrogenase activity, *Biochimie* 68:157–166.

Whitehead, J. P., Colpas, G. J., Bagyinka, C., and Maroney, M. J., 1991. X-ray absorption spectroscopic study of the reductive activation of *Thiocapsa roseopersicina* hydrogenase, *J. Am. Chem. Soc.* 113:6288–6289.

Wu, L. F., 1992. Putative nickel-binding sites of microbial proteins, *Res. Microbiol.* 143:347–351.

Wu, L.-F., and Mandrand, M. A., 1993. Microbial hydrogenases: Primary structure, classification, signatures and phylogeny, *FEMS Microbiol. Rev.* 104:243–270.

Wu, L. F., and Mandrand-Berthelot, M.-A., 1986. Genetic and physiological characterization of new *Escherichia coli* mutants impaired in hydrogenase activity, *Biochimie* 68:167–179.

Wu, L.-F., Mandrand-Berthelot, M.-A., Waugh, R., Edmonds, C. J., Holt, S. E., and Boxer, D. H., 1989. Nickel deficiency gives rise to the defective phenotype of *hydC* and *fnr* mutants in *Escherichia coli, Mol. Microbiol.* 3:1709–1718.

Wu, L.-F., Navarro, C., and Mandrand-Berthelot, M.-A., 1991. The *hydC* region contains a multicistronic operon (*nik*) involved in nickel transport in *Escherichia coli, Gene* 107:37–42.

Xiankong, Z., Tabita, F. R., and van Baalen, C., 1984. Nickel control of hydrogen production and uptake in *Anabaena* spp. strains CA and 1F, *J. Gen. Microbiol.* 130:1815–1818.

Xu, H.-W., and Wall, J. D., 1991. Clustering of genes necessary for hydrogen oxidation in *Rhodobacter capsulatus, J. Bacteriol.* 173:2401–2405.

Yamazaki, S., 1982. A selenium-containing hydrogenase from *Methanococcus vannielii, J. Biol. Chem.* 257:7926–7929.

Zaborosch, C., Schneider, K., Schlegel, H. G., and Kratzin, H., 1989. Comparison of the NH$_2$-terminal amino acid sequences of the four non-identical subunits of the NAD-linked hydrogenases from *Nocardia opaca* 1b and *Alcaligenes eutrophus* H16, *Eur. J. Biochem.* 181:175–180.

Zimmer, M., Schulte, G., Luo, X.-L., and Crabtree, R. H., 1991. Functional modeling of Ni,Fe hydrogenases: A nickel complex in an *N,O,S* environment, *Angew. Chem. Int. Ed. Engl.* 30:193–194.

Zinoni, F., Beier, A., Pecher, A., Wirth, R., and Böck, A., 1984. Regulation of the synthesis of hydrogenase (formate hydrogen-lyase) of *E. coli, Arch. Microbiol.* 139:299–304.

Zirngibl, C., van Dongen, W., Schwörer, B., von Bünau, R., Richter, M., Klein, A., and Thauer, R. K., 1992. H$_2$-forming methylenetetrahydromethanopterin dehydrogenase, a novel type of hydrogenase without iron–sulfur clusters in methanogenic archaea, *Eur. J. Biochem.* 208:511–520.

Zorin, N. A., 1986. Redox properties and active center of phototrophic bacterial hydrogenases, *Biochimie* 68:97–101.

Carbon Monoxide Dehydrogenase   5

5.1 Introduction

Carbon monoxide (CO) is metabolized by a wide range of microorganisms according to the following reversible reaction:

$$CO + H_2O \leftrightarrow CO_2 + 2H^+ + 2e^- \qquad (5\text{-}1)$$

Various aerobic bacteria including species of *Pseudomonas, Alcaligenes, Bacillus, Arthrobacter, Azotobacter,* and *Azotomonas* are capable of CO oxidation by using a molybdopterin–iron–sulfur–flavin-containing enzyme termed carbon monoxide:acceptor oxidoreductase (Meyer and Schlegel, 1983). Because this enzyme does not possess nickel, these carboxydotrophic microbes will not be further discussed here. Rather, this chapter focuses on anaerobic microorganisms that metabolize CO by using a nickel-containing enzyme that is often referred to as CO dehydrogenase. In many cases, anaerobic CO metabolism is only a side reaction for an enzyme complex which is normally involved in the cellular biosynthesis or degradation of acetate. Thus, CO dehydrogenase plays a central role in the growth of acetogenic bacteria and selected other autotrophs, in the physiology of aceticlastic methanogens, and in CO-dependent growth of several other microorganisms. The properties and roles of the nickel-dependent enzymes from each of these classes of microbes are discussed below.

5.2 CO Dehydrogenases Involved in Autotrophic Growth

The major mechanism of carbon dioxide fixation under anaerobic conditions makes use of a nickel-containing CO dehydrogenase. Several excellent reviews have summarized aspects of the recently identified autotrophic pathway (Diekert, 1988; Fuchs, 1986; Ljungdahl, 1986; Ragsdale, 1991; Ragsdale *et al.*, 1988, 1990; Wood and Ljungdahl, 1991; Wood *et al.*, 1986a,b,c). This section will describe studies demonstrating the presence of nickel in CO dehydrogenase of acetogenic bacteria, detail the evidence that this enzyme is an acetyl coenzyme A synthase that functions in a novel autotrophic pathway in acetogens, review the molecular biological studies that have been carried out with this system, summarize the properties of the nickel active site for the best studied acetyl coenzyme A synthase, that from *Clostridium thermoaceticum,* describe chemical models related to the mechanism of coenzyme A synthase, and discuss similar autotrophic pathways in nonacetogenic bacteria.

5.2.1 Acetogen CO Dehydrogenases Are Nickel-Containing Enzymes

The acetogenic bacteria are strict anaerobes that produce acetate either as a product of sugar fermentation or by autotrophic growth on hydrogen and carbon dioxide. Diekert and Thauer (1978) reported that cultures of two such organisms, *Clostridium thermoaceticum* and *Clostridium formicoaceticum,* are able to oxidize carbon monoxide to carbon dioxide. Cell extracts of these bacteria possess highly active CO dehydrogenase activities that couple CO oxidation to the reduction of methyl viologen, a synthetic dye. In these two species (Diekert and Thauer, 1980) and in the closely related but nonacetogenic microbe *Clostridium pasteurianum* (Diekert *et al.*, 1979), CO dehydrogenase activity was shown to be dependent on the presence of nickel in the medium. Using ^{63}Ni radioisotope, Drake *et al.* (1980) demonstrated that 14-fold-purified CO dehydrogenase from *C. thermoaceticum* possesses nickel, a result later corroborated by Diekert and Ritter (1983a). This metal ion was shown to be also a component of the enzyme from *Acetobacterium woodii* (Diekert and Ritter, 1982; Ragsdale *et al.*, 1983c), *C. pasteurianum* (Drake, 1982), and *Acetogenium kivui* (Yang *et al.*, 1989). Similarly, *Butyribacterium methylotrophicum* (Lynd *et al.*, 1982), *Eubacterium limosum* (Genthner *et al.*, 1982), *Peptostreptococcus productus* (Lorowitz and Bryant, 1984), and numerous other less well characterized acetogenic bacteria (Fuchs, 1986) are known to contain active CO dehydrogenase activity and are likely to possess analogous nickel-containing enzymes.

Despite its extreme sensitivity to oxygen, active CO dehydrogenase has been isolated from *A. woodii* and *C. thermoaceticum* by using strictly anaerobic procedures (Ragsdale *et al.*, 1983a,c). Several properties of these proteins are summarized in Table 5-1. Acetogen CO dehydrogenases are apparent trimers of dimers, each comprised of two subunits of M_r 78,000–80,000 and 68,000–71,000. A more recent report (Ramer *et al.*, 1989) suggested that the *C. thermoaceticum* enzyme possesses a third subunit (M_r 50,000) in equivalent stoi-

Table 5-1. Properties of Purified Nickel-Containing CO Dehydrogenases

Microorganism	Subunit M_r	Metal content[a]	Reference(s)
Acetobacterium woodii	80,000, 68,000	1.4 Ni, 9 Fe, 1.0 Mg or Zn/153,000	Ragsdale et al. (1983a)
Clostridium thermoaceticum	78,000, 71,000	1.7 Ni, 10.8 Fe, 1.3 Zn/155,000	Ragsdale et al. (1983c)
Desulfovibrio desulfuricans	NR[b]	NR[c,d]	Meyer and Fiebig (1985)
Methanococcus vannielii	89,000, 21,000	1.0 Ni, 8 Fe, 0.2 Zn/110,000	DeMoll et al. (1987)
Methanosarcina barkeri	92,000, 18,000	0.75 Ni, 15 Fe, 0.5 Cu, 0.5 Zn/116,000	Krzycki and Zeikus (1984); Krzycki et al. (1989)
Methanosarcina barkeri	84,500, 19,700	0.65 Ni, 7.8 Fe/104,200	Grahame and Stadtman (1987a)
Methanosarcina barkeri	84,500, 63,200, 53,000, 51,400, 19,700	0.86 Co[e]/271,800	Grahame (1991)
Methanosarcina thermophila	89,000, 71,000, 60,000, 58,000, 19,000	3.6 Ni, 25 Fe, 1.2 Co, 6.1 Zn/297,000	Terlesky et al. (1986)
Methanosarcina thermophila	89,000, 19,000	0.21 Ni, 7.7 Fe, 2.7 Zn/108,000	Abbanat and Ferry (1991)
Methanothrix soehngenii	79,400, 19,400	2.0 Ni, 12.5 Fe/98,800	Jetten et al. (1989a, 1991b)
Rhodospirillum rubrum	62,000	0.93 Ni, 8.0 Fe/62,000	Bonam and Ludden (1987); Ensign and Ludden (1991)
Rhodospirillum rubrum	62,000, 22,000	1.0 Ni, 14.2 Fe/84,000	Ensign and Ludden (1991)

[a] Number of metal ions per M_r of the minimal form of the enzyme.
[b] NR, Not reported.
[c] Nickel was shown to be present by co-chromatography of ^{63}Ni and enzyme.
[d] Iron was shown to be present spectroscopically.
[e] Nickel and iron are present, but were not quantitated.

chiometry to the other two, but which readily dissociates during gel filtration chromatography. However, Lu and Ragsdale (1991) have summarized evidence that the third subunit is not required for any of the known functions for this enzyme. The acetogen enzymes appear to possess two nickel ions per $\alpha\beta$ dimer and additionally contain several iron–sulfur clusters and a zinc and/ or magnesium ion. One of the nickel ions is reversibly removed upon addition of 1,10-phenanthroline without loss of CO dehydrogenase activity (Shin and Lindahl (1992a,b) but with effects on other activities (see section 5.2.2). The requirement for an electron acceptor is not met by flavins, nicotinamides, cytochrome c_3, or several other potential electron carriers, whereas Drake et *al.* (1980) showed that ferredoxin and a membrane-bound *b*-type cytochrome isolated from *C. thermoaceticum* is reduced by the enzyme in the presence of carbon monoxide. The most efficient electron acceptor for CO dehydrogenase, however, was found to be rubredoxin (Ragsdale *et al.,* 1983a,c). Although the enzyme is capable of metabolizing carbon monoxide, this chemistry is only a side reaction, and the *in vivo* role for CO dehydrogenase in acetogenic bacteria has proven to be far more complex. As described below, this enzyme, more aptly named acetyl coenzyme A synthase, functions in a novel autotrophic pathway.

5.2.2 Acetogen CO Dehydrogenases Function as Acetyl Coenzyme A Synthases

Early work carried out with acetogenic bacteria fermenting glucose had demonstrated that each mole of substrate is converted into three moles of acetate [reviewed by Ljungdahl and Wood (1982)]. The overall pathway can be broken down to four multistep reactions in a simplified treatment that ignores ATP synthesis: the glucose is initially degraded by the Embden–Meyerhof pathway to form two molecules of pyruvate (Eq. 5-2), one-half of the pyruvate is converted to acetate and carbon dioxide (Eq. 5-3), the carbon dioxide is both reduced to the methyl level and bound to tetrahydrofolic acid (H_4folate; Eq. 5-4), and the methyl group reacts with the second pyruvate to give rise to two additional acetate molecules (Eq. 5-5).

$$C_6H_{12}O_6 \rightarrow 2 \text{ pyruvate} + 4H^+ + 4e^- \qquad (5\text{-}2)$$

$$\text{pyruvate} + H_2O \rightarrow \text{acetate} + CO_2 + 2H^+ + 2e^- \qquad (5\text{-}3)$$

$$CO_2 + H_4\text{folate} + 6H^+ + 6e^- \rightarrow CH_3\text{-}H_4\text{folate} + 2H_2O \qquad (5\text{-}4)$$

$$CH_3\text{-}H_4\text{folate} + \text{pyruvate} + H_2O \rightarrow 2 \text{ acetate} + H_4\text{folate} \qquad (5\text{-}5)$$

The most remarkable aspect of this pathway involves the last step; that is, pyruvate cleavage appears to yield acetate and a bound one-carbon group that subsequently joins to the H_4folate-bound methyl group to provide the final acetate molecule. Drake *et al.* (1981) purified five components from *Clostridium thermoaceticum* that together carry out the reaction shown in Eq. (5-5). These investigators showed that one of these components includes CO dehydrogenase activity. In a benchmark study that greatly simplified the system and enhanced our understanding of the pathway, Hu *et al.* (1982) used purified CO dehydrogenase and a methyltransferase to show that acetyl coenzyme A is synthesized from methyltetrahydrofolate, carbon monoxide, and coenzyme A. In the cell, acetyl coenzyme A is converted to acetate with the concomitant synthesis of ATP. Hu *et al.* (1982) also reported that the fraction containing CO dehydrogenase catalyzes an unprecedented reversible exchange between the carboxyl portion of acetyl coenzyme A and carbon monoxide. Pezacka and Wood (1984b) demonstrated that carbon monoxide-dependent synthesis of acetyl coenzyme A is related to the reaction shown in Eq. (5-5). Using [1-^{14}C]pyruvate, they showed that CO dehydrogenase binds the radiolabeled carbon atom and couples it with methyltetrahydrofolate to form [1-^{14}C]acetyl coenzyme A. Thus, CO dehydrogenase clearly participates in an important step during sugar fermentation by *C. thermoaceticum*.

The number of functions ascribed to CO dehydrogenase in acetogens was subsequently expanded to include an essential role in a novel mode of autotrophic growth. Diekert and Ritter (1982) and Kerby and Zeikus (1983) showed that *Acetobacterium woodii* and *C. thermoaceticum* can grow by using carbon dioxide and hydrogen as the source of carbon and energy. Diekert and Ritter (1982) demonstrated that the autotrophic growth requires nickel ion in the medium. Kerby and Zeikus (1983) found that carbon dioxide and hydrogen can be replaced by carbon monoxide. These observations hinted at the involvement of a nickel-containing CO dehydrogenase in autotrophic growth in acetogens. This suspicion was confirmed when Pezacka and Wood (1984a) demonstrated that acetyl coenzyme A is synthesized *in vitro* from carbon dioxide, hydrogen, methyltetrahydrofolate, coenzyme A, and ATP and that this synthesis requires CO dehydrogenase in addition to helper enzymes and other components. Several lines of supportive *in vivo* evidence for the participation of CO dehydrogenase in acetate synthesis include (i) the detection of carbon monoxide from acetogenic bacteria growing on fructose or glucose (Diekert *et al.*, 1984) or on hydrogen and carbon dioxide (Diekert *et al.*, 1986), (ii) the demonstration that trace levels of radiolabeled carbon monoxide are incorporated into the carboxyl group of acetate in *A. woodii* cells that are grown on hydrogen and carbon dioxide (Diekert and Ritter, 1983b; Diekert *et al.*, 1986) or in *A. woodii* and *C. thermoaceticum* cells that are growing by sugar fermentation (Diekert *et al.*, 1984), and (iii) nuclear magnetic resonance spectroscopic evidence for incorporation of carbon monoxide into the carboxyl group of acetate by *Butyribacterium methylotrophicum* and *A. woodii* cells (Kerby *et al.*, 1983). Thus, CO dehydrogenase was found to be an important enzyme for both heterotrophic and autotrophic growth in acetogens.

The *in vivo* role for CO dehydrogenase in acetogenesis has recently been elucidated by Pezacka and Wood (1984a,b) and Ragsdale and Wood (1985) using the enzyme isolated from *C. thermoaceticum*. A simplified scheme illustrating its function as an acetyl coenzyme A synthase is shown in Fig. 5-1. Many experiments consistent with this scheme have been duplicated using enzyme from *A. woodii* (Shanmugasundaram *et al.*, 1988b). CO dehydrogenase

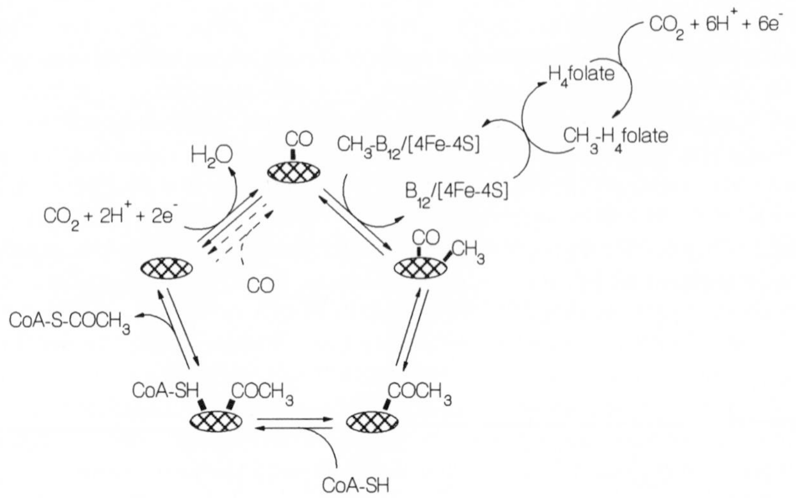

Figure 5-1. Simplified scheme depicting the role of acetogen CO dehydrogenase as an acetyl coenzyme A synthase. The autotrophic pathway involves catalysis of carbon–carbon bond formation between one-carbon units at the methyl and CO level. The CO-level one-carbon unit can be derived directly from CO gas or from reduction of carbon dioxide by CO dehydrogenase (represented by the cross-hatched ovals). The methyl unit is ultimately derived from a second molecule of carbon dioxide. This CO_2 is reduced to the methyl level and bound to tetrahydrofolate by other enzymes, transferred to a corrinoid/iron–sulfur protein (B_{12}/[4Fe-4S]), and then transferred to CO dehydrogenase. The nickel-containing acetogen enzyme catalyzes a carbonylation reaction to yield an acetylated enzyme intermediate. Coenzyme A (CoA-SH) additionally binds to the enzyme and reacts with the acetyl group to generate acetyl coenzyme A, which can be used in biosynthesis or for ATP generation with subsequent acetate release.

(represented by the cross-hatched oval symbol) can reversibly bind CO, or, in the presence of reduced ferredoxin, it can bind and reduce carbon dioxide to yield a bound one-carbon unit at the reduction level of CO. A second molecule of carbon dioxide is reduced to the methyl level in a series of reductive steps. The methyl group is transferred to a corrinoid/iron–sulfur protein (B_{12}/ [4Fe-4S]) that, in turn, transfers it to the CO dehydrogenase. The corrinoid/ iron–sulfur protein has been purified and extensively characterized (Hu *et al.,* 1984; Ragsdale *et al.,* 1987; Harder *et al.,* 1989). Figure 5-1 shows the CO binding first followed by the methyl group; however, there appears to be a random order of addition. Subsequent to the addition of the two one-carbon groups, the enzyme is proposed to carry out a carbonylation reaction in which a carbon–carbon bond is established between the bound CO and methyl groups. Another substrate, coenzyme A (CoA-SH), binds to a third site on the enzyme; both tryptophan and arginine residues have been implicated as being proximal to the coenzyme A binding site (Shanmugasundaram *et al.,*

1988a, 1989). Finally, coenzyme A is joined to the acetyl group, and acetyl coenzyme A is released from the enzyme to be used in biosynthesis or for ATP synthesis with acetate production.

The series of reactions illustrated in the pathway of Fig. 5-1 highlight the remarkable chemistry attributed to acetyl coenzyme A synthase. Below, the individual reactions carried out by the purified enzyme are detailed.

Clearly, the most important reaction is acetyl coenzyme A synthesis from CO, methylated corrinoid/iron–sulfur protein, and coenzyme A (Eq. 5-6; Lu and Ragsdale, 1991). Prior reaction of the enzyme (CODH) with methylated-corrinoid/iron–sulfur protein affords a methylated enzyme (CH_3-CODH; Eq. 5-7), that can generate acetyl coenzyme A upon addition of CO and coenzyme A (Eq. 5-8). In addition, several partial reactions are known to be catalyzed by the enzyme and provide further insight into its mechanism. Coenzyme A can exchange with the coenzyme A moiety of acetyl coenzyme A (Eq. 5-9; Pezacka and Wood, 1986), consistent with reversibility of the last two steps in the reaction. Also, CO can exchange with the carbonyl group of acetyl coenzyme A (Eq. 5-10; Ragsdale and Wood, 1985), providing clear evidence for the presence of distinct binding sites on the enzyme for the methyl, CO, and coenzyme A groups. Importantly, 1,10-phenanthroline-treated CO dehydrogenase, lacking one of the two nickel ions in the native protein, is incapable of catalyzing the reaction, but the activity can be recovered by subsequent addition of exogenous nickel ion (Shin and Lindahl, 1992a,b). Methyl exchange reactions, between methylated enzyme and methyl-corrinoid protein (Eq. 5-11) or methylated enzyme and the methyl group of acetyl coenzyme A (Eq. 5-12), are consistent with the presence of two methyl binding sites on the enzyme (Lu et al., 1990). Indeed, Pezacka and Wood (1988) have obtained evidence that a cysteine residue serves as the methyl acceptor on the enzyme, whereas Lu et al. (1990) argued on the basis of potentiometric analysis of the methyl exchange reaction that the methyl acceptor is most consistent with a metallocenter. Perhaps both groups are correct in that the methyl moiety may bind to two distinct sites; however, only one of these sites may be enzymatically significant, and the other may arise from a nonfunctional methyl-transfer reaction. Finally, acetyl coenzyme A synthase also carries out the apparently nonphysiological CO dehydrogenase reaction (Eq. 5-1).

$$CO + CH_3\text{-}B_{12}/[4Fe\text{-}4S] + CoA\text{-}SH \rightarrow$$

$$CoA\text{-}S\text{-}CO\text{-}CH_3 + B_{12}/[4Fe\text{-}4S] \quad (5\text{-}6)$$

$$CODH + CH_3\text{-}B_{12}/[4Fe\text{-}4S] \leftrightarrow CH_3\text{-}CODH + B_{12}/[4Fe\text{-}4S] \quad (5\text{-}7)$$

$$CH_3\text{-}CODH + CO + CoA\text{-}SH \rightarrow CoA\text{-}S\text{-}CO\text{-}CH_3 + CODH \quad (5\text{-}8)$$

$$CoA\text{-}S\text{-}CO\text{-}CH_3 + *CoA\text{-}SH \leftrightarrow *CoA\text{-}S\text{-}CO\text{-}CH_3 + CoA\text{-}SH \qquad (5\text{-}9)$$

$$CoA\text{-}S\text{-}CO\text{-}CH_3 + *CO \leftrightarrow CoA\text{-}S\text{-}*CO\text{-}CH_3 + CO \qquad (5\text{-}10)$$

$$CH_3\text{-}CODH + *CH_3\text{-}B_{12}/[4Fe\text{-}4S] \leftrightarrow$$

$$*CH_3\text{-}CODH + CH_3\text{-}B_{12}/[4Fe\text{-}4S] \qquad (5\text{-}11)$$

$$CH_3\text{-}CODH + CoA\text{-}S\text{-}CO\text{-}*CH_3 \leftrightarrow$$

$$*CH_3\text{-}CODH + CoA\text{-}S\text{-}CO\text{-}CH_3 \qquad (5\text{-}12)$$

Using highly purified preparations of acetyl coenzyme A synthase, corrinoid/iron–sulfur protein, methyltransferase, and ferredoxin, Roberts *et al.* (1992) demonstrated that acetyl coenzyme A is formed from methyltetrahydrofolate, CO, and coenzyme A at a rate that matches that observed for *in vivo* acetogenesis. Thus, *in vitro* studies carried out with these proteins are likely to be physiologically significant. Furthermore, these studies showed that the rate-limiting step in the reaction is methyl transfer from the methylated corrinoid/iron–sulfur protein to the acetyl coenzyme A synthase protein, rather than the series of steps that occur on the CO dehydrogenase itself.

In the acetogen cell, acetyl coenzyme A synthase is likely to interact with other bacterial components. A ferredoxin-dependent CO dehydrogenase disulfide reductase has been suggested to play a key role in the reaction cycle by reducing an internal disulfide of the enzyme (Pezacka and Wood, 1986). *In vitro,* this disulfide reductase can be replaced by dithioerythritol; however, *in vivo* it is tightly bound to the enzyme and may account for the third subunit described by Ramer *et al.* (1989). Furthermore, ferredoxin appears to form an electrostatically stabilized complex with CO dehydrogenase and is functional when covalently cross-linked to the enzyme (Shanmugasundaram and Wood, 1992). Thus, the *in vivo* form of the enzyme responsible for carrying out the reaction illustrated in Fig. 5-1 is probably a large acetyl coenzyme A synthase complex.

5.2.3 Molecular Biological Studies of Acetogen Acetyl Coenzyme A Synthase

A cluster of genes associated with *Clostridium thermoaceticum* acetyl coenzyme A synthase has been cloned and expressed in *Escherichia coli* (Roberts *et al.,* 1989). This gene cluster includes the coding region for the two CO dehydrogenase subunits, the genes encoding the two subunits of the corrinoid/iron–sulfur protein, and a methyltransferase gene (Fig. 5-2). The methyltransferase gene product, catalyzing the transfer of a methyl group from methyltetrahydrofolate to the corrinoid/iron–sulfur protein, is expressed as a fully

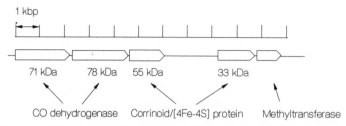

Figure 5-2. Acetyl coenzyme A synthase gene cluster. Several genes associated with acetyl coenzyme A synthase have been cloned and at least partially characterized in *Clostridium thermoaceticum* (Roberts *et al.*, 1989). These genes include those encoding the two subunits of the CO dehydrogenase, the two subunits of the corrinoid/iron–sulfur protein, and the methyltransferase.

functional dimeric protein in *E. coli,* whereas the CO dehydrogenase and corrinoid/iron–sulfur proteins are inactive.

Morton *et al.* (1991) have recently reported the DNA sequence for the region encoding the two subunits of CO dehydrogenase. The two subunits are predicted to possess 729 and 674 amino acids (calculated M_r = 81,730 and 72,928) and contain a total of 31 cysteine residues. The cysteine residues are clustered near the amino-terminal and carboxyl-terminal ends of each subunit in a pattern reminiscent of that found in iron–sulfur proteins. The genes for the two subunits are transcribed in the same orientation (small subunit, then large subunit) and separated by only 23 base pairs, consistent with an operon structure. The sequence for one subunit of the acetogen CO dehydrogenase is related to CO dehydrogenase sequences from other microbes, as noted in Sections 5.3.3 and 5.4.1, but neither peptide exhibits significant similarity to sequences of nickel-containing hydrogenases or to other enzymes.

5.2.4 Properties of the Acetogen Acetyl Coenzyme A Synthase Nickel Active Site

The detailed mechanism of *Clostridium thermoaceticum* acetyl coenzyme A synthase has been analyzed by numerous approaches. Here, I review certain aspects of the enzyme chemistry that relate to the mechanism; later in this section, I summarize results from spectroscopic characterization of the CO dehydrogenase metallocenters.

Stereochemical analysis has demonstrated that both CO exchange into acetyl coenzyme A (Raybuck *et al.*, 1987) and methyl transfer from methyltetrahydrofolate to the acetyl group (Lebertz *et al.*, 1987) lead to retention of methyl group configuration. Kinetic analysis of the CO and coenzyme A exchange reactions with acetyl coenzyme A has been reported (Raybuck *et al.*, 1988; Ramer *et al.*, 1989); the coenzyme A exchange rate exceeds the CO exchange rate as expected for an enzyme which is not designed to use free CO. The CO exchange reaction is highly sensitive to oxygen compared to the CO dehydrogenase activity. Furthermore, the CO dehydrogenase activity is unaffected by 1,10-phenanthroline-induced removal of the labile nickel ion, whereas the exchange activity is completely abolished (Shin and Lindahl, 1992a,b). These results, plus pronounced differences in inhibition by coenzyme A analogs for the two reactions (Raybuck et al., 1988), indicate that the CO dehydrogenase activity occurs at a site distinct from that involved in acetyl coenzyme A synthesis. However, both the coenzyme A/acetyl coenzyme A exchange reaction and methyl transfer reactions associated with this enzyme are dependent on the reduction potential (Lu and Ragsdale, 1991; Lu et al., 1990), and the measured reduction potential associated with these reactions matches that observed for carbon dioxide reduction (Lindahl et al., 1990a). Thus, all of the chemistry related to this enzyme may occur at different sites on or near the same metallocenter.

As summarized in Table 5-1, acetogen acetyl coenzyme A synthase $\alpha\beta$ dimers contain two nickel ions, several iron atoms, and at least one zinc ion. One of the nickel sites has been shown to reversibly dissociate as the nickel(II) ion in the presence of the chelator 1,10-phenanthroline (Shin and Lindahl, 1992a,b), but the identity of the second species remains unclear. Ragsdale et al. (1983a) reported that when nickel is extracted from the [63]Ni-labeled enzyme with perchloric acid, it chromatographs on a column of Sephadex G-25 as a complex ($M_r > 1350$) larger than the free cation, consistent with the presence of a nickel cofactor. This conclusion apparently has not stood the test of time as no subsequent mention of a nickel cofactor has appeared in the literature. In contrast, however, numerous spectroscopic studies have probed the protein-bound metallocenters.

Electron paramagnetic resonance (EPR) spectroscopy of anaerobic, dithionite-free enzyme was used to reveal the presence of two [4Fe-4S] centers (exhibiting g values of 2.04, 1.94, 1.90 and 2.01, 1.86, 1.75; Ragsdale et al., 1982) and a weak signal (g values of 2.21, 2.11, and 2.02; Ragsdale et al., 1983b) similar to the nickel(III) EPR signal observed in several hydrogenases (see Chapter 4). In the presence of CO or carbon dioxide, the enzyme exhibits a novel resonance (g values of 2.07 and 2.02; Ragsdale et al., 1982) that can be observed at 95 K. The latter signal accounts for 0.4 moles of spin per mole of enzyme and was attributed to a species that included Ni(III) (Ragsdale et al., 1982). Acetyl coenzyme A synthase was purified from both C. thermoaceticum and Acetobacterium woodii cells that were grown in the presence of [61]Ni-enriched medium. The observation of [61]Ni hyperfine-induced broadening confirms that the enzyme-bound nickel contributes, at least in part, to the novel CO-dependent signal and to the $g = 2.21, 2.11, 2.02$ signal (Ragsdale et al., 1983b). Broadening of the $g = 2.07, 2.02$ signal was also observed with

^{13}CO (nuclear spin, $I = \frac{1}{2}$) (Ragsdale *et al.*, 1983b) and with ^{57}Fe-enriched ($I = \frac{1}{2}$) enzyme (Ragsdale *et al.*, 1985), demonstrating that the CO carbon and the enzyme-bound iron and nickel are all spin-coupled in a spin $\frac{1}{2}$ complex which is responsible for the so-called NiFeC EPR spectrum. Some preparations of the CO-treated enzyme additionally contain a $g = 2.05$ signal that can be simulated with g values of 2.062, 2.047, and 2.028 (Ragsdale and Wood, 1985; Ragsdale *et al.*, 1985). The binding of coenzyme A converts the $g = 2.05$ signal to the original NiFeC signal, which can be simulated with g values of 2.074 and 2.028. These results and additional EPR spectroscopic changes observed in the presence of acetyl coenzyme A are consistent with coenzyme A and acetyl coenzyme A binding near the active-site nickel (Ragsdale *et al.*, 1985). The NiFeC signal is not observed in 1,10-phenanthroline-treated enzyme that contains only one nickel ion per catalytic unit (Shin and Lindahl, 1992a). The low spin quantitation noted for this signal (Ragsdale *et al.*, 1982) is not due to partial loss of the labile nickel ion during purification of the enzyme. Spin intensities on the order of 0.2 spin/mole of enzyme are present in cell extracts and in highly purified preparations that contain nearly a full complement of 2 Ni per molecule (Shin and Lindahl, 1993). These investigators suggest that the low spin value may reflect sample heterogeneity prior to purification. More recently, a variety of other methods have been used to further characterize the acetyl coenzyme A synthase metallocenters. Below, I discuss results obtained by X-ray absorption spectroscopy (XAS), controlled-potential EPR spectroscopy, Mössbauer analysis, electron-nuclear double-resonance (ENDOR) spectroscopy, and infrared spectroscopy for this nickel-containing enzyme.

Two groups have carried out XAS analysis of acetyl coenzyme A synthase: Cramer *et al.* (1987) characterized the rubredoxin-oxidized enzyme, and Bastian *et al.* (1988) the reduced enzyme. The former authors found that the nickel K absorption edge features are inconsistent with octahedral or tetrahedral geometries but consistent with either a square-pyramidal or a distorted square-planar site. XAS fine-structure analysis was used to obtain a best fit of approximately two Ni — S bonds of 2.21 Å and approximately two Ni — N/O bonds of 1.97 Å. Furthermore, reduction of the enzyme led to a shift in the absorption edge consistent with reduction of the nickel center (Cramer *et al.*, 1987). Bastian *et al.* (1988) examined the reduced, CO-free (Ni-EPR silent) form of the enzyme and reached similar conclusions; that is, their best fit was to 3.8 Ni — S bonds at 2.16 Å. They proposed that the nickel coordination was most consistent with square-planar geometry and pointed out that such a geometry leaves available the axial coordination sites for binding and chemical transformation of the substrates. Bastian *et al.* (1988) also found that the fit can be improved by inclusion of a metal scatterer at 3.25 Å, consistent with a possible Ni-S-Fe complex, but inconsistent with simple substitution of nickel for an iron atom in a [4Fe-4S] cluster. Iron XAS analysis is consistent with the presence of typical [4Fe-4S] clusters but would not be able to distinguish the scattering of one Fe—Ni bond in the large excess of Fe—S and Fe—Fe scatterers. All of the published XAS studies have presumably made use of the highest activity preparations available; that is, those containing nearly 2 Ni per molecule. However, it would be useful to compare results from these studies with those obtained on 1,10-phenanthroline-treated enzyme that lacks one of the two nickel sites yet retains full CO dehydrogenase activity.

In order to further characterize the various oxidation–reduction states of acetyl coenzyme A synthase, Lindahl *et al.* (1990a) combined EPR spectroscopy with controlled-potential coulometric reductive titrations. Under argon gas, oxidized samples of the acetogen protein require five to eight electrons per $\alpha\beta$ dimer for reduction. Four distinct EPR signals are detected: a $g_{ave} = 1.82$ signal, with a reduction potential of -220 mV, two $g_{ave} = 1.94$ signals differing in line width, with reduction potentials of -440 mV, and a $g_{ave} = 1.86$ signal associated with a reduction potential of -530 mV. These $S = \frac{1}{2}$ signals each account for only 0.2 to 0.3 spin per dimer, and additional features, perhaps corresponding to $S = \frac{3}{2}$ centers, were also observed at $g = 4$–6. In the presence of carbon dioxide, the previously described NiFeC signal was observed and found to have a potential of approximately -430 mV at pH 6.3 (Lindahl *et al.*, 1990a). This signal was shown by Gorst and Ragsdale (1991) to be catalytically competent; that is, the signal is formed at rates exceeding the CO/acetyl coenzyme A exchange rate. Furthermore, the NiFeC signal can be formed from acetyl coenzyme A, albeit with a very low spin concentration of 0.04 spin per mole of $\alpha\beta$ dimer, and is associated with a one-electron reduction potential of -541 mV (Gorst and Ragsdale, 1991). Additional reductive and oxidative titrations of the *C. thermoaceticum* enzyme have recently been reported by Shin *et al.* (1992). These workers used CO to reduce the protein and thionin to oxidize the protein while monitoring the metallocenters by UV-visible and EPR spectroscopies. They partially characterized three groups of redox centers: (i) four one-electron sites that include at least two iron–sulfur clusters, (ii) one or two one-electron sites that have absorbance properties that are distinct from those of typical iron–sulfur clusters, and (iii) sites involved in irreversible oxidation of the protein. Both of the former two groups are likely to be important in catalysis.

Mössbauer spectroscopy has proven to be a very powerful tool to study the metallocenters in acetyl coenzyme A synthase (Lindahl *et al.*, 1990b). All of the iron in anaerobically oxidized enzyme exhibits properties consistent with diamagnetic $[4Fe-4S]^{2+}$ clusters; since there are approximately 12 iron atoms per $\alpha\beta$ protein molecule, this result is consistent with the presence of three [4Fe-4S] clusters or metallocenters that are spectroscopically indistinguishable from this type. In the reduced, CO-bound form of the enzyme, several components were observed: approximately 40% of the iron corresponds to a $[4Fe-4S]^{1+}$ cluster, $\sim 18\%$ of the iron is associated with a possible [2Fe] cluster giving rise to the $g_{ave} = 1.82$ signal described above, $\sim 9\%$ of the iron corresponds to a reduced rubredoxin-type center, and $\sim 20\%$ of the iron is associated with the CO-induced EPR signal. The authors proposed that the latter signal may arise from a nickel site exchange-coupled to a $[4Fe-4S]^{2+}$ cluster where the metallocenters may be coupled through a bridging atom (X), perhaps CO. Similar to earlier EPR spectroscopic studies in which low spin concentrations were noted for several species, this study found that only ~ 0.35 spin per 12 iron atoms is associated with the NiFeC signal. The persistently low spin quantitation forced the authors to conclude that their enzyme preparation was spectroscopically heterogeneous despite being homogenous based on protein purity.

Another tool that has provided insight into the structure of the unique nickel-containing metallocenter in *C. thermoaceticum* acetyl coenzyme A synthase is ENDOR spectroscopy. Fan *et al.* (1991) used Q-band ENDOR spectroscopy to characterize the center giving rise to the NiFeC signal by isolating enzyme from cells grown in medium enriched in ^{57}Fe or ^{61}Ni and exposing these samples to CO and by incubating nonisotopically enriched enzyme in the presence of ^{13}CO. The ^{57}Fe hyperfine coupling results allowed the authors to infer that the center giving rise to the NiFeC signal contains three or four iron atoms. The ^{61}Ni ENDOR data, the first for any nickel-containing enzyme, are consistent with a single nickel site in the complex. The ^{13}C ENDOR data are unable to distinguish whether the CO is bound to the nickel or the iron. On the basis of the ENDOR studies and the Mössbauer data described above (Lindahl *et al.,* 1990b), Fan *et al.* (1991) proposed that the nickel-containing center in acetyl coenzyme A synthase has the following stoichiometry: $Ni_1Fe_{3-4}S_{\geq 4}C_1$.

Another spectroscopic method that has recently been used to probe the acetyl coenzyme A synthase metallocenter is Fourier transform infrared spectroscopy. In addition to the amide I and amide II bands at 1546 cm^{-1} and 1454 cm^{-1} that are typical of proteins, Kumar and Ragsdale (1992) were able to observe a spectral peak at 1995 cm^{-1} for the CO-bound form of the protein. Furthermore, substitution of ^{12}CO by ^{13}CO led to the expected 44-cm^{-1} isotopic shift in the band to 1951 cm^{-1}. The transitions giving rise to the observed CO-dependent bands were assigned to a terminally bound carbonyl group. These results exclude the possibility that CO forms a bridge between the nickel site and the iron–sulfur cluster, mentioned above. Rather, the center giving rise to the NiFeC signal is best described as $[NiXFe_{3-4}S_4] — C\equiv O$.

Hypothetical models for the NiFeC-signal-associated nickel center in acetyl coenzyme A synthase from acetogenic bacteria are shown in Fig. 5-3.

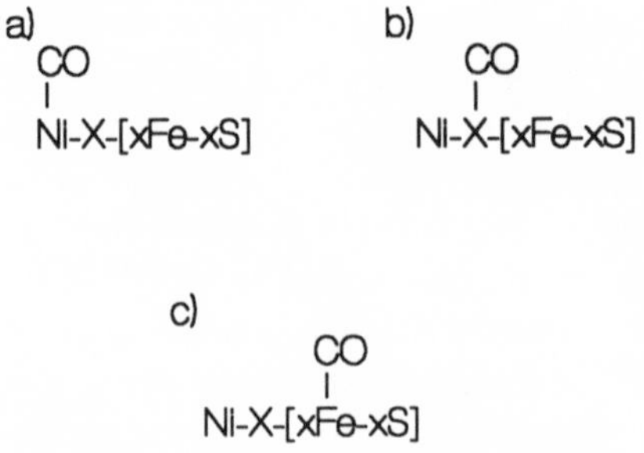

a)
CO
|
Ni-X-[xFe-xS]

b)
CO
|
Ni-X-[xFe-xS]

c)
CO
|
Ni-X-[xFe-xS]

Figure 5-3. Speculative models of the acetogen CO dehydrogenase active site. In the presence of CO, it is possible to observe an electron paramagnetic resonance spectroscopic signal, termed the NiFeC signal, that arises from a free electron that is distributed on nickel, an iron–sulfur center, and the carbon of CO (Ragsdale *et al.*, 1985). ENDOR spectroscopic measurements have demonstrated that the NiFeC signal arises from a metallocenter with an Ni_1-$Fe_{3-4}S_4C_1$ stoichiometry (Fan *et al.*, 1991); however, participation of the nickel in an [1Ni-3Fe-4S] cluster was precluded on the basis of X-ray absorption spectroscopic measurements (Bastian *et al.*, 1988). Hence, the spin-coupled system must involve a nickel site bridged to an [Fe-S] cluster. Infrared spectroscopic analysis was used to deduce terminal ligation of the CO (Kumar and Ragsdale, 1992); thus, the NiFeC signal is likely to arise from a bridged assembly where the CO binds to the nickel (a), to the bridging ligand, X (b), or to the iron–sulfur cluster (c).

5.2.5 Chemical Modeling of Acetogen Acetyl Coenzyme A Synthase

An important chemical approach to better understanding the biological reactions described above is to characterize analogous reactions carried out by model compounds. Functional modeling of CO dehydrogenase activity by a nickel catalyst has been reported (Lu *et al.*, 1993); however, it is now clear that microbial CO dehydrogenases actually catalyze far more sophisticated reactions than this simple chemistry. Exciting progress in modeling these complex reactions has been reported. For example, the acetyl coenzyme A condensation reaction between a methyl group and a carbonyl group is highly reminiscent of the Monsanto process for metal-promoted synthesis of acetate (Forster, 1979). In the industrial reaction, a rhodium catalyst facilitates the carbonylation of methanol to yield acetylated metal ion, which is subsequently hydrolyzed. As detailed below, more relevant and very elegant nickel-based chemical mimics of the acetyl coenzyme A synthase recently have been described by Holm and colleagues.

Stavropoulos *et al.* (1990, 1991) examined the acetyl coenzyme A synthase-related chemistry for nickel complexes containing tripodal ligands that provide a sulfur-rich environment for the metal ion. As shown in Fig. 5-4, the chloride salt of the nickel(II) species can be methylated to form the trigonal bipyramidal methyl-nickel complex. Subsequent addition of CO leads to instantaneous formation of the acetyl-nickel complex, which can undergo reaction with thiols to form acetyl thiolesters. In addition, the starting material was found to be capable of reduction with sodium borohydride to provide both the Ni-H and the nickel(I) species. The latter compound was shown to react with CO to yield the nickel–CO complex. Structures for each of the five

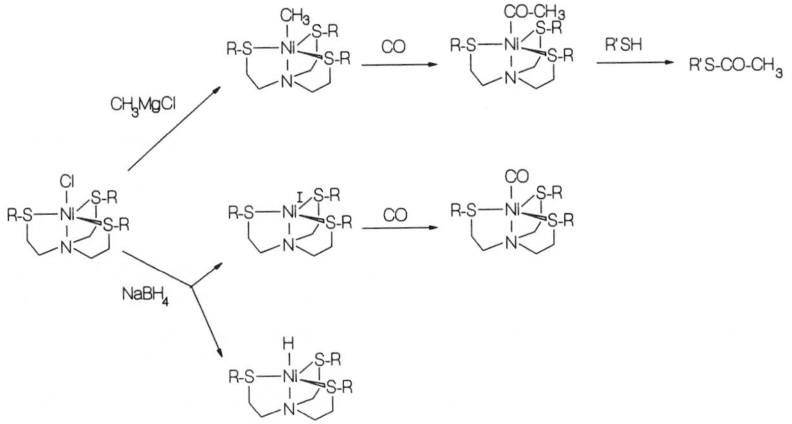

Figure 5-4. Chemical modeling of the acetyl coenzyme A synthase. Stavropoulos *et al.* (1990, 1991) have described the use of a nickel(II) complex with the indicated tripodal ligand (R = isopropyl or *t*-butyl group) to generate the methyl, acetyl, one-electron reduced, CO, and hydride species (all complexes have a 1+ charge that is not shown). Reaction of the acetylated ligand with thiol yields the acetyl thiolester.

nickel species have been determined by using X-ray crystallography. These complexes serve as excellent chemical models for possible intermediates in the acetyl coenzyme A synthase reaction.

Although the tripodal nickel complexes shown above carry out an impressive range of acetyl coenzyme A synthase-like reactions, these model compounds clearly do not mimic all aspects of the enzyme. For example, the thiolester formation step results in reduction of nickel to the nickel(0) state and precipitation of the ligand. Furthermore, this model nickel center is not associated with iron, as found in the enzyme. In this regard, Ciurli *et al.* (1990) have reported the synthesis of a model compound containing a [1Ni-3Fe-4S] cluster, and the same center has been produced in a ferredoxin (Conover *et al.*, 1990). This geometry, however, is unlikely to be pertinent to acetyl coenzyme A synthase as shown by the XAS measurements of Ni—Fe bond distances that are inappropriate for such a mixed metal cluster (Bastian *et al.*, 1988). Future model studies will work to overcome these limitations, for example, by analysis of model compounds containing bridged nickel/iron–sulfur-cluster assemblies.

5.2.6 CO Dehydrogenase-Dependent Autotrophic Growth in Nonacetogens

Autotrophic growth utilizing the acetyl coenzyme A synthase pathway not only occurs in acetogenic bacteria, as described above, but also in certain methanogenic and sulfate-reducing bacteria. A brief summary of the evidence for the presence of nickel-containing acetyl coenzyme A synthases in selected methanogens and sulfate reducers is provided below.

Many methanogenic bacteria are capable of growth in mineral medium by using carbon dioxide and hydrogen gases to yield methane:

$$CO_2 + 4H_2 \rightarrow CH_4 + 2H_2O \qquad\qquad (5\text{-}13)$$

In addition, these members of the archaea (formerly termed archaebacteria) must also use carbon dioxide as their sole carbon source for biosynthesis. Fuchs and Stupperich (1980) demonstrated that exogenous acetate can be incorporated into *Methanobacterium thermoautotrophicum* cellular material and proposed that the early steps in the autotrophic pathway lead to the synthesis of acetyl coenzyme A. *In vivo* radiolabeling studies clearly demonstrate that acetyl coenzyme A is, in fact, a rapidly synthesized intermediate in carbon assimilation and show that the two carbons in the acetyl group are labeled at different rates (Rühlemann *et al.*, 1985). Some of the acetyl moiety is released as acetate by *Methanosarcina* species (Westermann *et al.*, 1989), but at greatly reduced amounts compared to that observed in the acetogens. Stupperich and Fuchs (1983, 1984a) demonstrated that *in vitro* synthesis of acetyl coenzyme A from carbon dioxide in cell extracts of *M. thermoautotrophicum* is stimulated by methyl coenzyme M, an intermediate in the methane biosynthesis pathway, and is inhibited by methanogenesis inhibitors. More detailed analysis indicated that the methyl moiety of the acetyl group is derived from carbon dioxide by a pathway that shares the same initial intermediates as found in methanogenesis from carbon dioxide, whereas the carboxyl moiety is derived from a nonmethanogenesis pathway (Stupperich and Fuchs, 1984b). Holder *et al.* (1985) provided evidence, from *in vivo* and *in vitro* labeling studies, that the methyl groups of acetate and methane share the same one-carbon pathway at least through the formation of methylenetetrahydromethanopterin (methanopterin is a folate analog that is found in methanogenic bacteria). In contrast, as described below, the pathway for fixing the carboxyl group of acetyl coenzyme A is distinct and involves methanogen CO dehydrogenase.

Several species of methanogenic bacteria were shown to oxidize CO to carbon dioxide with concomitant reduction of the methanogen electron carrier, coenzyme F_{420} (Daniels *et al.*, 1977). The role for this CO dehydrogenase activity began to be clarified when Stupperich *et al.* (1983) demonstrated that CO is also incorporated specifically into the carboxyl group of acetyl coenzyme A during autotrophic growth of *M. thermoautotrophicum*. These results are consistent with the presence of an autotrophic system analogous to the clostridial acetyl coenzyme A pathway; that is, carbon dioxide appears to be reduced to form a protein-bound one-carbon unit equivalent to CO that subsequently joins to a methyl unit to form the acetyl group. The demonstration that CO is produced in autotrophic methanogen cultures provides further support for this pathway (Conrad and Thauer, 1983; Eikmanns *et al.*, 1985). Additional evidence was obtained by Länge and Fuchs (1987), who demonstrated that a partially purified protein fraction containing CO dehydrogenase activity and a corrinoid protein is capable of acetyl coenzyme A synthesis from coenzyme A, CO, and methylenetetrahydromethanopterin or methyltetrahydromethanopterin. Genetic evidence for the requirement for CO dehydrogenase in autotrophy has been obtained by Ladapo and Whitman (1990). Acetate auxotrophic mutants of *Methanococcus maripaludis* were isolated and shown to have greatly diminished levels of CO dehydrogenase activity. Perhaps related to this, methanogenic species that are unable to grow autotrophically (*Methanobrevibacter ruminantium, Methanobrevibacter smithii, Methanococcus voltae,* and *Methanospirillum hungatei* strain GP1) appear to be natural mutants that lack CO dehydrogenase activity (Bott *et al.*, 1985). Thus, autotrophic growth of methanogenic archaea appears to be a slight variation of the pathway shown in Fig. 5-1: the central role of a CO dehydrogenase as an acetyl coenzyme A synthase is retained, but reduction of carbon dioxide to the methyl level and transfer to a corrinoid protein uses a pathway unique to methanogens.

Several studies have begun to characterize the properties of purified methanogen CO dehydrogenases involved in autotrophic growth. Hammel *et al.* (1984) demonstrated that the enzyme activity from *Methanobrevibacter arboriphilicus* grown on hydrogen and carbon dioxide requires nickel ion in the medium. Furthermore, by using cells grown in the presence of [63]Ni, they showed that 21-fold-purified CO dehydrogenase comigrates with nickel. DeMoll *et al.* (1987) purified CO dehydrogenase from formate-grown *Methanococcus vannielii* and showed that the enzyme is comprised of two subunits (M_r = 89,000 and 21,000 in an $\alpha_2\beta_2$ stoichiometry) that contain two nickel ions per mole of enzyme (Table 5-1). Section 5.3 describes the properties of methanogen CO dehydrogenases that have been purified from acetate-grown cells. It is appropriate to mention here, however, that acetate-grown *Methanosarcina barkeri* cell suspensions synthesize acetate (Laufer *et al.*, 1987) from methyl iodide, carbon dioxide, and reducing equivalents. Furthermore, purified CO dehydrogenase from acetate-grown *Methanosarcina thermophila* catalyzes acetyl coenzyme A synthesis from methyl iodide, CO, and coenzyme A (Abbanat and Ferry, 1990).

In addition to the acetogens and the autotrophic methanogens, certain sulfate-reducing bacteria appear to utilize the acetyl coenzyme A synthase pathway for autotrophic growth (Fuchs, 1986). For example, Jansen *et al.* (1984, 1985) demonstrated that *Desulfovibrio baarsii* possesses many of the appropriate enzyme activities and exhibits the proper biosynthetic labeling patterns that one expects for an organism that uses this carbon dioxide fixation pathway. Other sulfate-reducing and sulfur-reducing bacteria also possess CO dehydrogenase and CO_2/acetyl coenzyme A exchange activities (Schauder *et al.*, 1986); however, these functions often appear to be involved in the degradation of acetate rather than its synthesis (see Section 5.3). *Desulfobacterium autotrophicum,* in contrast, appears to utilize CO dehydrogenase-dependent pathways both for reductive growth on carbon dioxide and for oxidative growth on acetate (Schauder *et al.*, 1989). As in the methanogens but not the acetogens, a tetrahydropterin derivative is used rather than tetrahydrofolate (Länge *et al.*, 1989). The only reported attempt to purify and characterize CO dehydrogenase from a sulfate-reducing bacterium is that for pyruvate-grown *Desulfovibrio desulfuricans* (Meyer and Fiebig, 1985): the enzyme was purified 585-fold and shown to possess nickel, based on co-chromatography of enzyme and [63]Ni, and iron–sulfur centers, based on UV-visible and EPR spectroscopic analysis. Remarkably, the *D. desulfuricans* enzyme is stable to oxygen, unlike most other reported CO dehydrogenases (see Section 5.3.1 for another exception). The stability of the protein makes this an attractive experimental system, and, because the spectroscopic studies have indicated clear differences from the *C. thermoaceticum* enzyme, the *D. desulfuricans* CO dehydrogenase needs to be characterized in greater detail.

5.3 CO Dehydrogenases Involved in Aceticlastic Methanogenesis

Most biologically produced methane is derived from methanogenic metabolism of acetate [reviewed by Ferry (1992a,b) and Jetten *et al.* (1992)]. Acetate-degrading methanogens carry out an aceticlastic reaction in which the carbon–carbon bond of acetate is split. The carboxyl group is subsequently oxidized to carbon dioxide, and the methyl group is reduced to methane.

$$CH_3\text{-}COOH \rightarrow CO_2 + CH_4 \qquad (5\text{-}14)$$

Although this reaction yields only a small amount of energy ($\Delta G = -36$ kJ/mol), these archaea are able to couple acetate degradation to ATP synthesis. Characterization of the aceticlastic reaction has been hampered by the slow growth rates of these microbes and by the poorly understood requirement for substrate adaptation exhibited by these cultures. With regard to the adaptation process, Krzycki *et al.* (1982) made the important observation that acetate-adapted *Methanosarcina barkeri* possesses significantly increased levels of CO dehydrogenase activity compared to cultures grown on methanol or carbon dioxide and hydrogen. This finding was the first indication that methanogen

CO dehydrogenase may function in the pathway for acetate degradation. This section will recount properties of CO dehydrogenases from aceticlastic methanogens, summarize evidence demonstrating that CO dehydrogenase plays a central acetate-cleaving role in these cells, review the molecular biological studies that have been carried out with this system, and detail aspects of the aceticlastic methanogen CO dehydrogenase nickel active site.

5.3.1 General Properties of Aceticlastic Methanogen CO Dehydrogenase

CO dehydrogenase is widely represented among the methanogenic bacteria (Daniels *et al.*, 1977). The enzyme has been purified from autotrophically grown *Methanococcus vannielii* (DeMoll *et al.*, 1987), as described in Section 5.2.6, and from acetate-grown *Methanosarcina barkeri* (Krzycki and Zeikus, 1984; Grahame and Stadtman, 1987a; Grahame, 1991), *Methanosarcina thermophila* (Terlesky *et al.*, 1986; Abbanat and Ferry, 1991), and *Methanothrix soehngenii* (Jetten *et al.*, 1989a). As summarized in Table 5-1, each of these enzymes, and partially purified enzyme from autotrophically growing *Methanobrevibacter arboriphilicus* (Hammel *et al.*, 1984), contains nickel. Moreover, each of these CO dehydrogenases has been isolated as $\alpha_2\beta_2$ species where the size of the two methanogen subunits ($M_r \approx 80,000-90,000$ and $18,000-21,000$) differs greatly from that observed in the enzyme isolated from acetogenic bacteria ($M_r = 78,000-80,000$ and $68,000-71,000$). The *M. thermophila* protein was initially isolated as a complex of five subunits (Terlesky *et al.*, 1986). Subsequent treatment with detergent yielded an active two-subunit form of the enzyme (Abbanat and Ferry, 1991). An additional component of the *M. thermophila* five-subunit complex is a dimeric corrinoid/iron–sulfur protein. This protein resembles the corrinoid/iron–sulfur protein associated with acetyl coenzyme A synthase, but again the subunit sizes differ significantly [$M_r = 60,000$ and $58,000$ for the methanogen protein (Abbanat and Ferry, 1991) versus $M_r = 55,000$ and $33,000$ for the protein isolated from acetogens (Hu *et al.*, 1984)]. A five-subunit, corrinoid-containing complex of the *M. barkeri* CO dehydrogenase also has recently been purified (Grahame, 1991), consistent with the possibility that an enzyme complex exists *in vivo* for all methanogen CO dehydrogenases.

The importance of CO dehydrogenase to methanogenic bacteria growing on acetate is evident from the amount of this enzyme which is produced. In acetate-grown *M. thermophila, M. barkeri,* and *M. soehngenii,* the enzyme accounts for ~10%, 5%, and 4% of the total soluble protein, whereas the enzyme only amounts to 0.2% of the protein in autotrophically growing *M. vannielii*. CO dehydrogenases from *M. vannielii, M. barkeri,* and *M. thermophila* are oxygen-labile, whereas the *M. soehngenii* enzyme is remarkably

stable to oxygen (Jetten *et al.*, 1989a), as previously noted for partially purified
CO dehydrogenase activity from *Desulfovibrio desulfuricans* (Meyer and Fiebig, 1985). The reason for the enhanced stability in the latter two enzymes is
unclear. Each of the methanogen CO dehydrogenases possesses millimolar
K_m values for CO, clearly indicating that CO is not a physiologically significant
substrate. Similar to the acetogen CO dehydrogenases that actually function
as acetyl coenzyme A synthases, the CO dehydrogenases from acetate-
degrading methanogens have been shown to have a role in acetate metabolism.

5.3.2 CO Dehydrogenase Functions in Aceticlastic Methanogenesis

A simplified current view for the role of CO dehydrogenase in aceticlastic
methanogens is illustrated in Fig. 5-5. Briefly, acetate is separately activated
to form acetyl coenzyme A, CO dehydrogenase reacts with the acetyl coenzyme
A to release coenzyme A and form an acetylated enzyme, and the acetyl group
is split to form enzyme-bound methyl and carbonyl groups; the methyl group
is transferred first to tetrahydromethanopterin (H_4MPT) with the participation
of a corrinoid/iron–sulfur protein (B_{12}/[Fe-S]) and then to coenzyme M
(CoM); and the protein-bound carbonyl group can exchange with CO or it

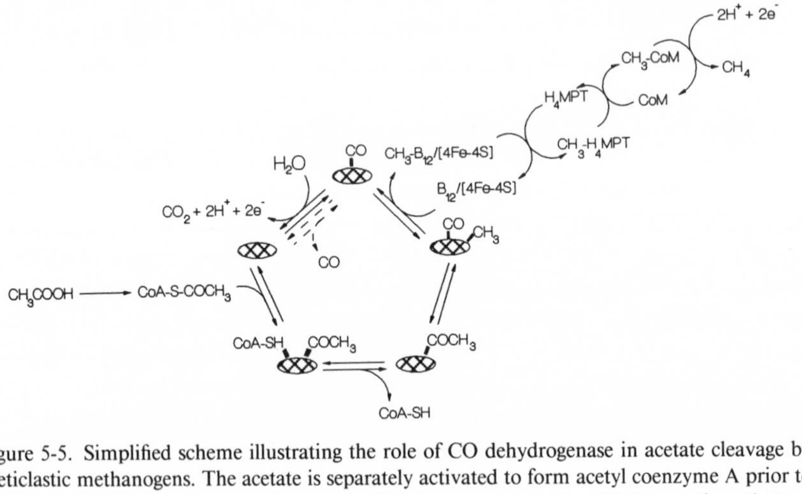

Figure 5-5. Simplified scheme illustrating the role of CO dehydrogenase in acetate cleavage by
aceticlastic methanogens. The acetate is separately activated to form acetyl coenzyme A prior to
binding to the CO dehydrogenase (represented by the cross-hatched ovals). By reactions that are
the reverse of the acetogen pathway, the acetyl coenzyme A is split to form coenzyme A, a bound
CO-level one-carbon unit, and a bound methyl-level one-carbon unit. The bound CO can reversibly
dissociate or it can be oxidized to carbon dioxide. Electrons generated in this oxidative process
are used to reduce the methyl group to form methane after methyl transfer to a corrinoid/[4Fe-
4S] protein, tetrahydromethanopterin (H_4MPT), and coenzyme M (CoM).

can be oxidized to form carbon dioxide, and the electrons generated in this reaction are eventually used to reduce the coenzyme M-bound methyl group to form methane. This pathway roughly corresponds to the reverse of the acetogen coenzyme A synthase pathway. Experiments that provide support for this pathway are described in more detail below.

Activation of acetate to form acetyl coenzyme A occurs either through use of acetyl coenzyme A synthetase (Eq. 5-15) or through a sequence of two reactions: phosphorylation of acetate by using acetate kinase (Eq. 5-16) and displacement of phosphate by coenzyme A in a reaction catalyzed by phosphotransacetylase (Eq. 5-17). The former enzyme has been purified from *Methanothrix soehngenii* (Jetten *et al.*, 1989b), and both of the latter enzymes have been purified from *Methanosarcina thermophila* (Aceti and Ferry, 1988; Lundie and Ferry, 1989). Consistent with a role in acetate catabolism, *M. soehngenii* and *M. thermophila* cells produce higher levels of these proteins when grown on acetate than when grown on methanol. Among many other methanogen species in which the latter two activities are present at high levels is *M. barkeri,* where acetyl phosphate has been shown to be an excellent substrate for methanogenesis in cell extracts in a coenzyme A-dependent manner (Grahame and Stadtman, 1987b; Fischer and Thauer, 1988).

$$CH_3\text{-}COO^- + ATP + CoA\text{-}S^- \rightarrow CoA\text{-}S\text{-}CO\text{-}CH_3 + AMP + PP_i \quad (5\text{-}15)$$

$$CH_3\text{-}COO^- + ATP \rightarrow CH_3\text{-}CO\text{-}OPO_3H^- + ADP \quad (5\text{-}16)$$

$$CH_3\text{-}CO\text{-}OPO_3H^- + CoA\text{-}S^- \rightarrow CoA\text{-}S\text{-}CO\text{-}CH_3 + HOPO_3^{2-} \quad (5\text{-}17)$$

The most convincing evidence for an acetate-cleaving role for methanogen CO dehydrogenase is derived from exchange studies. Purified preparations of CO dehydrogenase from *M. thermophila* (Raybuck *et al.*, 1991) and *M. soehngenii* (Jetten *et al.*, 1991a) catalyze an exchange between acetyl coenzyme A and CO (Eq. 5-10). An analogous exchange between carbon dioxide and the carboxyl group of acetate has been noted for whole cells of acetate-grown *M. barkeri* (Eikmanns and Thauer, 1984). This exchange is inhibited by cyanide, a potent inhibitor of CO dehydrogenase (Smith *et al.*, 1985). Additionally, exchange between coenzyme A and acetyl coenzyme A (Eq. 5-9) has been characterized for the purified *M. thermophila* enzyme (Raybuck *et al.*, 1991). Finally, as already noted, the purified *M. thermophila* CO dehydrogenase is capable of acetyl coenzyme A synthesis from methyl iodide, CO, and coenzyme A (Abbanat and Ferry, 1990). The principle of microscopic reversibility requires that the enzyme be capable of degrading acetyl coenzyme A to CO and coenzyme A in the presence of a suitable methyl acceptor.

The methyl transfer steps associated with the aceticlastic reaction are probably best understood in *M. barkeri*. Early evidence consistent with the involvement of a corrinoid enzyme in methanogenesis from acetate was provided by Eikmanns and Thauer (1985), who showed that propyl iodide is a potent, but light-reversible, inhibitor of the reaction. Using cell extracts of this microorganism, Fischer and Thauer (1989) identified a derivative of methyltetrahydromethanopterin as an intermediate in the reaction. More recently, Grahame (1991) used the purified five-subunit, corrinoid-containing CO dehydrogenase complex from this microbe to couple cleavage of acetyl coenzyme A to methylation of the tetrahydromethanopterin derivative, called tetrahydrosarcinapterin, that is found in this species. Also detected as products of the reaction were free coenzyme A and carbon dioxide. The tetrahydropterin-bound methyl group is subsequently transferred to coenzyme M, perhaps by action of a second corrinoid protein that also functions during growth on methanol or hydrogen and carbon dioxide. *In vivo* (Lovely *et al.,* 1984) and *in vitro* (Krzycki *et al.,* 1985) isotope labeling studies demonstrated that methyl coenzyme M is an intermediate in aceticlastic methanogenesis. Methyl coenzyme M reductase, a nickel-containing enzyme, is then used to reduce the methyl coenzyme M to produce methane (see Chapter 6).

Methane formation from methyl coenzyme M is a reductive process; that is, an electron donor is required for this reaction. Nelson and Ferry (1984) demonstrated that cell extracts of *M. thermophila* can reduce methyl coenzyme M by using CO. Thus, the ultimate source of electrons in the aceticlastic system is thought to be enzyme-bound CO. The detailed electron-transfer pathway may, however, be very complex. Kemner *et al.* (1987) performed optical spectroscopic analysis of a membrane preparation from *M. barkeri* cells and found results that are consistent with the presence of a *b*-type cytochrome that can be reduced in the presence of hydrogen gas and oxidized upon addition of methyl coenzyme M. In an elaboration of an earlier model (Zeikus *et al.,* 1985), the authors argued that the cytochrome and hydrogenase may function in energy conservation during aceticlastic methanogenesis. Consistent with such a model, hydrogen is both produced and consumed during growth of *M. thermophila* on acetate (Lovely and Ferry, 1985). If two hydrogenases are properly situated in the cell, this activity could be coupled to the development of a proton motive force. Furthermore, Terlesky and Ferry (1988) have shown that *M. thermophila* CO dehydrogenase can couple CO oxidation to production of hydrogen gas via ferredoxin, a *b*-type cytochrome, and a membrane-bound hydrogenase. A second hydrogenase may then be present in the cytoplasm to take up this hydrogen and couple its oxidation to methyl coenzyme M reduction (Ferry, 1992). Alternatively, Kemner (1993) provided evidence that CO oxidation in *M. barkeri* is coupled to release of hydrogen from a soluble hydrogenase and that a second, membrane-bound hydrogenase participates in electron transfer for methanogenesis. Despite the attractiveness of these hypotheses, hydrogen cycling has not been conclusively demonstrated to operate in aceticlastic methanogenesis.

CO dehydrogenases in aceticlastic methanogens appear to function in an acetate-degrading role by a pathway that is basically the reverse of that catalyzed by the acetogen acetyl coenzyme A synthase enzyme complex. Experiments with several acetate-degrading sulfate-reducing bacteria are consistent with a similar catabolic role for CO dehydrogenase in those organisms (Schauder *et al.*, 1986). High levels of CO dehydrogenase are present, and the cells catalyze an exchange between carbon dioxide and acetate. However, the only CO dehydrogenase from a sulfate reducer that has been partially purified and studied is that from *Desulfovibrio desulfuricans* (Meyer and Fiebig, 1985), for which no role in acetate metabolism has been demonstrated.

It is clear that CO dehydrogenases in anaerobes are utilized for both autotrophic synthesis and catabolic degradation of acetate, and one might expect that the methanogen and acetogen enzymes are highly related. As detailed below, however, the DNA-derived protein sequences of a methanogen CO dehydrogenase exhibit little similarity to the acetyl coenzyme A synthase proteins, and the properties of the nickel-containing active sites for the two complexes exhibit significant differences.

5.3.3 Molecular Biological Studies of Aceticlastic Methanogen CO Dehydrogenase

The *Methanothrix soehngenii* CO dehydrogenase structural genes have been cloned, sequenced, and expressed in *Escherichia coli* and *Desulfovibrio vulgaris* (Eggen *et al.*, 1991b). The two genes, *cdhA* and *cdhB,* encode proteins of 805 and 187 amino acids (calculated M_r of 89,461 and 21,008 compared to their apparent mobility during sodium dodecyl sulfate-polyacrylamide gel electrophoresis of 79,000 and 19,000). They are arranged in an operon-like structure with *cdhA* preceding *cdhB* by 20 base pairs, and both genes have consensus ribosome-binding sites. The small subunit is free of cysteine residues, whereas the large subunit contains 32 cysteines, including several that are in ferredoxin-like motifs. Furthermore, a stretch of 110 residues exhibits 24% identity to acyl-coenzyme A oxidase from *Candida tropicalis*. Remarkably, however, the methanogen-derived CO dehydrogenase sequences exhibit little similarity to sequences of the two subunits for the acetogen CO dehydrogenase (Morton *et al.*, 1991). The only stretches of identity noted are WATG and GSCV sequences (residues 495–500 and 548–551 in the *Clostridium thermoaceticum* β subunit and residues 545–550 and 583–586 in the *M. soehngenii* large subunit). Interestingly, both sequences also exhibit His-X-X-His-X-X-His-X-X-His motifs (residues 113–122 and 108–117, respectively). Kerby *et al.* (1992) have proposed an alignment of the sequences for these two proteins (and for the sequence of the *Rhodospirillum rubrum* enzyme) that exhibits relatively little overall similarity and requires numerous gaps in each sequence. Although the methanogen-derived peptides are expressed in both heterologous hosts, no CO dehydrogenase activity could be detected.

In addition to the two structural genes for CO dehydrogenase, the gene for the acetate-activating enzyme, acetyl coenzyme A synthetase, has been cloned and sequenced for *M. soehngenii* (Eggen *et al.,* 1991a). This gene is not located immediately adjacent to the CO dehydrogenase genes. Unlike the latter genes, the acetyl coenzyme A synthetase is expressed in a functional form in *E. coli.* Cloning and sequencing of methanogen genes encoding the corrinoid/iron–sulfur protein or the methyltransferase have not been reported.

5.3.4 Properties of the Aceticlastic Methanogen CO Dehydrogenase Nickel Active Site

Mechanistic characterization of methanogen CO dehydrogenase requires a detailed analysis of the enzyme metallocenters, particularly the nickel site. As in acetyl coenzyme A synthase, methanogen CO dehydrogenase participates in the reversible oxidation of CO, the CO/acetyl coenzyme A exchange reaction (Jetten *et al.,* 1991a; Raybuck *et al.,* 1991), and the coenzyme A/acetyl coenzyme A exchange reaction (Raybuck *et al.,* 1991). In addition, Bhatnagar *et al.* (1987) reported that the purified *M. barkeri* enzyme catalyzes a unidirectional hydrogen production activity, a finding that has not been verified by other researchers. Because the methanogen enzyme presumably catalyzes the reversal of the biosynthetic reaction observed in the acetogen enzyme, many of the same chemical and spectroscopic intermediates may be expected. This section will summarize the spectroscopic properties of the metallocenters that are found in CO dehydrogenases from different methanogen species.

Electron paramagnetic resonance spectroscopy of the five-subunit *Methanosarcina thermophila* CO dehydrogenase in the presence of CO reveals a spectrum with g values of 2.073, 2.049, and 2.028 (Terlesky *et al.,* 1986). This spectrum is similar to the NiFeC signal and the $g = 2.05$ signal observed in some preparations of CO dehydrogenase from *Clostridium thermoaceticum* (Ragsdale *et al.,* 1985). As in the acetogen enzyme case, the spectral features are broadened when CO dehydrogenase is purified from *M. thermophila* cells that are grown in the presence of ^{61}Ni, consistent with the spin density being localized at least partly on this metal ion (Terlesky *et al.,* 1986). A more detailed examination of the same enzyme demonstrated that the CO-dependent signal is actually comprised of two distinct species with g values of 2.090, 2.078, 2.030 and 2.057, 2.049, 2.027 (Terlesky *et al.,* 1987). Isotopic substitutions with ^{61}Ni, ^{57}Fe, or ^{13}CO led to spectral broadening of both signals, demonstrating that each is derived from a complex of nickel, iron, and the carbon of carbon monoxide. Furthermore, the former signal was quantitatively converted into the latter signal in the presence of acetyl coenzyme A, consistent with binding of acetyl coenzyme A near the metallocenter associated with the NiFeC signal (Terlesky et al., 1987). Again, this behavior is highly reminiscent of that observed in the C. thermoaceticum enzyme (Ragsdale et al., 1985).

Methanosarcina barkeri CO dehydrogenase has been examined by EPR spectroscopy both in vitro (Krzycki et al., 1989) and in vivo (Krzycki and Prince, 1990). Spectra of the purified $\alpha_2\beta_2$ form of the enzyme are consistent with the presence of two [4Fe-4S] clusters that generate signals in their reduced forms with g values of 2.05, 1.94, and 1.90 (with a reduction potential of -390 mV) and 2.01, 1.91, and 1.76 (reduction potential of -35 mV). Thionin oxidation generated two signals in the enzyme: a $g = 2.06$ species that may arise from copper and a $g = 2.016$ species that is consistent with the presence of a [3Fe-4S] cluster. The signals arising from iron–sulfur centers in the methanogen enzyme are very similar to signals previously described in Section 5.2.4 for the acetogen enzyme. However, upon addition of CO to the reduced enzyme, no signal analogous to the NiFeC signal in the C. thermoaceticum or M. thermophila enzymes was detected (Krzycki et al., 1989). Whole-cell EPR studies of M. barkeri (Krzycki and Prince, 1990) also failed to identify a NiFeC signal when CO was provided; rather, the only spectral perturbation was a shift in g value from 1.76 to 1.73 for the second [4Fe-4S] center. The reason for the lack of NiFeC signal in this enzyme is not clear.

EPR spectroscopy has been carried out on the Methanothrix soehngenii enzyme (Jetten et al., 1991a,b). The aerobically purified enzyme exhibits a single EPR signal at $g = 2.014$ that is characteristic of a [3Fe-4S] cluster. This signal is not affected by addition of CO or acetyl coenzyme A but disappears in the presence of dithionite. The addition of reductant also led to the generation of two new signals consistent with the presence of two [4Fe-4S] centers that are weakly coupled. Although the enzyme is oxygen-stable, it was also isolated anaerobically and characterized. Four distinct spectral signals were observed: one $S = \frac{1}{2}$ signal possesses g values of 2.048, 1.943, and 1.894 and is associated with a reduction potential of -410 mV, a second $S = \frac{1}{2}$ signal exhibits g values of 2.005, 1.894, and 1.733 with a reduction potential of -261 mV, and two further signals were attributed to $S = \frac{5}{2}$ and $S = \frac{9}{2}$ systems. None of the signals exhibit broadening from ^{61}Ni, and no NiFeC signal was observed. As in the case of C. thermoaceticum acetyl coenzyme A synthase, the signals associated with the M. soehngenii enzyme exhibit very low spin quantitation.

The active sites for methanogen CO dehydrogenases have not been as extensively studied as the C. thermoaceticum enzyme; however, it seems clear that significant differences exist between the aceticlastic system and the acetyl coenzyme A synthase. Further spectroscopic analysis and other types of enzyme characterization studies of both enzyme types are necessary to understand how these enzymes function in acetate synthesis and degradation. As described in the following section, a separate set of CO dehydrogenases appears to have no role in acetate metabolism.

5.4 CO Dehydrogenases Not Involved in Acetate Metabolism

CO dehydrogenase activity is present at high levels in certain photosynthetic bacteria. For example, Uffen (1976) reported that *Rubrivivax gelatinosus* (formerly *Rhodopseudomonas* or *Rhodocyclus gelatinosus*) can grow anaerobically in the dark under an atmosphere of 100% CO by carrying out the following reaction:

$$CO + H_2O \rightarrow CO_2 + H_2 \qquad (5\text{-}18)$$

The CO dehydrogenase activity in this purple, nonsulfur phototroph exhibits high affinity for CO ($K_m = 12.5 \ \mu M$), and the activity is induced by this substrate (Uffen, 1983). The membrane-associated *R. gelatinosus* enzyme has not been purified and the metal content is unknown; however, as described below, a similar, though readily solubilized, enzyme has been characterized from *Rhodospirillum rubrum*. Similar to the case of these phototrophs, high levels of CO dehydrogenase activity are present in *Clostridium pasteurianum* and certain strains of sulfate-reducing bacteria that are thought not to be capable of synthesis or degradation of acetate. Each of these cases is discussed in more detail, and alternative roles for CO dehydrogenase are considered.

5.4.1 *Rhodospirillum rubrum* CO Dehydrogenase

The nickel-containing *Rhodospirillum rubrum* CO dehydrogenase was partially purified and characterized by Bonam *et al.* (1984). The enzyme was enriched 600-fold to a specific activity of 95.4 units/mg from cells that were grown with glutamate under anaerobic, phototrophic conditions and shown to be heat-stable, oxygen-labile, and resistant to proteases. Optimal *in vitro* activity was observed at pH 10, where the K_m for CO is 110 μM (Bonam *et al.*, 1984).

A major breakthrough in characterizing the *R. rubrum* CO dehydrogenase was achieved by following up the observation of Uffen (1981) that CO induced high levels of the enzyme in this microbe. Bonam and Ludden (1987) grew

R. rubrum cells in the presence of CO and observed a 200-fold increase in specific activity compared to that measured previously (Bonam *et al.,* 1984). They were then able to readily purify CO dehydrogenase to homogeneity by a 212-fold enrichment process. In fact, two forms of the enzyme were isolated: the major form possesses ~50% higher activity (1079 versus 694 units/mg) and contains a higher metal content. Later studies by Ensign and Ludden (1991) suggested that the two forms represent a monomeric species of M_r 61,800 and a dimeric species in which the large peptide is associated with a second iron–sulfur-containing peptide of M_r 22,000. EPR spectroscopic analysis of either enzyme species revealed the development of a spectrum with g values of 2.042, 1.939, and 1.888 upon reduction with CO or dithionite (Bonam *et al.,* 1984). This signal was thought to be due to iron–sulfur centers, and no signal due to nickel could be identified in this study. Nickel is clearly present in both species as shown by atomic absorption analysis, and ^{63}Ni was incorporated into CO dehydrogenase when provided in the growth medium.

Another milestone in *R. rubrum* CO dehydrogenase research was the demonstration that the nickel-free enzyme could undergo both *in vivo* and *in vitro* activation by exogenous nickel (Bonam *et al.,* 1988). When cells are grown under nickel-depleted conditions in the presence of CO, a nickel-deficient form of CO dehydrogenase accumulates and can be purified as a monomeric species of $M_r = 61,800$. The nearly inactive sample is not an apoprotein because its iron–sulfur centers are intact, as demonstrated by UV-visible and EPR spectra that closely match analogous spectra of the holoenzyme. Whereas the iron–sulfur centers in the holoenzyme are rapidly reduced by CO, they fail to be reduced by this gas in the nickel-free protein (Ensign *et al.,* 1989a). These results are consistent with nickel serving to mediate electron flow from CO to the iron–sulfur centers in the protein. Direct electrochemical studies of holoenzyme and nickel-free protein demonstrated that both species possess equivalent one-electron, pH-independent reduction potentials of −418 mV (Smith *et al.,* 1992). The Ni-free CO dehydrogenase can be fully activated by addition of nickel to the cells, or purified protein can be activated *in vitro.* The final specific activity for fully active protein reported by Bonam *et al.* (1988) was 2640 units/mg, but Ensign *et al.* (1989a) later demonstrated even greater activity of 4400 units/mg. The kinetics and requirements for *in vitro* activation were further detailed by Ensign *et al.* (1990). Nickel incorporation is a two-phase process involving an initial binding step ($K_d = 755$ μM) followed by an irreversible association with the enzyme. For the binding to occur, an iron–sulfur center with a midpoint potential of −475 mV must be reduced. A scheme illustrating the steps involved in nickel incorporation is shown in Fig. 5-6.

A nickel and iron-containing metallocenter in *R. rubrum* CO dehydrogenase was identified by comparison of the EPR spectra for anaerobically

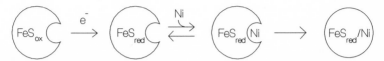

Figure 5-6. Mechanism of nickel incorporation into *Rhodospirillum rubrum* CO dehydrogenase. The nickel-free form of the enzyme, purified from cells grown in the absence of nickel, can be reconstituted to full activity only after a one-electron reduction of an iron–sulfur center (Ensign *et al.*, 1990). Nickel ion can then reversibly associate with the reduced, nickel-free form of the enzyme followed by an irreversible binding step.

oxidized holoenzyme and nickel-free protein (Stephens *et al.*, 1989). The methyl viologen-oxidized holoenzyme exhibits a signal near $g = 4.3$ and two overlapping signals: (i) $g = 2.04$, 1.90, and 1.71 and (ii) $g = 1.95$, 1.88, and 1.79. The latter signals can be distinguished by their power saturation and temperature dependence behavior. In contrast, the nickel-deficient protein lacks the $g = 4.3$ signal and the 2.04, 1.90, 1.71 signal, but retains the low-g-value signal. Although there are several spectral similarities to CO dehydrogenases from other microbes, no NiFeC signal, as noted in *Clostridium thermoaceticum* acetyl coenzyme A synthase or in the *Methanosarcina thermophila* acetate-degrading enzyme, is observed in the CO-treated *R. rubrum* enzyme. When holoenzyme was isolated from cells that were grown in the presence of [57]Fe, all three signals were broadened, consistent with each of these signals arising from a center that includes the participation of iron. Analogous studies with holoenzyme purified from cultures grown in the presence of [61]Ni demonstrated broadening of the 2.04 feature of the second signal, demonstrating that nickel contributes to the center that gives rise to this signal. Addition of cyanide, a slow-binding inhibitor of CO dehydrogenase that has been proposed to bind to the nickel (Ensign *et al.*, 1989b), led to the abolishment of the $g = 4.3$ signal and a shift in the second signal. These data are consistent with assignment of the $g = 4.3$ and $g = 2.04$ signals to the same metallocenter species.

Extended XAS analysis of the nickel center in the *R. rubrum* enzyme was used to identify atoms in the first coordination shell: two sulfur atoms at 2.23 Å and two to three nitrogen/oxygen atoms at 1.87 Å (Tan *et al.*, 1992). As in the *Clostridium thermoaceticum* acetyl coenzyme A synthase, the data preclude the nickel from being in a [Ni-3Fe-4S] cubane cluster. Near-edge data were interpreted as indicating either a distorted tetrahedral or five-coordinate geometry in both the oxidized and reduced states.

The *R. rubrum* enzyme possesses a single subunit when isolated from heat-treated cells (Bonam and Ludden, 1987) but can be isolated as a two-subunit species by using detergent or ethanol treatment of membranes (Ensign and Ludden, 1991). The second peptide (M_r 22,000) is an iron–sulfur protein that is thought to function in electron transfer from CO dehydrogenase to hydrogenase. In either case, the subunit structure of the phototroph-derived enzyme differs from that for both the acetogen and methanogen enzymes as shown in Table 5-1. As described above, the metallocenters in this protein are also clearly distinct from the complex array of centers in the acetogen and methanogen cases. Further biophysical studies of this simple system will distinguish whether the *R. rubrum* enzyme is a variation of the more complicated systems or whether it is structurally and functionally unique.

Kerby *et al.* (1992) have reported the cloning and sequencing of the genes in the *R. rubrum* CO dehydrogenase system. The structural gene for CO dehydrogenase (*cooS*) encodes a protein comprised of 639 amino acid residues. This sequence is 46% identical to the β subunit of *Clostridium thermoaceticum* CO dehydrogenase and 23% identical to the α subunit of the *Methanothrix soehngenii* enzyme. The *cooS* gene is immediately preceded by the *cooF* gene that encodes a 190-residue peptide. This peptide is predicted to contain 16 cysteines in four 4-cysteine motifs and is identical to the iron–sulfur protein that transfers electrons from CO dehydrogenase to hydrogenase. No significant sequence similarity was found between this electron-transfer peptide and either subunit of the acetogen or methanogen enzymes. Based on the *R. rubrum* CO dehydrogenase single subunit, one can reasonably conclude that the nickel binding site is localized to the β and α subunits of the acetogen and methanogen proteins. Approximately 450 bp upstream of *cooF* is a partial open reading frame (*cooH*) that exhibits a high degree of similarity to the large subunit of nickel-containing hydrogenases. Though not part of the same transcriptional unit as *cooF* and *cooS,* the *cooH* gene is likely to be the hydrogenase that is specifically induced by CO in this microbe (Bonam *et al.,* 1989). Finally, downstream of *cooS* is the start of an open reading frame that encodes a protein containing an ATP-binding motif. Although its role is unknown, this gene is expressed in a CO-dependent manner. Based on the precedence of accessory genes in the urease and hydrogenase gene clusters, one could speculate that the product of the downstream gene may function in maturation of CO dehydrogenase. The ability to easily manipulate *R. rubrum* DNA makes this an attractive system for detailed analysis of the CO dehydrogenase mechanism by site-directed mutagenesis.

The function of the *R. rubrum* CO dehydrogenase remains unclear. No evidence for a role in acetate metabolism has been identified. The high affinity for CO and the CO-specific induction of this enzyme are consistent with a role as a true CO dehydrogenase. The existence of an iron–sulfur electron-transfer peptide and a CO-inducible hydrogenase is consistent with *R. rubrum* cells possessing a system designed to use CO to reduce protons as a mode of energy generation. However, few naturally occurring environments are likely to possess CO concentrations that would make this a viable method for growth.

5.4.2 Other CO Dehydrogenases

Although closely related to acetogenic bacteria, *Clostridium pasteurianum* apparently does not synthesize acetate, yet this microbe possesses a nickel-containing CO dehydrogenase (Drake, 1982; Diekert *et al.,* 1979). It may be that growth conditions to enhance acetate synthesis have simply not been identified or that, under appropriate conditions, acetyl coenzyme A is made but only used for cell growth. Alternatively, it is possible that *C. pasteurianum* is a naturally occurring acetogen mutant that is defective in the acetyl co-enzyme A synthase complex while retaining a functional CO dehydrogenase. Finally, it remains possible that the enzyme plays an as yet unrecognized role in this bacterium. It has not been possible to purify the enzyme because of the extreme lability of the nickel center. Perhaps, the generation of CO dehydrogenase mutants of *C. pasteurianum* would help to distinguish the true role among the suggested possibilities.

The first evidence for anaerobic oxidation of CO was reported by Yagi (1958), using cell extracts of *Desulfovibrio desulfuricans.* The high levels of CO dehydrogenase activity in this microorganism are not associated with acetate degradation, as in many other sulfate reducers (Schauder *et al.,* 1986), because it lacks the ability to oxidize acetyl coenzyme A. Furthermore, no autotrophic role has been identified for CO dehydrogenase in this microbe. Partial purification of the enzyme from cells grown in the presence of ^{63}Ni did, however, demonstrate that nickel is present (Meyer and Fiebig, 1985). Similarly to the *D. desulfuricans* example, *D. vulgaris* possesses a highly active CO dehydrogenase that does not appear to function in acetate metabolism (Lupton *et al.,* 1984). As in the case of *C. pasteurianum,* appropriate growth conditions for acetate degradation or acetyl coenzyme A synthase activity may not have been identified in these microbes. Alternatively, these *Desulfovibrio* species may represent natural mutants that have a defective acetate-metabolizing system while retaining a functional CO dehydrogenase. Lastly, one may speculate that the enzyme could participate in a totally novel cellular role. For example, Lupton *et al.* (1984) suggested, without any experimental evidence, that the *D. vulgaris* CO dehydrogenase may function in the degradation of pyruvate to acetate. Again, further physiological and genetic approaches may help to discriminate among these possibilities.

5.5 Perspective

This chapter has described the properties of nickel-containing CO dehydrogenases from several types of microorganisms, summarized the evidence that this enzyme participates as an acetyl coenzyme A synthase in acetogenic bacteria and in selected autotrophs, discussed the acetate-degrading function of CO dehydrogenase in methanogens and certain sulfur and sulfate-reducing bacteria, and indicated that the enzyme functions in nonacetate metabolism in several other microbes. Many questions remain to be answered regarding the detailed mechanisms of catalysis for these enzymes and the structures of the nickel active sites. Further biophysical and spectroscopic comparison of these enzymes may facilitate our understanding of their functional similarities and differences. Crystallization and structural determination would clarify many aspects of the novel chemistries associated with CO dehydrogenases. Molecular biological studies of CO dehydrogenases have begun to provide sequence and gene organization information, but much remains to be done to dissect the regulation of the pathways and the interaction between components. Site-directed mutagenesis offers the hope of revealing the essential residues that participate in metallocenter ligation and catalysis. Excellent progress has been made in developing chemical mimics of the acetyl coenzyme A synthase pathway, but further modeling studies are required to fully understand the unique chemistry of these enzymes. Finally, one can consider the biotechnological possibilities of harnessing CO dehydrogenase for detoxification of CO or other uses. This nickel-containing enzyme clearly offers multiple exciting opportunities for future research.

References

Abbanat, D. R., and Ferry, J. G., 1990. Synthesis of acetyl coenzyme A by carbon monoxide dehydrogenase complex from acetate-grown *Methanosarcina thermophila, J. Bacteriol.* 172: 7145–7150.

Abbanat, D. R., and Ferry, J. G., 1991. Resolution of component proteins in an enzyme complex from *Methanosarcina thermophila* catalyzing the synthesis or cleavage of acetyl-CoA, *Proc. Natl. Acad. Sci. USA* 88:3272–3276.

Aceti, D. J., and Ferry, J. G., 1988. Purification and characterization of acetate kinase from acetate-grown *Methanosarcina thermophila.* Evidence for regulation of synthesis, *J. Biol. Chem.* 263:15444–15448.

Bastian, N. R., Diekert, G., Niederhoffer, E. C., Teo, B.-T., Walsh, C. T., and Orme-Johnson, W. H., 1988. Nickel and iron EXAFS of carbon monoxide dehydrogenase from *Clostridium thermoaceticum* strain DSM, *J. Am. Chem. Soc.* 110:5581–5582.

Bhatnagar, L., Krzycki, J. A., and Zeikus, J. G., 1987. Analysis of hydrogen metabolism in *Methanosarcina barkeri:* Regulation of hydrogenase and role of CO-dehydrogenase in H_2 production, *FEMS Microbiol. Lett.* 41:337–343.

Bonam, D., and Ludden, P. W., 1987. Purification and characterization of carbon monoxide dehydrogenase, a nickel, zinc, iron–sulfur protein, from *Rhodospirillum rubrum, J. Biol. Chem.* 262:2980–2987.

Bonam, D., Murrel, S. A., and Ludden, P. W., 1984. Carbon monoxide dehydrogenase from *Rhodospirillum rubrum, J. Bacteriol.* 159:693–699.

Bonam, D., McKenna, M. C., Stephens, P. J., and Ludden, P. W., 1988. Nickel-deficient carbon monoxide dehydrogenase from *Rhodospirillum rubrum: In vivo* and *in vitro* activation by exogenous nickel, *Proc. Natl. Acad. Sci. USA* 85:31–35.

Bonam, D., Lehman, L., Roberts, G. P., and Ludden, P. W., 1989. Regulation of carbon monoxide dehydrogenase and hydrogenase in *Rhodospirillum rubrum:* Effects of CO and oxygen on synthesis and activity, *J. Bacteriol.* 171:3102–3107.

Bott, M. H., Eikmanns, B., and Thauer, R. K., 1985. Defective formation and/or utilization of carbon monoxide in H_2/CO_2 fermenting methanogens dependent on acetate as carbon source, *Arch. Microbiol.* 143:266–269.

Ciurli, S., Yu, S.-B., Holm, R. H., Srivastava, K. K. P., and Münck, E., 1990. Synthetic $NiFe_3Q_4$ cubane-type clusters ($S = \frac{3}{2}$) by reductive rearrangement of linear $[Fe_3Q_4(SEt)_4]^{3-}$ (Q = S, Se), *J. Am. Chem. Soc.* 112:8169–8171.

Conover, R. C., Park, J.-B., Adams, M. W. W., and Johnson, M. K., 1990. Formation and properties of a $NiFe_3S_4$ cluster in *Pyrococcus furiosus* ferredoxin, *J. Am. Chem. Soc.* 112: 4562–4564.

Conrad, R., and Thauer, R. K., 1983. Carbon monoxide production by *Methanobacterium thermoautotrophicum, FEMS Microbiol. Lett.* 20:229–232.

Cramer, S. P., Eidsness, M. K., Pan, W.-H., Morton, T. A., Ragsdale, S. W., DerVartanian, D. V., Lungdahl, L. G., and Scott, R. A., 1987. X-ray absorption spectroscopic evidence for a unique nickel site in *Clostridium thermoaceticum* carbon monoxide dehydrogenase, *Inorg. Chem.* 26:2477–2479.

Daniels, L., Fuchs, G., Thauer, R. K., and Zeikus, J. G., 1977. Carbon monoxide oxidation by methanogenic bacteria, *J. Bacteriol.* 132:118–126.

DeMoll, E., Grahame, D. A., Harnly, J. M., Tsai, L., and Stadtman, T. C., 1987. Purification and properties of carbon monoxide dehydrogenase from *Methanococcus vannielii, J. Bacteriol.* 169:3916–3920.

Diekert, G., 1988. Carbon monoxide dehydrogenase of acetogens, in *The Bioinorganic Chemistry of Nickel* (J. R. Lancaster, Jr., ed.), VCH Publishers, New York, pp. 299–309.

Diekert, G., and Ritter, M., 1982. Nickel requirement of *Acetobacterium woodii, J. Bacteriol.* 151:1043–1045.

Diekert, G., and Ritter, M., 1983a. Purification of the nickel protein carbon monoxide dehydrogenase of *Clostridium thermoaceticum, FEBS Lett.* 151:41–44.

Diekert, G., and Ritter, M., 1983b. Carbon monoxide fixation into the carboxyl group of acetate during growth of *Acetobacterium woodii* on H_2 and CO_2, *FEMS Microbiol. Lett.* 17:299–302.

Diekert, G. B., and Thauer, R. K., 1978. Carbon monoxide oxidation by *Clostridium thermoaceticum* and *Clostridium formicoaceticum, J. Bacteriol.* 136:597–606.

Diekert, G., and Thauer, R. K., 1980. The effect of nickel on carbon monoxide dehydrogenase formation in *Clostridium thermoaceticum* and *Clostridium formicoaceticum, FEMS Microbiol. Lett.* 7:187–189.

Diekert, G. B., Graf, E. G., and Thauer, R. K., 1979. Nickel requirement for carbon monoxide dehydrogenase formation in *Clostridium pasteurianum, Arch. Microbiol.* 122:117–120.

Diekert, G., Hansch, M., and Conrad, R., 1984. Acetate synthesis from 2 CO_2 in acetogenic bacteria: Is carbon monoxide an intermediate?, *Arch. Microbiol.* 138:224–228.

Diekert, G., Schrader, E., and Harder, W., 1986. Energetics of CO formation and CO oxidation in cell suspensions of *Acetobacterium woodii, Arch. Microbiol.* 144:386–392.

Drake, H. L., 1982. Occurence of nickel in carbon monoxide dehydrogenase from *Clostridium pasteurianum* and *Clostridium thermoaceticum, J. Bacteriol.* 149:561–566.

Drake, H. L., Hu, S.-I., and Wood, H. G., 1980. Purification of carbon monoxide dehydrogenase, a nickel enzyme from *Clostridium thermoaceticum, J. Biol. Chem.* 255:7174–7180.

Drake, H. L., Hu, S.-I., and Wood, H. G., 1981. Purification of five components from *Clostridium thermoaceticum* which catalyze synthesis of acetate from pyruvate and methyltetrahydrofolate. Properties of phosphotransacetylase, *J. Biol. Chem.* 256:11137–11144.

Eggen, R. I. L., Geerling, A. C. M., Boshoven, A. B. P., and de Vos, W. M., 1991a. Cloning, sequence analysis, and functional expression of the acetyl coenzyme A synthetase gene from *Methanotrix soehngenii* in *Escherichia coli, J. Bacteriol.* 173:6383–6389.

Eggen, R. I. L., Geerling, A. C. M., Jetten, M. S. M., and de Vos, W. M., 1991b. Cloning, expression, and sequence analysis of the genes for carbon monoxide dehydrogenase of *Methanothrix soehngenii, J. Biol. Chem.* 266:6883–6887.

Eikmanns, B., and Thauer, R. K., 1984. Catalysis of an isotopic exchange between CO_2 and the carboxyl group of acetate by *Methanosarcina barkeri* grown on acetate, *Arch. Microbiol.* 138:365–370.

Eikmanns, B., and Thauer, R. K., 1985. Evidence for the involvement and role of a corrinoid enzyme in methane formation from acetate in *Methanosarcina barkeri, Arch. Microbiol.* 142:175–179.

Eikmanns, B., Fuchs, G., and Thauer, R. K., 1985. Formation of carbon monoxide from CO_2 and H_2 by *Methanobacterium thermoautotrophicum, Eur. J. Biochem.* 146:149–154.

Ensign, S. A., and Ludden, P. W., 1991. Characterization of the CO oxidation/H_2 evolution system of *Rhodospirillum rubrum.* Role of a 22-kDa iron–sulfur protein in mediating electron transfer between carbon monoxide dehydrogenase and hydrogenase, *J. Biol. Chem.* 266: 18395–18403.

Ensign, S. A., Bonam, D., and Ludden, P. W., 1989a. Nickel is required for the transfer of electrons from carbon monoxide to the iron–sulfur center(s) of carbon monoxide dehydrogenase from *Rhodospirillum rubrum, Biochemistry* 28:4968–4973.

Ensign, S. A., Hyman, M. R., and Ludden, P. W., 1989b. Nickel-specific, slow-binding inhibition of carbon monoxide dehydrogenase from *Rhodospirillum rubrum* by cyanide, *Biochemistry* 28:4973–4979.

Ensign, S. A., Campbell, M. J., and Ludden, P. W., 1990. Activation of the nickel-deficient carbon monoxide dehydrogenase from *Rhodospirillum rubrum:* Kinetic characterization and reductant requirement, *Biochemistry* 29:2162–2168.

Fan, C., Gorst, C. M., Ragsdale, S. W., and Hoffman, B. M., 1991. Characterization of the Ni-Fe-C complex formed by reaction of carbon monoxide dehydrogenase from *Clostridium thermoaceticum* by Q-band ENDOR, *Biochemistry* 30:431–435.

Ferry, J. G., 1992a. Methane from acetate, *J. Bacteriol.* 174:5489–5495.

Ferry, J. G., 1992b. Biochemistry of methanogenesis, *Crit. Rev. Biochem. Mol. Biol.* 27:473–503.

Fischer, R., and Thauer, R. K., 1988. Methane formation from acetyl phosphate in cell extracts of *Methanosarcina barkeri.* Dependence of the reaction on coenzyme A, *FEBS Lett.* 228: 249–253.

Fischer, R., and Thauer, R. K., 1989. Methyltetrahydromethanopterin as an intermediate in methanogenesis from acetate in *Methanosarcina barkeri, Arch. Microbiol.* 151:459–465.

Forster, D., 1979. Mechanistic pathways in the catalytic carbonylation of methanol by rhodium and iridium complexes, *Adv. Organomet. Chem.* 17:255–267.

Fuchs, G., 1986. CO_2 fixation in acetogenic bacteria: Variations on a theme, *FEMS Microbiol. Rev.* 39:181–213.

Fuchs, G., and Stupperich, E., 1980. Acetyl CoA, a central intermediate of autotrophic CO_2 fixation in *Methanobacterium thermoautotrophicum, Arch. Microbiol.* 127:267–272.

Genthner, B. R. S., and Bryant, M. P., 1982. Growth of *Eubacterium limosum* with carbon monoxide as the energy source, *Appl. Environ. Microbiol.* 43:70–74.

Gorst, C. M., and Ragsdale, S. W., 1990. Characterization of the NiFeCO complex of carbon monoxide dehydrogenase as a catalytically competent intermediate in the pathway of acetyl-coenzyme A synthesis, *J. Biol. Chem.* 266:20687–20693.

Grahame, D. A., 1991. Catalysis of acetyl-CoA cleavage and tetrahydrosarcinapterin methylation by a carbon monoxide dehydrogenase–corrinoid enzyme complex, *J. Biol. Chem.* 266:22227–22233.

Grahame, D. A., and Stadtman, T. C., 1987a. Carbon monoxide dehydrogenase from *Methanosarcina barkeri*. Disaggregation, purification, and physicochemical properties of the enzyme, *J. Biol. Chem.* 262:3706–3712.

Grahame, D. A., and Stadtman, T. C., 1987b. *In vitro* methane and methyl coenzyme M formation from acetate: Evidence that acetyl-CoA is the required intermediate activated form of acetate, *Biochem. Biophys. Res. Commun.* 147:254–258.

Hammel, K. E., Cornwell, K. L., Diekert, G. B., and Thauer, R. K., 1984. Evidence for a nickel-containing carbon monoxide dehydrogenase in *Methanobrevibacter arboriphilicus*, *J. Bacteriol.* 157:975–978.

Harder, S. R., Lu, W.-P., Feinberg, B. A., and Ragsdale, S. W., 1989. Spectroelectrochemical studies of the corrinoid/iron–sulfur protein involved in acetyl coenzyme A synthesis by *Clostridium thermoaceticum*, *Biochemistry* 28:9080–9087.

Holder, U., Schmidt, D.-E., Stupperich, E., and Fuchs, G., 1985. Autotrophic synthesis of activated acetic acid from two CO_2 in *Methanobacterium thermoautotrophicum*. III. Evidence for common one-carbon precursor pool and the role of corrinoid, *Arch. Microbiol.* 141:229–238.

Hu, S.-I., Drake, H. L., and Wood, H. G., 1982. Synthesis of acetyl coenzyme A from carbon monoxide, methyltetrahydrofolate, and coenzyme A by enzymes from *Clostridium thermoaceticum*, *J. Bacteriol.* 149:440–448.

Hu, S.-I., Pezacka, E., and Wood, H. G., 1984. Acetate synthesis from carbon monoxide by *Clostridium thermoaceticum*. Purification of the corrinoid protein, *J. Biol. Chem.* 259:8892–8897.

Jansen, K., Thauer, R. K., Widdel, F., and Fuchs, G., 1984. Carbon assimilation pathways in sulfate reducing bacteria. Formate, carbon dioxide, carbon monoxide, and acetate assimilation by *Desulfovibrio baarsii*, *Arch. Microbiol.* 138:257–262.

Jansen, K., Fuchs, G., and Thauer, R. K., 1985. Autotrophic CO_2 fixation by *Desulfovibrio baarsii*: Demonstration of enzyme activities characteristic for the acetyl-CoA pathway, *FEMS Microbiol. Lett.* 28:311–315.

Jetten, M. S. M., Stams, A. J. M., and Zehnder, A. J. B., 1989a. Purification and characterization of an oxygen-stable carbon monoxide dehydrogenase of *Methanothrix soehngenii*, *Eur. J. Biochem.* 181:437–441.

Jetten, M. S. M., Stams, A. J. M., and Zehnder, A. J. B., 1989b. Isolation and characterization of acetyl-coenzyme A synthetase from *Methanothrix soehngenii*, *J. Bacteriol.* 171:5430–5435.

Jetten, M. S. M., Hagen, W. R., Pierik, A. J., Stams, A. J. M., and Zehnder, A. J. B., 1991a. Paramagnetic centers and acetyl-coenzyme A/CO exchange activity of carbon monoxide dehydrogenase from *Methanothrix soehngenii*, *Eur. J. Biochem.* 195:385–391.

Jetten, M. S. M., Pierik, A. J., and Hagen, W. R., 1991b. EPR characterization of a high-spin system in carbon monoxide dehydrogenase from *Methanothrix soehngenii*, *Eur. J. Biochem.* 202:1291–1297.

Jetten, M. S. M., Stams, A. J. M., and Zehnder, A. J. B., 1992. Methanogenesis from acetate: A comparison of the acetate metabolism in *Methanothrix soehngenii* and *Methanosarcina* spp., *FEMS Microbiol. Rev.* 88:181–198.

Kemner, J. M., 1993. Characterization of electron transfer activities associated with acetate dependent methanogenesis by *Methanosarcina barkeri* MS, Ph.D. thesis, Michigan State University.

Kemner, J. M., Krzycki, J. A., Prince, R. C., and Zeikus, J. G., 1987. Spectroscopic and enzymatic evidence for membrane-bound electron transport carriers and hydrogenase and their relation to cytochrome *b* function in *Methanosarcina barkeri, FEMS Microbiol. Lett.* 48:267–272.

Kerby, R., and Zeikus, J. G., 1983. Growth of *Clostridium thermoaceticum* on H_2/CO_2 or CO as energy source, *Curr. Microbiol.* 8:27–30.

Kerby, R., Niemczura, W., and Zeikus, J. G., 1983. Single-carbon catabolism in acetogens; analysis of carbon flow in *Acetobacterium woodii* and *Butyribacterium methylotrophicum* by fermentation and ^{13}C nuclear magnetic resonance measurement, *J. Bacteriol.* 155:1208–1218.

Kerby, R. L., Hong, S. S., Ensign, S. A., Coppoc, L. G., Ludden, P. W., and Roberts, G. P., 1992. Genetic and physiological characterization of the *Rhodospirillum rubrum* carbon monoxide dehydrogenase system, *J. Bacteriol.* 174:5284–5294.

Krzycki, J. A., and Prince, R. C., 1990. EPR observation of carbon monoxide dehydrogenase, methylreductase and corrinoid in intact *Methanosarcina barkeri* during methanogenesis from acetate, *Biochim. Biophys. Acta* 1015:53–60.

Krzycki, J. A., and Zeikus, J. G., 1984. Characterization and purification of carbon monoxide dehydrogenase from *Methanosarcina barkeri, J. Bacteriol.* 158:231–237.

Krzycki, J. A., Wolkin, R. H., and Zeikus, J. G., 1982. Comparison of unitrophic and mixotrophic substrate metabolism by an acetate-adapted strain of *Methanosarcina barkeri, J. Bacteriol.* 149:247–254.

Krzycki, J. A., Lehman, L. J., and Zeikus, J. G., 1985. Acetate catabolism by *Methanosarcina barkeri:* Evidence for involvement of carbon monoxide dehydrogenase, methyl coenzyme M, and methylreductase, *J. Bacteriol.* 163:1000–1006.

Krzycki, J. A., Mortenson, L. E., and Prince, R. C., 1989. Paramagnetic centers of carbon monoxide dehydrogenase from aceticlastic *Methanosarcina barkeri, J. Biol. Chem.* 264:7217–7221.

Kumar, M., and Ragsdale, S. W., 1992. Characterization of the CO binding site of carbon monoxide dehydrogenase from *Clostridium thermoaceticum* by infrared spectroscopy, *J. Am. Chem. Soc.* 114:8713–8715.

Ladapo, J., and Whitman, W. B., 1990. Method for isolation of auxotrophs in the methanogenic archaebacteria: Role of the acetyl-CoA pathway of autotrophic CO_2 fixation in *Methanococcus maripaludis, Proc. Natl. Acad. Sci. USA* 87:5598–5602.

Länge, S., and Fuchs, G., 1987. Autotrophic synthesis of activated acetic acid from CO_2 in *Methanobacterium thermoautotrophicum.* Synthesis from tetrahydromethanopterin-bound C_1 units and carbon monoxide, *Eur. J. Biochem.* 163:147–154.

Länge, S., Scholtz, R., and Fuchs, G., 1989. Oxidative and reductive acetyl CoA/carbon monoxide dehydrogenase pathway in *Desulfobacterium autotrophicum.* I. Characterization and metabolic function of the cellular tetrahydropterin, *Arch. Microbiol.* 151:77–83.

Laufer, K., Eikmanns, B., Frimmer, U., and Thauer, R. K., 1987. Methanogenesis from acetate by *Methanosarcina barkeri:* Catalysis of acetate formation from methyl iodide, CO_2, and H_2 by the enzyme system involved, *Z. Naturforsch. C* 42:360–372.

Lebertz, H., Simon, H., Courtney, L. F., Benkovic, S. J., Zydowsky, L. D., Lee, K., and Floss, H. G., 1987. Stereochemistry of acetic acid formation from 5-methyltetrahydrofolate by *Clostridium thermoaceticum, J. Am. Chem. Soc.* 109:3173–3174.

Lindahl, P. A., Münck, E., and Ragsdale, S. W., 1990a. CO dehydrogenase from *Clostridium thermoaceticum.* EPR and electrochemical studies in CO_2 and argon atmospheres, *J. Biol. Chem.* 265:3873–3879.

Lindahl, P. A., Ragsdale, S. W., and Münck, E., 1990b. Mössbauer study of CO dehydrogenase from *Clostridium thermoaceticum, J. Biol. Chem.* 265:3880–3888.

Ljungdahl, L. G., 1986. The autotrophic pathway of acetate synthesis in acetogenic bacteria, *Annu. Rev. Microbiol.* 40:415–450.

Ljungdahl, L. G., and Wood, H. G., 1982. Acetate synthesis, in *Vitamin B_{12}* (D. Dolphin, ed.), Wiley Interscience, New York, pp. 165–202.

Lorowitz, W. H., and Bryant, M. P., 1984. *Peptostreptococcus productus* strain that grows rapidly with CO as the energy source, *Appl. Environ. Microbiol.* 47:961–964.

Lovely, D. R., and Ferry, J. G., 1985. Production and consumption of H_2 during growth of *Methanosarcina* spp. on acetate, *Appl. Environ. Microbiol.* 49:247–249.

Lovely, D. R., White, R. H., and Ferry, J. G., 1984. Identification of methyl coenzyme M as an intermediate in methanogenesis from acetate in *Methanosarcina* spp., *J. Bacteriol.* 160:521–525.

Lu, W.-P., and Ragsdale, S. W., 1991. Reductive activation of the coenzyme A/acetyl-CoA isotopic exchange reaction catalyzed by carbon monoxide dehydrogenase from *Clostridium thermoaceticum* and its inhibition by nitrous oxide and carbon monoxide, *J. Biol. Chem.* 266:3554–3564.

Lu, W.-P., Harder, S. R., and Ragsdale, S. W., 1990. Controlled potential enzymology of methyl transfer reactions involved in acetyl-CoA synthesis by CO dehydrogenase and the corrinoid/iron–sulfur protein from *Clostridium thermoaceticum, J. Biol. Chem.* 265:3124–3133.

Lu, Z., White, C., Rheingold, A. L., and Crabtree, R. H., 1993. Functional modeling of CO dehydrogenase: Catalytic reduction of methylviologen by CO/H_2O with an N, O, S-ligated nickel catalyst, *Angew. Chem. Int. Ed. Engl.* 32:9294.

Lundie, L. L., Jr., and Ferry, J. G., 1989. Activation of acetate by *Methanosarcina thermophila.* Purification and characterization of phosphotransferase, *J. Biol. Chem.* 264:18392–18396.

Lupton, F. S., Conrad, R., and Zeikus, J. G., 1984. CO metabolism *Desulfovibrio vulgaris* strain Madison: Physiological function in the absence or presence of exogeneous substrates, *FEMS Microbiol. Lett.* 23:263–268.

Lynd, L., Kerby, R., and Zeikus, J. G., 1982. Carbon monoxide metabolism of the methylotrophic acidogen *Butyribacterium methylotrophicum, J. Bacteriol.* 149:255–263.

Meyer, O., and Fiebig, K., 1985. Enzymes oxidizing carbon monoxide, in *Gas Enzymology* (H. Degn, R. P. Cox, and H. Toftlund, eds.), D. Reidel, Dordrecht, The Netherlands, pp. 147–168.

Meyer, O., and Schlegel, H. G., 1983. Biology of aerobic carbon monoxide oxidizing bacteria, *Annu. Rev. Microbiol.* 37:277–310.

Morton, T. A., Runquist, J. A., Ragsdale, S. W., Shanmugasundaram, T., Wood, H. G., and Ljungdahl, L. G., 1991. The primary structure of the subunits of carbon monoxide dehydrogenase/acetyl coenzyme A synthase from *Clostridium thermoaceticum, J. Biol. Chem.* 266:23824–23828.

Nelson, M. J. K., and Ferry, J. G., 1984. Carbon monoxide-dependent methylcoenzyme M reductase in acetotrophic *Methanosarcina* spp., *J. Bacteriol.* 160:526–532.

Pezacka, E., and Wood, H. G., 1984a. The synthesis of acetyl-CoA by *Clostridium thermoaceticum* from carbon dioxide, hydrogen, coenzyme A and methyltetrahydrofolate, *Arch. Microbiol.* 137:63–69.

Pezacka, E., and Wood, H. G., 1984b. Role of carbon monoxide dehydrogenase in the autotrophic pathway used by acetogenic bacteria, *Proc. Natl. Acad. Sci. USA* 81:6261–6265.

Pezacka, E., and Wood, H. G., 1986. The autotrophic pathway of acetogenic bacteria. Role of CO dehydrogenase disulfide reductase, *J. Biol. Chem.* 261:1609–1615.

Pezacka, E., and Wood, H. G., 1988. Acetyl-CoA pathway of autotrophic growth. Identification of the methyl-binding site of the CO dehydrogenase, *J. Biol. Chem.* 263:16000–16006.

Ragsdale, S. W., 1991. Enzymology of the acetyl-CoA pathway of CO_2 fixation, *Crit. Rev. Biochem. Mol. Biol.* 26:261–300.

Ragsdale, S. W., and Wood, H. G., 1985. Acetate biosynthesis by acetogenic bacteria: Evidence that carbon monoxide dehydrogenase is the condensing enzyme that catalyzes the final steps of the synthesis, *J. Biol. Chem.* 260:3970–3977.

Ragsdale, S. W., Ljungdahl, L. G., and DerVartanian, D. V., 1982. EPR evidence for nickel–substrate interaction in carbon monoxide dehydrogenase from *Clostridium thermoaceticum,* *Biochem. Biophys. Res. Commun.* 108:658–663.

Ragsdale, S. W., Clarke, J. E., Ljungdahl, L. G., Lundie, L. L., and Drake, H. L., 1983a. Properties of purified carbon monoxide dehydrogenase from *Clostridium thermoaceticum,* a nickel, iron–sulfur protein, *J. Biol. Chem.* 258:2364–2369.

Ragsdale, S. W., Ljungdahl, L. G., and DerVartanian, D. V., 1983b. ^{13}C and ^{61}Ni isotope substitutions confirm the presence of a nickel (III)-carbon species in acetogenic CO dehydrogenase, *Biochem. Biophys. Res. Commun.* 115:658–665.

Ragsdale, S. W., Ljungdahl, L. G., and DerVartanian, D. V., 1983c. Isolation of carbon monoxide dehydrogenase from *Acetobacterium woodii* and comparison of its properties with those of the *Clostridium thermoaceticum* enzyme, *J. Bacteriol.* 155:1224–1237.

Ragsdale, S. W., Wood, H. G., and Antholine, W. E., 1985. Evidence that an iron–nickel–carbon complex is formed by reaction of CO with the CO dehydrogenase from *Clostridium thermoaceticum,* *Proc. Natl. Acad. Sci. USA* 82:6811–6814.

Ragsdale, S. W., Lindahl, P. A., and Münck, E., 1987. Mössbauer, EPR, and optical studies of the corrinoid/iron–sulfur protein involved in the synthesis of acetyl coenzyme A by *Clostridium thermoaceticum,* *J. Biol. Chem.* 262:14289–14297.

Ragsdale, S. W., Wood, H. G., Morton, T. A., Ljungdahl, L. G., and DerVartanian, D. V., 1988. Nickel in CO dehydrogenase, in *The Bioinorganic Chemistry of Nickel* (J. R. Lancaster, Jr., ed.), VCH Publishers, New York, pp. 311–332.

Ragsdale, S. W., Baur, J. R., Gorst, C. M., Harder, S. R., Lu, W.-P., Roberts, D. L., Rundquist, J. A., and Schiau, I., 1990. The acetyl-CoA synthase from *Clostridium thermoaceticum:* From gene cluster to active-site metal clusters, *FEMS Microbiol. Rev.* 87:397–402.

Ramer, S. E., Raybuck, S. A., Orme-Johnson, W. H., and Walsh, C. T., 1989. Kinetic characterization of the [3-^{32}P]coenzyme A/acetyl coenzyme A exchange catalyzed by a three-subunit form of the carbon monoxide dehydrogenase/acetyl-CoA synthase from *Clostridium thermoaceticum,* *Biochemistry* 28:4675–4680.

Raybuck, S. A., Bastian, N. R., Zydowsky, L. D., Kobayashi, K., Floss, H., Orme-Johnson, W. H., and Walsh, C. T., 1987. Nickel-containing CO dehydrogenase catalyzes reversible decarbonylation of acetyl CoA with retention of stereochemistry at the methyl carbon, *J. Am. Chem. Soc.* 109:3171–3173.

Raybuck, S. A., Bastian, N. R., Orme-Johnson, W. H., and Walsh, C. T., 1988. Kinetic characterization of the carbon monoxide-acetyl-CoA exchange activity of the acetyl-CoA synthesizing CO dehydrogenase from *Clostridium thermoaceticum,* *Biochemistry* 27:7698–7702.

Raybuck, S. A., Ramer, S. E., Abbanat, D. R., Peters, J. W., Orme-Johnson, W. H., Ferry, J. G., and Walsh, C. T., 1991. Demonstration of carbon–carbon bond cleavage of acetyl coenzyme A by using isotopic exchange catalyzed by the CO dehydrogenase complex from acetate-grown *Methanosarcina thermophila,* *J. Bacteriol.* 173:929–932.

Roberts, D. L., James-Hagstrom, J. E., Garvin, D. K., Gorst, C. M., Runquist, J. A., Baur, J. R., Haase, F. C., and Ragsdale, S. W., 1989. Cloning and expression of the gene cluster encoding key proteins involved in acetyl-CoA synthesis in *Clostridium thermoaceticum:* CO dehydrogenase, the corrinoid/Fe-S protein, and methyltransferase, *Proc. Natl. Acad. Sci. USA* 86: 32–36.

Roberts, J. R., Lu, W.-L., and Ragsdale, S. W., 1992. Acetyl-coenzyme A synthesis from methyltetrahydrofolate, CO, and coenzyme A by enzymes purified from *Clostridium thermoace-*

ticum: Attainment of *in vivo* rates and identification of rate-limiting steps, *J. Bacteriol.* 174: 4667–4676.

Rühlemann, M., Ziegler, K., Stupperich, E., and Fuchs, G., 1985. Detection of acetyl coenzyme A as an early CO_2 assimilation intermediate in *Methanobacterium, Arch. Microbiol.* 141: 399–406.

Schauder, R., Eikmanns, B., Thauer, R. K., Widdel, F., and Fuchs, G., 1986. Acetate oxidation to CO_2 in anaerobic bacteria via a novel pathway not involving reactions of the citric acid cycle, *Arch. Microbiol.* 145:162–172.

Schauder, R., Preuß, A., Jetten, M., and Fuchs, G., 1989. Oxidative and reductive acetyl CoA/ carbon monoxide dehydrogenase pathway in *Desulfobacterium autotrophicum.* 2. Demonstration of the enzymes of the pathway and comparison of CO dehydrogenase, *Arch. Microbiol.* 151:84–89.

Shanmugasundaram, T., and Wood, H. G., 1992. Interaction of ferredoxin with carbon monoxide dehydrogenase from *Clostridium thermoaceticum, J. Biol. Chem.* 267:897–900.

Shanmugasundaram, T., Kumar, G. K., and Wood, H. G., 1988a. Involvement of tryptophan residues at the coenzyme A binding site of carbon monoxide dehydrogenase from *Clostridium thermoaceticum, Biochemistry* 27:6499–6503.

Shanmugasundaram, T., Ragsdale, S. W., and Wood, H. G., 1988b. Role of carbon monoxide dehydrogenase in acetate synthesis by the acetogenic bacterium, *Acetobacterium woodii, Biofactors* 1:147–152.

Shanmugasundaram, T., Kumar, G. K., Shenoy, B. C., and Wood, H. G., 1989. Chemical modification of the functional arginine residues of carbon monoxide dehydrogenase from *Clostridium thermoaceticum, Biochemistry* 28:7112–7116.

Shin, W., and Lindahl, P. A., 1992a. Function and CO binding properties of the NiFe complex in carbon monoxide dehydrogenase from *Clostridium thermoaceticum, Biochemistry* 31: 12870–12875.

Shin, W., and Lindahl, P. A., 1992b. Discovery of a labile nickel ion required for CO/acetyl-CoA exchange activity in the NiFe complex of carbon monoxide dehydrogenase from *Clostridium thermoaceticum, J. Am. Chem. Soc.* 114:9718–9719.

Shin, W., and Lindahl, P. A., 1993. Low spin quantitation of NiFeC EPR signal from carbon monoxide dehydrogenase is not due to damage incurred during protein purification, *Biochim. Biophys. Acta* 1161:317–322.

Shin, W., Stafford, P. R., and Lindahl, P. A., 1992. Redox titrations of carbon monoxide dehydrogenase from *Clostridium thermoaceticum, Biochemistry* 31:6003–6011.

Smith, M. J., Lequerica, J. L., and Hart, M. R., 1985. Inhibition of methanogenesis and carbon metabolism in *Methanosarcina* sp. by cyanide, *J. Bacteriol.* 162:67–71.

Smith, E. T., Ensign, S. A., Ludden, P. W., and Feinberg, B. A., 1992. Direct electrochemical studies of hydrogenase and CO dehydrogenase, *Biochem. J.* 285:181–185.

Stavropoulos, P., Carrié, M., Muetterties, M. C., and Holm, R. H., 1990. Reaction sequence related to that of carbon monoxide dehydrogenase (acetyl coenzyme A synthase): Thioester formation mediated at structurally defined nickel centers, *J. Am. Chem. Soc.* 112:5385–5387.

Stavropoulos, P., Muetterties, M. C., Carrié, M., and Holm, R. H., 1991. Structure and reaction chemistry of nickel complexes in relation to carbon monoxide dehydrogenase: A reaction system simulating acetyl-coenzyme A synthase activity, *J. Am. Chem. Soc.* 113:8485–8492.

Stephens, P. J., McKenna, M.-C., Ensign, S. A., Bonam, D., and Ludden, P. W., 1989. Identification of a Ni- and Fe-containing cluster in *Rhodospirillum rubrum* carbon monoxide dehydrogenase, *J. Biol. Chem.* 264:16347–16350.

Stupperich, E., and Fuchs, G., 1983. Autotrophic acetyl coenzyme A synthesis *in vitro* from two CO_2 in *Methanobacterium, FEBS Lett.* 156:345–348.

Stupperich, E., and Fuchs, G., 1984a. Autotrophic synthesis of activated acetic acid from two CO_2 in *Methanobacterium thermoautotrophicum*. I. Properties of *in vivo* system, *Arch. Microbiol.* 139:8–13.

Stupperich, E., and Fuchs, G., 1984b. Autotrophic synthesis of activated acetic acid from two CO_2 in *Methanobacterium thermoautotrophicum*. II. Evidence for different origins of acetate carbon atoms, *Arch. Microbiol.* 139:14–20.

Stupperich, E., Hammel, K. E., Fuchs, G., and Thauer, R. K., 1983. Carbon monoxide fixation into the carboxyl group of acetyl coenzyme A during autotrophic growth of *Methanobacterium*, *FEBS Lett.* 152:21–23.

Tan, G. O., Ensign, S. A., Ciurli, S., Scott, M. J., Hedman, B., Holm, R. H., Ludden, P. A., Korszun, Z. R., Stephens, P. J., and Hodgson, K. O., 1992. On the structure of the nickel/iron/sulfur center of the carbon monoxide dehydrogenase from *Rhodospirillum rubrum:* An X-ray absorption spectroscopic study, *Proc. Natl. Acad. Sci. USA* 89:4427–4431.

Terlesky, K. C., and Ferry, J. G., 1988. Ferredoxin requirement for electron transport from the carbon monoxide dehydrogenase complex to a membrane-bound hydrogenase in acetate-grown *Methanosarcina thermophila*, *J. Biol. Chem.* 263:4075–4079.

Terlesky, K. C., Nelson, M. J. K., and Ferry, J. G., 1986. Isolation of an enzyme complex with carbon monoxide dehydrogenase activity containing corrinoid and nickel from acetate-grown *Methanosarcina thermophila*, *J. Bacteriol.* 168:1053–1058.

Terlesky, K. C., Barber, M. J., Aceti, D. J., and Ferry, J. G., 1987. EPR properties of the Ni-Fe-C center in an enzyme complex with carbon monoxide dehydrogenase activity from acetate-grown *Methanosarcina thermophila*. Evidence that acetyl-CoA is a physiological substrate, *J. Biol. Chem.* 262:15392–15395.

Uffen, R. L., 1976. Anaerobic growth of a *Rhodopseudomonas* species in the dark with carbon monoxide as sole carbon and energy source, *Proc. Natl. Acad. Sci. USA* 73:3298–3302.

Uffen, R. L., 1981. Metabolism of carbon monoxide, *Enzyme Microb. Technol.* 3:197–206.

Uffen, R. L., 1983. Metabolism of carbon monoxide by *Rhodopseudomonas gelatinosa:* Cell growth and properties of the oxidation system, *J. Bacteriol.* 155:956–965.

Wakin, B. T., and Uffen, R. L., 1983. Membrane association of the carbon monoxide oxidation system in *Rhodopseudomonas gelatinosa*, *J. Bacteriol.* 153:571–573.

Westermann, P., Ahring, B. K., and Mah, R. A., 1989. Acetate production by methanogenic bacteria, *Appl. Environ. Microbiol.* 55:2257–2261.

Wood, H. G., and Ljungdahl, L. G., 1991. Autotrophic character of the acetogenic bacteria, in *Variations in Autotrophic Life*, Academic Press, New York pp. 201–250.

Wood, H. G., Ragsdale, S. W., and Pezacka, E., 1986a. A new pathway of autotrophic growth utilizing carbon monoxide or carbon dioxide and hydrogen, *Biochem. Int.* 12:421–440.

Wood, H. G., Ragsdale, S. W., and Pezacka, E., 1986b. The acetyl-CoA pathway: A newly discovered pathway of autotrophic growth, *Trends Biochem. Sci.* 11:14–18.

Wood, H. G., Ragsdale, S. W., and Pezacka, E., 1986c. The acetyl-CoA pathway of autotrophic bacteria, *FEMS Microbiol. Rev.* 39:345–362.

Yagi, T., 1958. Enzymatic oxidation of carbon monoxide, *Biochim. Biophys. Acta* 30:194–195.

Yang, H., Daniel, S. L., Hsu, T., and Drake, H. L., 1989. Nickel transport by the thermophilic acetogen *Acetogenium kivui*, *Appl. Environ. Microbiol.* 55:1078–1081.

Zeikus, J. G., Kerby, R., and Krzycki, J. A., 1985. Single-carbon chemistry of acetogenic and methanogenic bacteria, *Science* 227:1167–1173.

Methyl Coenzyme M Reductase ⑥

6.1 Introduction

The biogenesis of methane is carried out by a group of strictly anaerobic archaea (formerly termed archaebacteria), referred to as methanogens, that obtain energy from one- or two-carbon substrates according to the following reactions [reviewed by Daniels et al. (1984), Ferry (1992), Jones et al. (1987), and Thauer (1990)]:

$$4H_2 + CO_2 \rightarrow CH_4 + 2H_2O \qquad (6\text{-}1)$$

$$4HCOOH \rightarrow CH_4 + 3CO_2 + 2H_2O \qquad (6\text{-}2)$$

$$4CH_3OH \rightarrow 3CH_4 + CO_2 + 2H_2O \qquad (6\text{-}3)$$

$$4CH_3NH_3^+ + H_2O \rightarrow 3CH_4 + CO_2 + 4NH_4^+ \qquad (6\text{-}4)$$

$$4CO + 2H_2O \rightarrow CH_4 + 3CO_2 \qquad (6\text{-}5)$$

$$CH_3COOH \rightarrow CH_4 + CO_2 \qquad (6\text{-}6)$$

Using distinct pathways, these substrates and other substrates are all converted to a common methylated intermediate that is subsequently converted to methane. Each pathway involves a series of novel coenzymes. The structures and functions of these singular compounds have been reviewed (DiMarco et al., 1990; Rouvière and Wolfe, 1988), and only an overview is presented here. Carbon dioxide is reduced to the formyl level with the participation of a furan-containing compound, termed methanofuran. The formyl group is transferred to a special pterin, named tetrahydromethanopterin, where it is converted to the methenyl derivative and sequentially reduced to the methylene and methyl levels. Reduction of the methenyltetrahydromethanopterin requires an unusual 5-deazaflavin, denoted coenzyme F_{420}, as the electron

147

donor. The methyl group is subsequently transferred to 2-mercaptoethane-sulfonate (coenzyme M, HS-CoM) to form 2-(methylthio)ethanesulfonate (methyl coenzyme M, methyl-S-CoM), which is the direct precursor of methane. The ultimate source of electrons in each of these reactions is hydrogen, requiring the participation of nickel-containing hydrogenases (discussed in Chapter 4). Formate metabolism requires the action of formate dehydrogenase (containing an atypical molybdopterin cofactor) and a nickel-containing hydrogenase; the combined action of these enzymes results in the release of carbon dioxide and hydrogen gas. Formate-derived carbon dioxide is reduced to methane by the same reactions as just described. Methanol or methylamines are transformed directly into the common methylated intermediate methyl-S-CoM by action of a corrinoid protein that contains a modified benzimidazole base compared to that found in vitamin B_{12}. Carbon monoxide is metabolized by a nickel-containing CO dehydrogenase (discussed in Chapter 5). Oxidation of four moles of CO to CO_2 provides electrons that are used to reduce one mole of CO_2 to methane by the same pathway as described above. Finally, acetate cleavage similarly requires participation of the nickel-containing CO dehydrogenase. In this case, the enzyme forms a methylated intermediate and a bound carbonyl group. Oxidation of the carbonyl group is coupled to methyl group reduction to form methane. This chapter focuses on the last step in methanogenesis that is common to all pathways: conversion of the methylated intermediate into methane.

The common intermediate for each pathway, methyl-S-CoM, is a coenzyme found uniquely in all methanogens. The methyl group of methyl-S-CoM is released as methane by a methyl coenzyme M reductase-catalyzed reaction (Fig. 6-1) that requires the participation of another distinctive methanogen coenzyme, 7-mercaptoheptanoylthreonine phosphate (HS-HTP). The second product of this reaction is the CoM-S-S-HTP disulfide (Bobik *et al.*, 1987; Ellermann *et al.*, 1988). Reduction of the disulfide is carried out by a separate enzyme using reduced coenzyme F_{420} (Deppenmeier *et al.*, 1990; Hedderich and Thauer, 1988; Hedderich *et al.*, 1990).

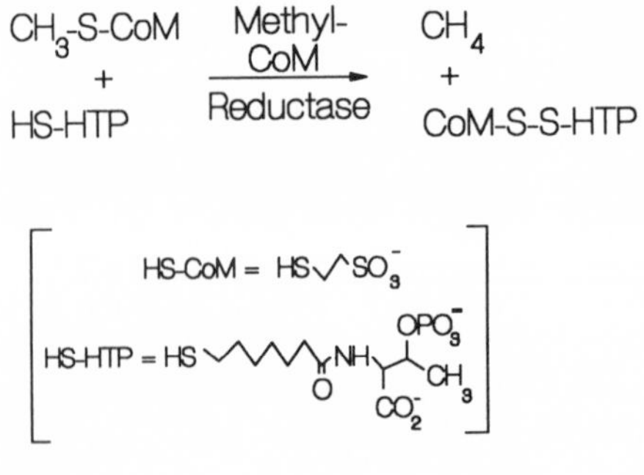

Figure 6-1. The reaction carried out by methyl coenzyme M reductase. Regardless of the bacterial growth substrate (carbon dioxide plus hydrogen, formate, methanol, methylamine, carbon monoxide, or acetic acid), the ultimate step in methanogenesis is conserved. This reaction involves the conversion of methyl-S-CoM to methane with concomitant formation of the HTP-S-S-CoM disulfide.

Methyl coenzyme M reductase is a nickel-containing enzyme found in all methanogenic bacteria. Unlike the other nickel-containing enzymes discussed in Chapters 3–5, the methyl coenzyme M reductase nickel is found in a tetrapyrrolic structure, designated coenzyme F_{430}. Below, I will review the evidence that methyl coenzyme M reductase contains nickel, examine the structure of the coenzyme and enzyme, detail the spectroscopic, oxidation–reduction, and other properties of the F_{430} nickel center in comparison to related nickel model compounds, and summarize current understanding of methyl coenzyme M reductase catalysis.

6.2 Evidence for Nickel in Methyl Coenzyme M Reductase

Stimulated by Jean LeGall's unpublished observations, Gunsalus and Wolfe (1978) isolated a yellow, nonfluorescent compound from heat-treated cell extracts of *Methanobacterium thermoautotrophicum* strain ΔH. At the time, no role could be assigned to the chromophore, and it was designated factor F_{430} because it possesses an intense absorbance maximum at 430 nm. Shortly thereafter, Thauer and co-workers (Schönheit *et al.*, 1979) observed a nickel requirement for growth of *M. thermoautotrophicum* strain Marburg. These two experimental observations converged when both groups independently demonstrated that factor F_{430} contains nickel. The Illinois researchers established the presence of nickel in *Methanobacterium bryantii* factor F_{430} by subjecting the purified compound to neutron activation analysis (Whitman and Wolfe, 1980), whereas the German group documented that 70% of the [63]Ni radioactivity taken up by *M. thermoautotrophicum* cells can be extracted and copurified with factor F_{430} (Diekert *et al.*, 1980c). Other methanogens, including *Methanobrevibacter smithii, Methanosarcina barkeri, Methanococcus vannielii,* and *Methanospirillum hungatei,* soon afterwards were found to require nickel for growth and to possess factor F_{430} (Diekert *et al.*, 1981). Indeed, all methanogens possess the nickel-containing factor F_{430}, and this chromophore is found only in methanogens.

The role of factor F_{430} in methanogens remained unknown until purified methyl coenzyme M reductase became available. Gunsalus and Wolfe (1980) had resolved *M. thermoautotrophicum* cell extracts into three fractions that are necessary to convert methyl-S-CoM to methane under hydrogen gas. The actual methyl coenzyme M reductase was demonstrated to be in a fraction called component C (Ellefson and Wolfe, 1980). The component C protein was purified, shown to possess three subunits (M_r = 68,000, 45,000, and 38,500) in an $\alpha_2\beta_2\gamma_2$ stoichiometry, and found to be yellow in color with an absorbance maximum at 425 nm and a shoulder at 455 nm (Ellefson and Wolfe, 1981). Upon treatment with 80% methanol, the yellow color was shown to be released from the protein, and concomitantly there was a shift in its absorbance spectrum to yield a chromophore with maximal absorbance at 430 nm (Ellefson *et al.,* 1982). This compound was identical to the nickel-containing factor, F_{430}. Moura *et al.* (1983) also reported the isolation from *Methanosarcina barkeri* and *Methanobacterium thermoautotrophicum* of yellow proteins that, when heated or acidified, yielded factor F_{430}; however, they were unable to assign a role to the proteins. In contrast, the Wolfe laboratory demonstrated that this chromophore, now called *coenzyme* F_{430}, is an essential component of methyl coenzyme M reductase (Ellefson *et al.,* 1982).

6.3 Structure of Coenzyme F_{430}

Early studies to unravel the structure of coenzyme F_{430} involved growth of methanogenic bacteria in the presence of various radioactively labeled compounds and examination of the extracted chromophore for incorporation of radioactivity. This highly successful approach immediately provided compelling evidence for a tetrapyrrolic structure; that is, eight molecules of succinate or δ-aminolevulinic acid are incorporated per F_{430} molecule (Diekert *et al.,* 1980a,b). Furthermore, uroporphyrinogen III can serve as a precursor for F_{430} biosynthesis, and the cellular uroporphyrinogen III levels increase for cells grown in the absence of nickel (Gilles and Thauer, 1983). Additional labeling studies indicated that two methionine-derived methyl groups are present in F_{430} (Jaenchen *et al.,* 1981), similar to the methylation content observed in siroheme and vitamin B_{12} tetrapyrroles. Indeed, sirohydrochlorin, a precursor in the synthesis of the latter compounds, is metabolized by cell-free extracts of *Methanobacterium thermoautotrophicum* to form F_{430} (Mucha *et al.,* 1985).

In an elegant series of collaborative studies [summarized by Pfaltz (1988)], the Thauer and Eschenmoser groups tentatively assigned the structure of the nickel-containing F_{430} macrocycle. *Methanobacterium thermoautotrophicum* strain Marburg was grown in the presence of specifically mono-^{13}C-labeled δ-aminolevulinic acid (with the label at carbons 2 through 5) or with [*methyl*-^{13}C]methionine. Factor F_{430} was extracted with perchloric acid, purified, and methylated to enhance solubility in apolar solvents. Finally, the chromophore was analyzed by ^{13}C-nuclear magnetic resonance (NMR) spectroscopy and other methods (Pfaltz *et al.,* 1982). The pentamethyl ester tetrapyrrolic structure (Fig. 6-2) possesses a novel biological ring system which combines elements of both the corrins and the porphyrins. Compounds with this framework previously had been synthesized, designated corphins, and extensively characterized [reviewed by Eschenmoser (1986)]. Factor F_{430} is a tetrahydrocorphin in which one acetamide side chain is fused with a ring carbon to form a lactam

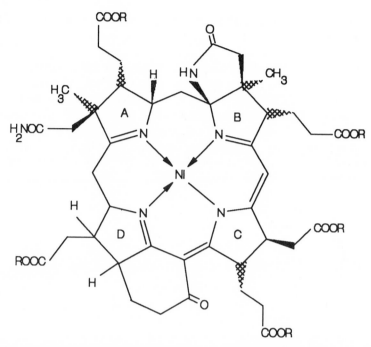

Figure 6-2. Tentative assignment of the F_{430} structure. Specifically ^{13}C-labeled precursors were provided to methanogen cells, F_{430} was extracted at low temperature with the use of perchloric acid, the purified chromophore was subjected to methanolysis, and the structure of the corphin derivative was deduced from NMR spectroscopic data (Pfaltz et al., 1982). The study could not eliminate ambiguity in the stereochemistry of groups in the D ring. The pentamethyl ester illustrated (R = methyl) is often referred to as F_{430}M.

and a propionic acid side chain is cyclized to form a six-membered carbocyclic ring. The NMR spectroscopic studies, combined with comparisons of the circular dichroism spectra for F_{430} derivatives and sirohydrochlorin (Fässler et al., 1985; Pfaltz et al., 1982), were used to establish the absolute configuration of rings A, B, and C; however, ambiguity remained concerning the stereochemical assignments in the D ring.

Further structural characterization of this molecule (Keltjens et al., 1982, 1983a,b) suggested that several native forms of F_{430} exist in the cell and that these species possess additional peripheral components including coenzyme M and 6,7-dimethyl-8-ribityl-5,6,7,8-tetrahydrolumazine. These adducts and supposed structural variants, however, were later shown to be contaminants and artifacts generated by extracting F_{430} from whole cells under harsh conditions such as high temperature or low pH. For example, several thermal denaturation products of F_{430}, with absorbance maxima at 430 nm but exhibiting slightly perturbed peak shape when compared to that of the native coenzyme, and an oxidation product, termed F_{560} for its absorbance maximum, have been observed (Diekert *et al.*, 1981) and structurally characterized (Pfaltz *et al.*, 1985). The denatured species arise from epimerization at one or two positions in the C ring or from oxidation within this ring. Partial structures of the resulting 13-epi-F_{430}, 12,13-diepi-F_{430}, and F_{560} derivatives are illustrated in Fig. 6-3 and compared to that of the native chromophore.

Pfaltz *et al.* (1985) showed that native F_{430} is not the thermodynamically preferred epimer. Rather, at thermal equilibrium the relative populations of F_{430}, 13-epi-F_{430}, and 12,13-diepi-F_{430} are 4%, 8%, and 88%, respectively.

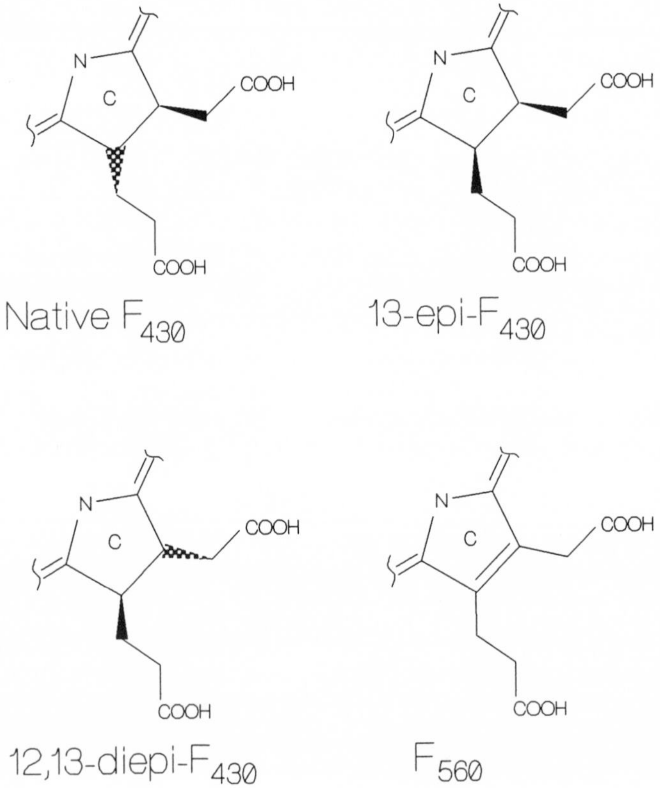

Figure 6-3. Structure of three F_{430} denaturation products. The partial structure of the native F_{430} molecule is compared with that of two compounds that form by epimerization upon heating: the 13-epi-F_{430} and 12,13-diepi-F_{430}. As in authentic F_{430}, these denatured compounds retain maximal absorbances at 430 nm. In addition, the partial structure is shown for an oxidized product of F_{430} that absorbs at 560 nm and is designated F_{560} (Pfaltz *et al.*, 1985).

This raises questions regarding how epimerization is prevented or corrected *in vivo*, especially for the thermophilic methanogens. Cellular oxidation of F_{430} is of less concern in the highly reducing environment of these strict anaerobes. Nevertheless, methanogen cells possess a system that is capable of reducing F_{560} to F_{430} (Keltjens *et al.*, 1988). One could speculate that correction of the epimeric denatured forms may involve an oxidation–reduction cycle. *In vitro*, oxidized F_{560} is readily reduced by zinc in acetic acid to yield a single species that, surprisingly, matches the native epimer isolated from whole cells or from purified protein by using mild extraction conditions. The single native form of F_{430} is free of peripheral components as established by fast-atom bombardment mass spectrometry and according to other criteria (Hausinger *et al.*, 1984; Livingston *et al.*, 1984). Further evidence excluding the presence of coenzyme M in F_{430} was obtained by growing *Methanobrevibacter ruminantium*, a coenzyme M-dependent methanogen, in the presence of [2-^{14}C]coenzyme M and demonstrating the absence of radioactivity in the purified chromophore (Hüster *et al.*, 1985). A distinct nickel-containing tetrapyrrole ($15,17^3$-seco-F_{430}-17^3-acid) is formed in *Methanobrevibacter arboriphilicus* or *Methanobacterium thermoautotrophicum* (Pfaltz *et al.*, 1987) when grown under conditions of low nickel concentration and in the presence of δ-aminolevulinic acid; however, this compound was shown to be an F_{430} biosynthetic precursor and not a second native form. Thus, the native coenzyme F_{430} does not possess peripheral components on the corphin ring and is present in a single structural form. Complete assignment of its configuration combined the use of recently developed two-dimensional NMR spectroscopic methods for F_{430} with X-ray crystallographic and NMR structural analysis of 12,13-diepi-F_{430} pentamethyl ester crystals (Färber *et al.*, 1991; Olson *et al.*, 1990; Won *et al.*, 1990, 1992). The complete coenzyme F_{430} tetrapyrrolic structure is the penta-acid corphin shown in Fig. 6-4.

6.4 General Properties of Methyl Coenzyme M Reductase

I have divided discussion of the general properties of the methyl coenzyme M reductase into four sections: protein characterization, coenzyme binding properties, molecular biological studies and sequence analysis, and mechanistic enzymology.

6.4.1 Protein Characterization

Methyl coenzyme M reductase is present at high levels in all methanogenic cells; for example, Ellefson and Wolfe (1981) estimated that this enzyme ac-

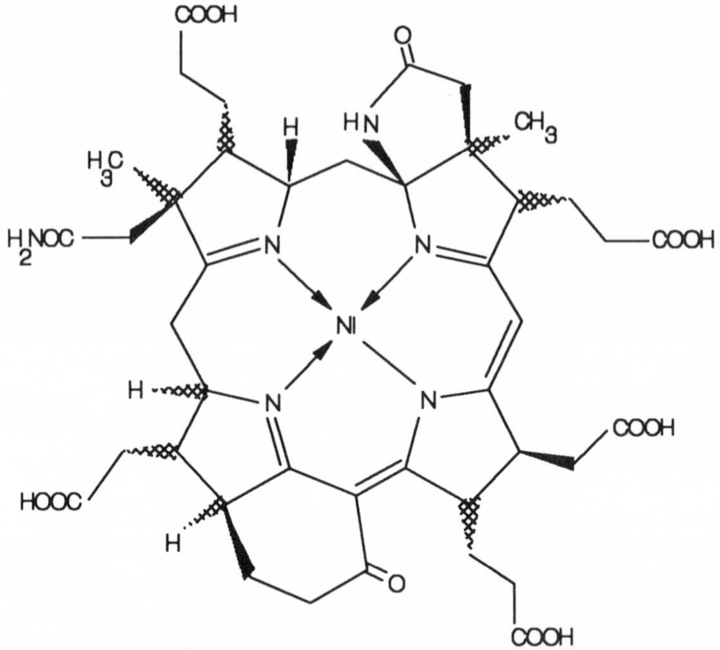

Figure 6-4. Complete structural assignment for the F_{430} tetrapyrrole. The structure of native, penta-acid coenzyme F_{430} is illustrated, including the stereochemical assignment for all chiral centers (Färber *et al.*, 1991).

counts for over 10% of methanogen cell protein. Most of the protein characterization studies involving methyl coenzyme M reductase have been carried out by using enzyme isolated from *Methanobacterium thermoautotrophicum*. This microbe served as the source for activity in preliminary cell fractionation efforts (Gunsalus and Wolfe, 1978), in the first successful methyl coenzyme M reductase purification (Ellefson and Wolfe, 1981), and in much of the subsequent methanogen research carried out in the Wolfe and other laboratories. Hartzell and Wolfe (1986b), however, demonstrated that the enzymes isolated from *Methanococcus voltae, Methanococcus jannaschii,* and *Methanosarcina barkeri* are all similar to this benchmark enzyme in possessing $\alpha_2\beta_2\gamma_2$ structures [total M_r of $\sim$300,000 comprised of subunits with M_r 66,000–73,000 (α), 41,500–45,000 (β), and 33,000–38,500 (γ)] and in containing F_{430}. Furthermore, they showed that the enzymes are interchangeable in their ability to reconstitute hydrogen-dependent methyl-S-CoM reduction using accessory protein fractions from various sources in their assays (Hartzell and Wolfe, 1986b). Rouvière and Wolfe (1987) extended this comparison of peptide subunit sizes to 18 species of methanogens and suggested that partial purification of methyl coenzyme M reductase and gel analysis can serve as a rapid initial taxonomic assessment for new methanogens. The enzyme, however, does not have the same quaternary structure in all methanogens. Jablonski and Ferry (1991) demonstrated that *Methanosarcina thermophila* methyl coenzyme M reductase possesses an $\alpha_1\beta_1\gamma_1$ stoichiometry, where the peptides are similar in size to those described above. Nevertheless, the basic structural unit for the enzyme appears to be a highly conserved trimer which is usually found dimerized. Electron microscopic analysis of the *M. thermoautotrophicum* enzyme revealed an oligomeric complex consistent with a pair of trimers in an eclipsed position (Wackett *et al.*, 1987a). No crystals of the enzyme suitable for X-ray crystallographic analysis have been reported.

Despite extensive efforts throughout the 1980s to characterize methyl coenzyme M reductase from *M. thermoautotrophicum,* a major aspect of its structure had been overlooked; namely, both the ΔH and Marburg strains of this microbe possess two genetically distinct forms of the enzyme (Bonacker *et al.*, 1992; Brenner *et al.*, 1992; Rospert *et al.*, 1990). One enzyme form predominates during growth under limiting hydrogen and carbon dioxide conditions, whereas the other form predominates during exponential-growth conditions. The relative expression levels of the two enzyme forms are also affected by temperature and pH. Earlier studies had been carried out using primarily the former, lower activity, species. The two forms possess distinct specific activities, amino-terminal sequences, subunit sizes, and charge properties. Nevertheless, they contain identical F_{430} coenzymes, and both possess an $\alpha_2\beta_2\gamma_2$ subunit structure. The need for two distinct enzyme species in the same cell is not clear; however, Brenner *et al.* (1992) proposed that methanogenesis from the more active species may be uncoupled from cell growth and that the cell may use this enzyme as a mechanism to rapidly reduce the local environmental concentration of hydrogen, an inhibitor of syntrophic microbes.

Localization of methyl coenzyme M reductase within the cell has yielded conflicting results. The *M. thermoautotrophicum* enzyme purifies as if it were a cytoplasmic protein, consistent with initial immunogold electron microscopic results which indicated a random distribution in the cell (Ossmer *et al.*, 1986). The enzyme from *Methanococcus voltae* similarly behaves like a soluble protein, but, in contrast to its properties during purification, immunocytochemical localization indicates an association with the cytoplasmic membrane (Ossmer *et al.*, 1986). Reexamination of the *M. thermoautotrophicum* enzyme under conditions of low protein biosynthesis indicated a shift in location to the vicinity of the cytoplasmic membrane (Aldrich *et al.*, 1987). Furthermore, on the basis of other electron microscopic evidence, Mayer *et al.* (1988) suggested that in the methanogenic bacterium strain Göl the methyl coenzyme M reductase is part of a large membrane-bound complex of several proteins that they denoted the methanoreductosome. Despite these apparent interstrain differences and the dependence on growth conditions, it now appears that there is no requirement for the enzyme to be membrane-associated in order for the cells to carry out electron transport phosphorylation. Rather, the coenzyme F_{420}: heterodisulfide oxidoreductase that reduces the CoM-S-S-HTP mixed disulfide appears to function in proton translocation (Deppenmeier *et al.*, 1990).

6.4.2 Coenzyme Binding Properties

Cellular coenzyme F_{430} exists in both protein-free and methyl coenzyme M reductase-bound forms (Hausinger *et al.*, 1984; Livingston *et al.*, 1984). Both the total chromophore content of the cells (Diekert *et al.*, 1980d) and the protein-free/protein-bound ratio of F_{430} (Ankel-Fuchs *et al.*, 1984) depend on the nickel concentration of the medium. At low nickel concentrations, F_{430} is nearly all protein-associated, whereas at high nickel concentrations ($>1 \mu M$) the protein-free form predominates. The protein-free species appears to be a biosynthetic precursor for methyl coenzyme M reductase synthesis: when cells are shifted from nickel-rich medium to nickel-deficient medium, the protein-free species is converted to the protein-bound form (Ankel-Fuchs *et al.*, 1984). Although the free and bound forms have distinct absorbance spectra (wavelength maximum at 430 nm vs. a peak at ~420 nm with a shoulder at 445 nm), the maximal extinction coefficients are nearly identical ($23,000 \ M^{-1} \ cm^{-1}$). (For further discussion of the optical spectrum of methyl coenzyme M reductase, see Sections 6.4.4 and 6.5.) Furthermore, when the protein-bound chromophore is dissociated from the protein either by a freeze-thaw cycle in $1 \ M$ NaCl, by 80% ethanol containing $2 \ M$ LiCl, or by perchloric acid extraction, its structure is identical to that of the protein-free molecule (Hausinger *et al.*, 1984; Livingston *et al.*, 1984). *In vitro* reconstitution of F_{430} into the denatured protein has also been reported: Hartzell and Wolfe (1986a) separated the three subunits of methyl coenzyme M reductase from each other and from the chromophore by sodium dodecyl sulfate (SDS) polyacrylamide gel electrophoresis, extracted the peptides from the gel, and were able to demonstrate reassociation to yield 70% active enzyme after adding F_{430}. These investigators suggested that F_{430} binds specifically to the large subunit of the enzyme, but further studies are necessary to verify this assignment.

Methyl coenzyme M reductase binds not only coenzyme F_{430}, but also stoichiometric amounts of HS-CoM (Hartzell *et al.*, 1987; Hausinger *et al.*, 1984; Keltjens *et al.*, 1982) and HS-HTP (Noll and Wolfe, 1986). Neither HS-CoM nor HS-HTP is covalently attached to the protein or to the F_{430} molecule. The protein-bound HS-CoM does not exchange with free HS-CoM, but it does exchange with the 2-mercaptoethanesulfonate portion of methyl-S-CoM, as shown by experiments in which ^{35}S but not ^{3}H is incorporated from [*methyl*-^{3}H, *thio*-^{35}S]2-methylthioethanesulfonate (Hartzell *et al.*, 1987). Importantly, this reaction occurs only during enzyme turnover and requires multiple turnovers for complete exchange. This exchange reaction has important mechanistic implications for the enzyme as will be discussed in Section 6.5.4.

6.4.3 Protein Sequence Analysis

Methyl coenzyme M reductase gene clusters have been cloned and sequenced from *Methanococcus vannielli* (Cram *et al.*, 1987), *Methanosarcina barkeri* (Bokranz and Klein, 1987), *Methanobacterium thermoautotrophicum* strain Marburg (Bokranz *et al.*, 1988), *Methanothermus fervidus* (Weil *et al.*, 1988), and *Methanococcus voltae* (Klein *et al.*, 1988). In each case, the cluster was shown to contain five contiguous open reading frames, named *mrcB, mrcD, mrcC, mrcG,* and *mrcA,* as shown in Fig. 6-5.

Allmansberger *et al.* (1989), Friedmann *et al.* (1990), Klein *et al.* (1988), and Weil *et al.* (1988, 1989) analyzed sequences of the methyl coenzyme M reductase genes in order to provide insight into the enzyme structure and mechanism. The α subunit (550–570 amino acids, $M_r \approx 61,000$), β subunit (434–443 amino acids, $M_r \approx 47,000$), and γ subunit (248–261 amino acids, $M_r \approx 29,000$) of the protein are encoded by the *mcrA, mcrB,* and *mcrG* genes, respectively. The corresponding subunit sequences share over 50% identity and contain several highly conserved domains. The binding sites for coenzyme F_{430}, HS-CoM, and HS-HTP are not revealed by the sequence, and the functions of the individual subunits are unknown. A similar degree of conservation was observed in the *mcrC* genes that are predicted to encode peptides of 189–206 amino acids having $M_r \approx 22,000$. It is not known, however, whether these peptides are synthesized in methanogens, and the function, if any, for this gene is unknown. The predicted peptide sequences (131–171 amino acids, M_r 15,100–19,400) encoded by the *mcrD* genes are much less conserved and exhibit less than 35% identity in some cases. The function, if any, for this

| | mcrB | mcrD | mcrC | mcrG | mcrA | |

Figure 6-5. The methyl coenzyme M reductase (*mcr*) gene cluster. All methyl coenzyme M reductase genes that have been sequenced occur in clusters containing five genes. The *mcrA, mcrB,* and *mcrG* genes encode the α, β, and γ subunits of the enzyme, whereas the function for the other two genes is unknown.

gene is also unknown, but the gene product was shown by antigenic methods to be synthesized in *Methanococcus vannielli* (Sherf and Reeve, 1990). Furthermore, when antibodies raised against the *mcrD* gene product were used to precipitate the protein from cell extracts, methyl coenzyme M reductase coprecipitated (Sherf and Reeve, 1990). The McrD protein–methyl coenzyme M reductase protein complex is not stable to nondenaturing gel electrophoresis or enzyme purification. These results demonstrate the presence of a weak association between the enzyme and the *mcrD* gene product, and they raise the possibility of similar interactions with other proteins. Such a multiple-protein–methyl coenzyme M reductase complex may give rise to the methanoreductosome structure observed in methanogenic bacterium strain Göl, as described in Section 6.4.1.

6.4.4 Enzymology

Early enzymological studies with methyl coenzyme M reductase were hampered by the need for a complex, strictly anaerobic assay that required three accessory protein fractions, hydrogen gas, HS-HTP, flavin adenine dinucleotide, magnesium, and ATP in addition to methyl-S-CoM (Gunsalus and Wolfe, 1980; Nagle and Wolfe, 1983). Corrins were shown to further activate the enzyme system three- to fivefold in extracts of *Methanobacterium bryantii,* but this phenomenon was not thought to be physiologically significant (Whitman and Wolfe, 1985). Nevertheless, this finding led to the development of a greatly simplified assay system for methyl coenzyme M reductase that eliminated the need for addition of the three accessory proteins, magnesium, and ATP from the Marburg strain of *Methanobacterium thermoautotrophicum* (Ankel-Fuchs and Thauer, 1986; Ankel-Fuchs *et al.,* 1986, 1987). The highly purified enzyme catalyzes conversion of methyl-S-CoM to methane in the presence of HS-HTP, cob(III)alamin, and dithiothreitol or $SnCl_2$. The corrinoid is reduced to cob(II)alamin, which disproportionates to yield cob(I)alamin, the probable reductant in the reaction. Hartzell *et al.* (1988) extended these studies to *M. thermoautotrophicum* strain ΔH and demonstrated that magnesium plus ATP stimulates the activity in the simplified assay, that titanium(III) citrate is a more effective electron donor, and that contaminating peptides are still required for activity to be present in this strain; that is, enzyme that is purified by using an affinity column does not exhibit activity under these conditions. Using the same strain, Rouvière *et al.* (1988) found that one of the accessory proteins is no longer needed using this reductive system, but two other accessory proteins must still be present to observe activity. Ellermann *et al.* (1989) anaerobically purified the enzyme from the Marburg strain to a point at which no contaminating peptides were able to be detected and still found excellent activity in the simplified assay. Thus, real differences appear to exist in the enzymes purified from the two strains of *M. thermoautotrophicum*. This result is not altogether surprising. Although these microbes share the same species assignment, these cultures are not closely related based on DNA hydridization analysis (Brandis *et al.,* 1981). Accessory protein-independent assay of homogeneous enzyme from the ΔH strain was shown to be possible, however, when samples incubated with HS-HTP and titanium(III) citrate are illuminated by light of wavelengths longer than 400 nm (Olson *et al.,* 1991). The photoactivation, in the case of the ΔH strain, or simple reductive activation, for the Marburg strain, probably transforms the nickel(II) F_{430} to a nickel(I) species. Evidence for the existence and function of this species will be detailed in Sections 6.5.3 and 6.5.4.

A major concern regarding much of the early enzymological studies of methyl coenzyme M reductase is that the activity in cell extracts represents, at most, only a few percent of that observed in the intact cells (typically, 3– 5 μmol min^{-1} mg protein^{-1} in *M. thermoautotrophicum;* Ellermann *et al.,* 1989). These results have been interpreted to suggest a requirement for a membrane, the presence of an unidentified requisite cofactor, or the requirement for labile protein–protein interactions in methanogenesis [reviewed by Daniels *et al.* (1984) and Wackett *et al.* (1988)]. Furthermore, even after use of the methods described above for reductive activation, the specific activity of purified enzyme (2.5 μmol min^{-1} mg protein^{-1}) is far lower than that expected (50 μmol min^{-1} mg protein^{-1}) based on whole-cell methanogenesis rates (Ellermann *et al.,* 1989). Recently, however, Rospert *et al.* (1991) demonstrated that preincubation of *M. thermoautotrophicum* cells with 100% hydrogen gas prior to cell disruption leads to the retention of activity at near the *in vivo* levels. They purified the enzyme tenfold in the presence of methyl-S-CoM, thought to stabilize the activity, and obtained a specific activity of 20 μmol min^{-1} mg protein^{-1}. This enzyme preparation possesses distinct spectral properties compared to the enzyme as normally prepared; for example, rather than the usual absorbance maximum at 420 nm and a shoulder at 445 nm, the highly active enzyme displays an absorbance maximum at 386 nm and a shoulder at 420 nm (Rospert *et al.,* 1991). A few other properties of this highly active enzyme form are discussed in Section 6.5.3. In contrast to the sparsity of information available for the high-activity form of methyl coenzyme M reductase, the less active form of the enzyme has been intensively examined with a goal of characterizing the enzyme mechanism. Although the results from studying low-activity enzyme must be viewed with caution, experiments involving analysis of substrate analogs and inhibitors can be revealing. The next few paragraphs summarize our current knowledge of these active-site ligands.

An early study demonstrated that ethyl-S-CoM is an alternate substrate for the enzyme and reacts at 20% of the rate of methyl-S-CoM to generate ethane (Gunsalus *et al.*, 1978). Using (*R*)- and (*S*)-[1-^{2}H,^{3}H]ethyl-S-CoM, Ahn *et al.* (1991) probed the stereospecificity of the methyl coenzyme M reductase reaction. They showed that the labeled ethane carbon has inverted configuration compared to the labeled substrate compounds (Fig. 6-6). In order to assess the stereochemistry of the released ethane, they illuminated the product in chlorine gas, treated the resulting chloroethane with basic permanganate, and used established methods to deduce the stereochemistry of the product acetic acid (Ahn *et al.*, 1991). By analogy, the enzyme is thought to react with methyl-S-CoM by a mechanism that leads to inversion of stereochemistry in the product methane. Any mechanistic scheme proposed for the enzyme must account for the observed stereochemistry.

A third compound that functions as a substrate for methyl coenzyme M reductase is methyl-Se-CoM (Wackett *et al.*, 1987b). The substitution of selenium for the thioether sulfur results in a threefold increase in both the k_{cat} and K_m, consistent with a weaker carbon–heteroatom dissociation energy and weaker binding to the enzyme (perhaps via a Ni–heteroatom bond). By contrast, the oxygen analog, 2-methoxyethanesulfonate, is not a substrate for the enzyme. This result is not surprising because this compound has a stronger carbon–heteroatom dissociation energy than methyl-S-CoM.

A fourth substrate for methyl coenzyme M reductase is difluoromethyl-S-CoM. This compound was shown to have a twofold higher turnover than the normal substrate (Wackett *et al.*, 1987b). In contrast, the trifluoromethyl-S-CoM analog is not a substrate. The addition of electron-withdrawing fluorine atoms would be expected to increase the energy required for carbon–sulfur

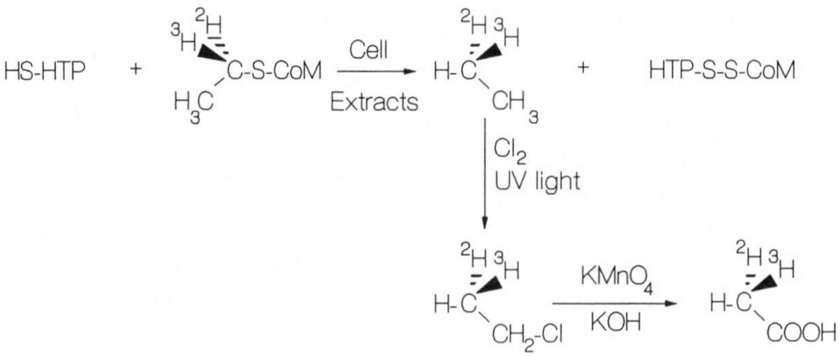

Figure 6-6. Stereochemistry of the methyl coenzyme M reductase reaction. Using stereospecifically labeled ethyl-S-CoM as an alternate substrate, the enzyme generated ethane that has an inverted configuration around the labeled carbon, as shown by conversion of the product ethane to acetic acid and subsequent analysis (Ahn *et al.*, 1991).

dissociation while at the same time stabilizing an alkyl-nickel hypothetical intermediate. Despite our present inability to clearly understand why certain compounds (such as the difluoromethyl-S-CoM) are substrates for the enzyme whereas very similar compounds (such as the trifluoromethyl-S-CoM) are not, it is clear that further studies involving these four substrate analogs may be invaluable in establishing the enzyme mechanism.

A historically important inhibitor of the enzyme is bromoethanesulfonate. This compound is a precursor used in the synthesis of methyl-S-CoM, and hence methyl-S-CoM must be carefully purified before use (Gunsalus *et al.*, 1978). The mode of action of this reversible inhibitor has not been demonstrated, but it does not involve alkylation of the enzyme or coenzyme. Other HS-CoM analogs that inhibit the enzyme include 3-bromopropanesulfonate and 2-azidoethanesulfonate (Ellermann *et al.*, 1989), 2-methoxyethanesulfonate, trifluoromethyl-S-CoM, trifluoromethyl-Se-CoM, allyl-S-CoM, cyano-S-CoM, and difluoromethyl sulfoxide-CoM (Wackett *et al.*, 1987b), and bromomethanesulfonic acid (Olson *et al.*, 1992).

The enzyme exhibits exquisite specificity for HS-HTP; the one-carbon longer or shorter methylene chains in 8-mercaptooctanoylthreonine phosphate and 6-mercaptohexanoylthreonine phosphate result only in inhibition (Ellermann *et al.*, 1988). Because methyl-S-HTP is an inhibitor (Ellermann *et al.*, 1987; Noll and Wolfe, 1987), mechanisms of catalysis involving methyl transfer via this intermediate have been greatly reduced in probability. Further studies are needed to detail the mechanisms by which the HS-CoM and HS-HTP analogs and other inhibitory compounds interact with the enzyme, especially those interactions involving coenzyme F_{430}.

6.5 Characterization of the Nickel Center

The planar nickel(II) structure of coenzyme F_{430} shown in Fig. 6-4 is deceptive. This figure does not illustrate either the bend in the tetrapyrrole to yield a saddle shape or the exchangeable axial ligands that are present both in protein-free and enzyme-bound species. Furthermore, the structure does not indicate that the F_{430} nickel apparently cycles through a nickel(I) redox state during catalysis. In this section, I will summarize the results from a variety of spectroscopic methods that have been used to probe the chemistry and geometry of coenzyme F_{430}, both free in solution and bound to methyl coenzyme M reductase.

6.5.1 Nonplanarity of the Corphin Ring

Low-spin nickel(II) ion is ideally suited for binding to corrin rings in a square-planar geometry to yield an unstrained nickel–nitrogen distance of

1.87 Å. In contrast, this metal ion is inappropriately small for analogous binding to porphyrin or corphin macrocycles. Rather, a phenomenon termed coordination hole contraction occurs (Eschenmoser, 1986; Pfaltz, 1988). Kratky *et al.* (1985) documented such tetrapyrrole deformations in X-ray crystal structures of 13 model compounds. Contraction of the ring retains coplanarity of the metal ion and the four coordinating pyrrole nitrogens, whereas the four *meso* carbon atoms are displaced alternately above and below the plane. This distortion results in a ruffled or saddle-shaped appearance for the tetrapyrroles, as is illustrated in Fig. 6-7.

The more pronounced the slope of the saddle, the shorter are the nickel–nitrogen distances (Kratky *et al.*, 1985). The steepest saddle shape yet observed by X-ray crystallography or NMR spectroscopy is that for the pentamethyl ester of 12,13-diepi-F_{430} (Färber *et al.*, 1991; Won *et al.*, 1992). Unfortunately, the crystal structure for native F_{430} is not available; thus, the extent of ruffling is unknown for the coenzyme. Nevertheless, some puckering of the ring must occur for F_{430} in polar solvents. By making use of molecular mechanical calculations, Zimmer and Crabtree (1990) proposed that low-spin nickel(II) F_{430} may exhibit less ruffling than might be anticipated from the results with model porphyrins. In addition, they posed the radical hypothesis that bending of the F_{430} ring could accommodate trigonal bipyramidal coordination of nickel. Model compounds consistent with this type of coordination in a nickel tetrapyrrole have not been reported. As discussed below, however, axial ligation of F_{430} would lead to a reduction of corphin ring saddle-shape distortion. The capability for ring distortion and axial ligation in the chemistry of nickel corphins has potential implications regarding the role for F_{430} in methyl co-enzyme M reductase catalysis.

6.5.2 Coenzyme F_{430} Axial Ligation

Nickel(II)-corrin complexes are uniformly diamagnetic and exhibit no reactivity at the axial position, as expected for square-planar nickel molecules. Nickel-corphin or nickel-porphyrin compounds are similarly diamagnetic in apolar solvents, but many of these compounds exhibit paramagnetism in

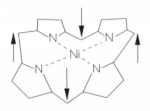

Figure 6-7. Coordination hole contraction of nickel-porphyrins or nickel-corphins. Binding of low-spin nickel(II) to many tetrapyrroles leads to a puckering or ruffling of the ring in which the *meso* carbons are distorted out of the plane of the ring, as evidenced by a series of X-ray crystal structures (Kratky *et al.*, 1985).

polar solvents or upon addition of nucleophiles. [This is the reason that the original NMR spectroscopic studies of F_{430} (Pfaltz *et al.*, 1982) were not carried out with the native compound but rather using the pentamethyl ester, which is more soluble in apolar solvents.] Strain energy generated by ring distortion during coordination hole contraction of low-spin ($S = 0$, diamagnetic) four-coordinate nickel(II) complexes in apolar solvents appears to be alternatively released, in some cases, by conversion to six-coordinate high-spin ($S = 1$, paramagnetic) complexes. Indeed, both crystallographic structure analyses (Kratky *et al.*, 1984) and spectroscopic observations (Fässler *et al.*, 1984) are consistent with axial ligation in model compounds of F_{430} (Fässler *et al.*, 1982). The addition of axial bonds leads to relaxation of the macrocyclic ring with concomitant lengthening of the nickel–nitrogen bonds; thus, measurement of bond length provides an indication of the extent of axial coordination.

Experiments to characterize axial ligation in F_{430} were compromised in early studies by substantial sample contamination with 13-epi-F_{430} and 12,13-diepi-F_{430}. For example, although resonance Raman spectroscopy can clearly discern the presence of axial ligands in nickel corphinoids (Shelnutt, 1987), the first resonance Raman spectroscopic studies of the methanogen coenzyme (Shiemke *et al.*, 1983) were carried out with the heat-extracted, diepimeric form. No evidence for axial ligation was observed, but this result left open the possibility of axial ligation in the native coenzyme. Problems with sample preparation have been especially prevalent in X-ray absorption spectroscopic (XAS) studies, a method that directly probes the number, size, and distance of the immediate nickel ligands. Diakun *et al.* (1985) carried out an extended XAS fine-structure analysis of F_{430} obtained from heat-treated *Methanobacterium thermoautotrophicum* cell extracts. They suggested that nickel is coordinated by two nitrogen atoms at 1.92 Å and two nitrogen atoms at 2.10 Å. Also using XAS, Scott *et al.* (1986) and Eidsness *et al.* (1986) reported similar nickel–nitrogen distances of 1.91 Å and 2.14 Å (with a coordination number of four to six) for heat-extracted chromophore. Furthermore, the spectra possessed a pre-edge feature that is consistent with four-coordinate geometry. In the latter study, the investigators probed the nickel site in methyl coenzyme M reductase and in F_{430} extracted from the purified protein by two different procedures. The enzyme-bound cofactor was found to be clearly distinct from that isolated from heat-treated cell extracts: the methyl coenzyme M reductase nickel was five- or six-coordinate with a single nickel–nitrogen distance of 2.09 Å, and no pre-edge feature was observed. Although not emphasized by the authors, no nickel–sulfur interaction was detected in the enzyme sample, thus ruling out axial coordination of HS-HTP or HS-CoM to coenzyme F_{430}. Extraction of the tetrapyrrole from the protein by heat treatment yielded a species that possesses spectral properties similar to those for heat-treated cell extracts, whereas salt-induced dissociation of the chromophore yielded a species that spectroscopically resembles that bound to the protein. Shiemke *et al.* (1988a) examined highly purified samples of non-heat-treated native F_{430} and the 12,13-diepi-F_{430} by XAS and found that under the experimental conditions examined the native molecule is six-coordinate with a single nickel–nitrogen (or oxygen) distance of ~2.1 Å whereas the diepimer is four-coordinate with a nickel–nitrogen distance of ~1.9 Å. They concluded that the purified samples from heat-treated cell extracts used by Diakun *et al.* (1985) and Eidsness *et al.* (1986) contained a mixture of species, that native F_{430} is pseudooctahedral, and that the diepimer is a distorted square-planar molecule. The short distances in the diepimer are consistent with conformational ring contraction to yield the saddle shape, as subsequently seen in the X-ray crystal structure of this compound (Färber *et al.*, 1991), whereas the longer distances in the native chromophore would be consistent with minimal ring contraction.

Further evaluation of the F_{430} species and methyl coenzyme M reductase by using optical, resonance Raman, and low-temperature magnetic circular dichroism spectroscopies, as well as additional XAS analysis, indicated that the situation is even more complex than described above; that is, both the native F_{430} and the F_{430} diepimer are in equilibrium between four- and six-coordinate forms, and this equilibrium is affected by several parameters. Shiemke *et al.* (1988b) found that the purified native F_{430} compound gives rise to a mix of two spectroscopically distinguishable species by resonance Raman spectroscopy. The major component is consistent with six-coordinate geometry, whereas the minor component appeared to be four-coordinate. The spectrum of the minor component is similar in some ways to that of the diepimer but differs significantly from that of the enzyme-bound species. The presence of cyanide, pyridine, or 1-methylimidazole led to changes in the optical and resonance Raman spectra of both F_{430} and the 12,13-diepi-F_{430} species, consistent with axial ligation by these compounds in both chromophores (Shiemke *et al.*, 1989a). In the same study, XAS analysis of the diepimer demonstrated loss of the pre-edge feature and conversion to six-coordinate geometry in the presence of these ligands. Using a combination of XAS, optical spectroscopy, and Raman spectroscopy, Shiemke *et al.* (1989b) showed that the equilibrium between four- and six-coordination is influenced by temperature, with the six-coordinate species predominating at lower temperatures. Similar observations were reported by Hamilton *et al.* (1989) using low-temperature magnetic circular dichroism and other spectroscopic methods to monitor the paramagnetic species. They found that the proportion of paramagnetic species for either F_{430} or 12,13-diepi-F_{430} increases with decreasing temperature and with increasing concentration of glycerol, consistent with conversion of four-coordinate nickel(II) species to six-coordinate nickel(II) compounds. Hexacoordinate F_{430} predominates at low temperature in 50%

glycerol, whereas the diepimer is still predominantly square planar under these conditions (which are similar to the conditions used for XAS samples). In contrast to these F_{430} samples, F_{560} is diamagnetic regardless of temperature or the addition of glycerol (Hamilton *et al.*, 1989), consistent with the four-coordinate geometry observed by XAS (Shiemke *et al.*, 1989b). The magnetic circular dichroism spectrum of methyl coenzyme M reductase is nearly identical to that of the native cofactor at low temperature (Cheeseman *et al.*, 1989), but, unlike the free chromophore, the protein-bound species does not exhibit temperature-dependent changes in coordination (Hamilton *et al.*, 1989). The stable axial ligands in the enzyme are thought not to be from histidine imidazoles, but rather may be oxygen ligands. An illustration of the methyl coenzyme M reductase active-site nickel center that is consistent with these data is provided in Fig. 6-8.

6.5.3 Redox Activity of F_{430}

To this point, we have only considered F_{430} in terms of possessing a divalent nickel ion. However, alternative redox states for nickel are well known in other biological systems as described for hydrogenases in Chapter 4 and carbon monoxide dehydrogenases in Chapter 5. This section details attempts to characterize other nickel oxidation states of F_{430} and similar corphins, and it examines the evidence for redox chemistry of methyl coenzyme M reductase.

The pentamethyl ester of F_{430} was shown to be capable of reduction to the nickel(I) level by cyclic voltammetry or by chemical reduction with sodium amalgam (Jaun and Pfaltz, 1986). The electrochemical reduction is completely reversible in tetrahydrofuran or dimethylformamide solvents, but the product could not be obtained in high yield by this method. In contrast, chemical reduction afforded large amounts of a stable product: the nickel(I) species possesses optical absorbance maxima at 382 nm [similar to the absorbance maximum in the highly active enzyme preparations of Rospert *et al.* (1991) described in Section 6.4.4] and 754 nm (extinction coefficients of 29,600 and 2500 M^{-1} cm^{-1}) and exhibits an electron paramagnetic resonance (EPR)

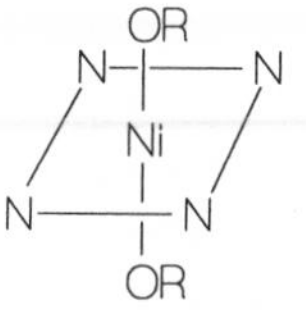

Figure 6-8. Methyl coenzyme M reductase nickel active-site geometry. This figure illustrates the geometry of the nickel site for F_{430} when it is bound to the low-activity form of the enzyme, as typically isolated. The corphin (previously shown in Fig. 6-4) is denoted here by the nickel ion and four nitrogens in a plane and probably adopts a slightly puckered shape. In addition to the tetrapyrrolic ligand complex, the nickel possesses one or two axial oxygen ligands that may arise from water or from protein amino acid side chains.

spectroscopic signal with g = 2.25, 2.074, and 2.065. The latter was assigned to a square-planar nickel(I) ion with a lone electron in the $d_{x^2-y^2}$ orbital.

Not only is F_{430} capable of being reduced, but Jaun (1990) has shown that the F_{430} pentamethyl ester can be oxidized electrolytically in high yield to form the nickel(III) complex. The oxidized compound absorbs strongly at 368 nm and weakly at 595, 890, and 1020 nm, whereas the esterified nickel(II) F_{430} absorbs at 274 and 439 nm in acetonitrile. The EPR spectrum of oxidized F_{430} is strongly isotropic with $g_\perp$ = 2.211 and $g_\parallel$ = 2.020. This spectrum was interpreted as arising from a lone electron in the d_{z^2} orbital of the nickel. These results demonstrate that either nickel(I) or nickel(III) states of the enzyme are plausible intermediates in the mechanism of methyl coenzyme M reductase.

Hamilton *et al.* (1991) extended F_{430} reduction studies to include analysis of various pentaalkylamide derivatives of this coenzyme and of its diepimer. Depending on the amine alkyl substituent, the series of F_{430} derivatives differ in their solubility in different aprotic solvents; however, their electrochemical properties are all similar. Furthermore, chemical reduction of the 12,13-diepi-F_{430} pentabutylamide derivative provided sufficient sample for spectroscopic analysis. The optical spectrum possesses peaks at 382 nm and 756 nm, and the EPR spectrum reveals g values of 2.244, 2.076, and 2.060. These properties of the diepimer (Hamilton *et al.*, 1991) are nearly identical to those for the native coenzyme described above (Jaun and Pfaltz, 1986).

Extended XAS fine-structure analysis has been used to probe the consequences of nickel reduction in the pentamethyl ester of F_{430} (Furenlid *et al.*, 1990a) and in model compounds (Furenlid *et al.*, 1990b, 1991; Renner *et al.*, 1991). The low-spin form of the methylated nickel(II) coenzyme in apolar solvent exhibits pre-edge features typically observed in tetracoordinate planar complexes and possesses nickel–nitrogen distances of 1.90 Å. These data are nearly identical to those previously reported for the diepimer (Shiemke *et al.*, 1989a). Reduction of the F_{430} compound leads to a 2–3-eV shift in the edge position, retention of the pre-edge feature, and the presence of two distinct nickel–nitrogen distances of 1.88 Å and 2.03 Å (Furenlid *et al.*, 1990a). Analogous pairing of nickel–nitrogen distances for nickel(I) model compounds has been observed by XAS and confirmed by X-ray crystal structure analyses (Furenlid *et al.*, 1991; Renner *et al.*, 1991). Thus, the F_{430} corphin framework is sufficiently flexible to accommodate the requisite perturbations that occur upon nickel reduction while maintaining distorted square-planar geometry in aprotic organic solvents.

The first direct evidence that methyl coenzyme M reductase can exist in a form detectable by EPR spectroscopy was reported by Albracht *et al.* (1986) and Ankel-Fuchs *et al.* (1986). They observed a new EPR signal (designated MCR-ox1) in thawed whole cells of *Methanobacterium thermoautotrophicum* strain Marburg, and, after growing cells on medium that was enriched in [61]Ni, they showed that the signal exhibits hyperfine interactions consistent with its assignment to a nickel species. The signal is distinct from those previously described for nickel-containing hydrogenases (see Chapter 4) and appears to arise from nickel in octahedral coordination comprised of four equivalent nitrogen atoms in a plane and two unidentified axial ligands. Suspecting that the source for the new signal was methyl coenzyme M reductase, Albracht *et al.* (1986) purified the enzyme and showed that it gives rise to the same EPR signal (g = 2.233, 2.168, and 2.154), albeit at a spin concentration that accounts for less than 25% of the amount of enzyme. Furthermore, treatment of purified F_{430} with dithionite yielded, at a spin concentration of only 1%, a species giving rise to a similar signal. The authors initially suggested that the nickel-associated signal arises from active enzyme and that much of the purified methyl coenzyme M reductase is in an inactive form. This preliminary conclusion was abandoned when further studies (Albracht *et al.*, 1988) indicated a lack of correlation between the intensity of this signal and enzyme activity. Brenner *et al.* (1992) found that the *M. thermoautotrophicum* strain ΔH methyl coenzyme M reductase that is usually isolated from slow-growing cells does not give rise to the MCR-ox1 signal, whereas the genetically distinct, less active enzyme from exponentially growing cells easily converts to this paramagnetic species. These results lead to a reasonable suggestion for the variability in MCR-ox1 signal intensity levels found by Albracht *et al.* (1988): different preparations of cells may have been harvested at variable times or may have been grown with slightly altered gassing rates, leading to different ratios of the two enzyme species. An analogous signal has also been observed in *Methanothrix soehngenii* (Jetten *et al.*, 1991). Although the MCR-ox1 signal now appears to be nonphysiological, it did demonstrate that the nickel in F_{430} bound to methyl coenzyme M reductase is capable of generating an EPR spectroscopically detectable species.

Further EPR spectroscopic analysis of intact cells and cell extracts from *M. thermoautotrophicum* identified several new paramagnetic species derived from methyl coenzyme M reductase and demonstrated that the enzyme is redox active (Albracht *et al.*, 1988). The key features in this very complex study are summarized in Fig. 6-9 and described below.

When intact cells are harvested under a nitrogen atmosphere, they do not exhibit an EPR signal from the enzyme. After addition of hydrogen gas to the cells, a strongly axial signal ($g_\perp$ 2.24 and $g_\parallel$ 2.052; denoted MCR-red1) develops, reaches a maximum, and converts to a rhombic signal (g = 2.284, 2.231, and 2.175; termed MCR-red2), consistent with a change in axial ligation. Analogous MCR-red1 and MCR-red2 signals have also been observed in whole cells of *Methanosarcina barkeri* (Krzycki and Prince, 1990). When the hydrogen-reduced *M. thermoautotrophicum* cells are exposed to air at low

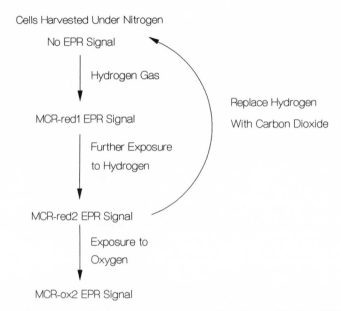

Figure 6-9. Redox activity of nickel in methyl coenzyme M reductase. Whole-cell EPR spectroscopic studies demonstrated an absence of any signal that could be attributed to methyl coenzyme
M reductase (except for a nonphysiological species denoted MCR-ox1) for cells harvested under
a stream of nitrogen. Upon exposure to hydrogen gas, two signals appear in the sequence MCR-
red1 and then MCR-red2. Exposure of the reduced sample to oxygen leads to the development
of a signal designated MCR-ox2, which is also thought not to be physiologically significant. When
the reduced cells are oxidized anaerobically by replacement of hydrogen with carbon dioxide, the
cycle is completed to form the EPR-silent state (Albracht *et al.,* 1988). For cells harvested after
gassing with 100% hydrogen, MCR-red1 and MCR-red2 species are present at high levels, and
the bacteria retain very high levels of activity (Rospert *et al.,* 1991).

temperature, a species is generated that possesses g values of 2.24 and 2.122.
This signal, called MCR-ox2, is thought not to be physiologically significant.
In contrast, addition of carbon dioxide to the hydrogen-reduced cells at their
growth temperature of 60°C leads to the disappearance of all signals arising
from methyl coenzyme M reductase, consistent with a return to the original
EPR-silent state of the enzyme. These results strongly support the proposal
that the methyl coenzyme M reductase mechanistic cycle involves reduction
of the resting nickel(II) enzyme form by hydrogen and reoxidization by transfer
of electrons to carbon dioxide. A major drawback to the proposal, however,
is that the MCR-red1 and MCR-red2 species were not observed in cell extracts
or in the purified enzyme. However, these experiments were carried out with
preparations that possess very low methanogenic activity compared to the
whole-cell rates so that trace amounts of active species may have been im-

possible to detect. The finding by Rospert *et al.* (1991) that highly active methyl coenzyme M reductase, with activity comparable to the *in vivo* level, can be isolated by treatment of cells with 100% hydrogen prior to cell disruption led to a more complete understanding of the functional redox states of enzyme-bound F_{430}. Cell extracts from these preparations were found to exhibit the MCR-red1 and MCR-red2 signals that had previously only been observed in intact cells. Furthermore, enzyme purified in the presence of methyl-S-CoM retains the MCR-red1 spectral signature. In more recent studies, Rospert *et al.* (1992) reported that (i) the MCR-red2 signal of purified enzyme can be converted to the MCR-red1 signal upon addition of methyl coenzyme M, (ii) addition of the potent inhibitor 3-bromopropanesulfonate to the MCR-red2 state yielded a novel axial EPR signal, which was identical to that observed when the fluoro- and iodo- analogs were examined, and (iii) 7-bromoheptanoylthreonine phosphate converted the MCR-red1 or MCR-red2 signals to yet another novel signal. All of these signals appear to represent the same nickel redox state. Importantly, the MCR-red1 signal is similar in appearance to the EPR spectra of reduced F_{430} pentaalkyl esters (Jaun and Pfaltz, 1986; Hamilton *et al.*, 1991). These results provide compelling evidence that nickel(I) is important in methyl coenzyme M reductase catalysis.

6.5.4 Mechanism of Catalysis

The mechanism of methyl coenzyme M reductase is beginning to be understood. As just described, there is compelling EPR spectroscopic evidence for the involvement of at least two nickel(I) species during catalysis, but the kinetic competence of these species has not been examined. The identity of the F_{430} nickel(I) axial ligands in the MCR-red1 and MCR-red2 species has not been determined. In contrast, for the nickel(II) species the axial ligands are thought to contain oxygen. It is possible that two axial ligands are not present in the nickel(I) species: on the basis of conformational calculations, Zimmer and Crabtree (1990) proposed that only a single nontetrapyrrole ligand need bind because the nickel(I) corphin should be able to accommodate a trigonal bipyramidal geometry. Further analysis of the spectroscopically distinguishable enzyme forms, as well as the resting EPR-silent state, is required before the mechanism can be completely understood.

Nearly all of methyl coenzyme M reductase studies in the literature have been carried out necessarily with enzyme that possesses very low activity compared to that found in the intact cells. Only since the report of Rospert *et al.* (1991) has highly active enzyme been available. Although mechanistic interpretations derived from the earlier work can be questioned, most of the experimental results are likely to remain valid. However, the low-activity form of the enzyme is probably not from a small amount of active enzyme amid a large contaminating amount of dead enzyme. Hartzell *et al.* (1987) showed that all of the enzyme-bound HS-CoM can exchange with the thioethanesulfonate portion of methyl-S-CoM, that the reaction requires enzyme turnover, and that multiple turnovers are required for complete exchange. If only a portion of the enzyme is active, the amount of exchange would be substoichiometric. The experimental result is most easily understood in terms of all of the enzyme being capable of methyl coenzyme M reductase activity but only a low proportion of the enzyme being active at any one time. Structural changes that occur upon shifting the equilibrium from the well-studied low-activity form to the poorly understood highly active enzyme are completely unknown. Despite these caveats, in the following paragraphs an attempt is made to use the experimental results from studies of methyl coenzyme M reductase and model systems to propose a mechanism for the enzyme.

Wackett *et al.* (1988) enumerated three plausible mechanistic intermediates for the enzyme, involving nickel-hydride, nickel-sulfonium, or nickel-methyl compounds. These structures and a fourth intermediate suggested by Berkessel (1991), involving a complex of nickel and the thiolate anion of HS-HTP, are shown in Fig. 6-10. Note that axial ligation to nickel occurs in each of these intermediates, consistent with the axial reactivity in F_{430}.

The nickel-hydride intermediate is strongly reminiscent of a postulated activated state for nickel-containing hydrogenases, as described in Chapter 4. An activated hydride could theoretically attack methyl-S-CoM, resulting in release of methane with inversion of stereochemistry as found by Ahn *et al.* (1991), but no chemical precedents are known for this reaction. Furthermore, this mechanism would lead to displacement of the thiolate anion of HS-CoM rather than formation of the HTP-S-S-CoM disulfide. A mechanism could be invoked in which disulfide formation is coupled to nickel reduction in order to activate F_{430} to form the hydride intermediate, but this would involve a two-electron reduction of the coenzyme. No experimental evidence for a nickel-hydride intermediate in catalysis has been observed, and a mechanism involving this intermediate appears unlikely.

The nickel-sulfonium ion intermediate shown in Fig. 6-10 would be expected to activate methyl-S-CoM to attack by a hydride ion. Again, the released

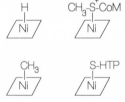

Figure 6-10. Potential intermediates in the methyl coenzyme M reductase mechanism. Wackett *et al.* (1988) proposed three possible intermediates involving nickel-hydride, nickel-thioether, or nickel-methyl species that could exist in the catalytic mechanism. A fourth potential intermediate, involving a nickel-thiolate, was proposed by Berkessel (1991). Each of these compounds involves axial coordination to the nickel tetrapyrrole, indicated by the plane, as is consistent with the chemical reactivity of F_{430}.

methane would possess the requisite inverted stereochemistry. Wackett *et al.* (1987b) suggested that such an intermediate is consistent with the decreased K_m observed for methyl-Se-CoM compared to methyl-S-CoM because nickel–heteroatom interaction is dependent on the softness of the heteroatom. Furthermore, chemical precedents for nickel(II)–thioether interactions are well known (Wackett *et al.*, 1988). Although only the low-activity form of the enzyme has been examined, no nickel–sulfur interaction has been observed by XAS or resonance Raman spectroscopy. At present, a mechanism involving this intermediate cannot be dismissed, but there is no compelling evidence for the presence of direct interaction between the sulfur atom of methyl-S-CoM and the nickel of coenzyme F_{430}.

The nickel-methyl intermediate is an attractive possibility for methyl coenzyme M reductase because there are clear precedents for nickel-methyl complexes in model chemistry and an analogous cobalt intermediate exists in vitamin B_{12}-dependent methionine biosynthesis. Wackett *et al.* (1988) have reviewed several examples in which methyl-nickel compounds undergo protonolysis to yield methane. Furthermore, Jaun and Pfaltz (1988) have shown that the pentamethyl ester of nickel(II) F_{430} reacts catalytically with iodomethane, methylsulfonium ion, and other methylated substrates (but *not* thioethers) under reducing conditions (zinc amalgam in dimethylformamide) to yield methane. These results are consistent with reduction of F_{430} to the nickel(I) state, nucleophilic attack by the corrinoid on the substrate to yield a nickel-methyl intermediate, and subsequent protonolysis. Lin and Jaun (1992) have more fully characterized the kinetics and properties of the reductive cleavage of methylsulfonium ion. They used isotope labeling studies to verify that the fourth hydrogen in the released methane is indeed from solvent proton. F_{430} can also catalyze reductive dechlorination of carbon tetrachloride under reducing conditions [titanium(III) citrate], presumably by an analogous reaction series (Krone *et al.*, 1989). Similarly, nickel(I) octaethylisobacteriochlorin was shown to be an exceptional nucleophile for reaction with alkyl halides (Helveston and Castro, 1992). Recently, Lin and Jaun (1991) directly demonstrated that the nickel of $F_{430}M$ can be methylated by $(CD_3)_2Mg$, as detected by deuterium nuclear magnetic resonance spectroscopy. In methanogens, nucleophilic attack of nickel(I) on the substrate to yield a nickel-methyl intermediate followed by protonolysis would be expected to result in inversion of stereochemistry, consistent with the results reported by Ahn *et al.* (1991). Not accounted for in a simple reduction, nucleophilic attack, protonolysis scheme is the formation of the second product, the HTP-S-S-CoM disulfide. Nevertheless, a nickel-methyl intermediate is very appealing for further consideration.

The last possibility we will consider is the nickel-thiolate intermediate shown in Fig. 6-10. Berkessel (1991) has synthesized a series of nickel compounds that possess a thioether appropriately positioned to coordinate axially to the metal ion. Electron transfer from thioether sulfur to nickel(II) occurs much more readily than for noncoordinated species. Analogous one-electron transfer from thiolate anion to nickel(II) of F_{430} was suggested to be a reasonable thermodynamic possibility.

An elegant mechanism for methyl coenzyme M reductase that involves two of the hypothetical intermediates was proposed by Berkessel (1991) and is shown in Fig. 6-11. The reaction initiates with electron transfer from the HS-HTP thiolate anion to F_{430}. The HTP-thiyl radical is suggested to react with methyl-S-CoM to yield a sulfuranyl radical by well-known chemistry. Methyl migration to the nickel(I) F_{430} affords the nickel-methyl intermediate, which can undergo protonolysis to yield methane. This mechanism is consistent with the stereochemical studies of Ahn *et al.* (1991), with the presence of two EPR-detectable nickel(I) species, and with the observed disulfide product formation. No spectroscopic evidence for the thiyl or sulfuranyl radicals has been reported; however, high-activity enzyme has not been available until very recently. The Berkessel proposal appears to serve as a useful working model and will certainly stimulate further mechanistic studies.

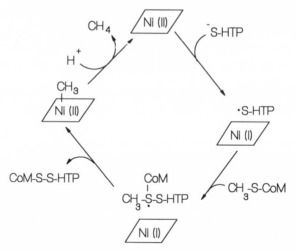

Figure 6-11. Hypothetical mechanism for methyl coenzyme M reductase. The catalytic cycle, modified from Berkessel (1991), is proposed to involve electron transfer from the thiolate anion of HS-HTP to nickel(II) F_{430}, yielding a thiyl radical and nickel(I) F_{430}. The thiyl radical is suggested to couple with methyl-S-CoM to yield a sulfuranyl radical. Methyl transfer occurs from the sulfuranyl radical to the corphin to yield methylated nickel(II) F_{430} and the HTP-S-S-CoM disulfide. Protonolysis of the methyl-nickel(II) F_{430} is then proposed to yield methane. This elegant mechanism incorporates most of the experimental results from the enzyme and from model chemistry and is a useful working model for the methyl coenzyme M reductase reaction.

6.6 Perspective

In this chapter, I have summarized the evidence that coenzyme F_{430} is an essential component of coenzyme M reductase, detailed the structure of the nickel-containing tetrapyrrole, discussed several general properties of the enzyme, assessed the reactivity of the nickel active site, and examined evidence related to the enzyme mechanism. Although much has been learned about F_{430} and this nickel-containing enzyme in the few brief years since their discoveries, many questions remain open for investigation. The coenzyme structure is quite well understood from NMR spectroscopy and model studies; however, an X-ray crystal structure of F_{430} is still desirable. Furthermore, a crystal structure of either the traditional low-activity protein or the highly active enzyme would be of great interest in understanding the roles for the three subunits and in elucidating the interactions among the enzyme-bound F_{430}, methyl-S-CoM, and HS-HTP cofactors. The inactivation phenomenon that, until very recently, resulted in isolation and characterization of low-level-activity enzyme (unless cells are harvested under 100% hydrogen) needs to be better understood. The results from many early enzymological studies demand reevaluation using the fully active enzyme. Continued characterization of the two genetically distinct methyl coenzyme M reductases that exist in the same cell is of paramount importance; that is, how are these two systems differentially regulated and what are their roles? The identity of and role for the two nonsubunit genes that are conserved in all methyl coenzyme M reductase gene clusters require further study. Perhaps these genes are related to formation of a labile protein complex such as the methanoreductosome. Finally, methods must be developed to further explore the mechanism of methyl coenzyme M reductase. For example, the kinetic competence of presumed intermediates possibly could be examined by freeze-quench EPR spectroscopic methods. Furthermore, the identities of these compounds could perhaps be probed by analysis of hyperfine interactions arising from [77]Se-substituted methyl-Se-CoM (or HSe-HTP) or ^{13}C-substituted methyl-S-CoM. This fascinating enzyme should continue to hold the interests of enzymologists, spectroscopists, and inorganic chemists for some time.

References

Ahn, Y., Krzycki, J. A., and Floss, H. G., 1991. Steric course of reduction of ethyl coenzyme M to ethane catalyzed by methyl coenzyme M reductase from *Methanosarcina barkeri, J. Am. Chem. Soc.* 113:4700–4701.

Albracht, S. P. J., Ankel-Fuchs, D., van der Zwaan, J. W., Fontijn, R. D., and Thauer, R. K., 1986. A new EPR signal of nickel in *Methanobacterium thermoautotrophicum, Biochim. Biophys. Acta* 870:50–57.

Albracht, S. P. J., Ankel-Fuchs, D., Böcher, R., Ellermann, J., Moll, J., van der Zwaan, J. W., and Thauer, R. K., 1988. Five new EPR signals assigned to nickel in methyl-coenzyme M reductase from *Methanobacterium thermoautotrophicum*, strain Marburg, *Biochim. Biophys. Acta* 955:86–102.

Aldrich, H. C., Beimborn, D. B., Bokranz, M., and Schönheit, P., 1987. Immunocytochemical localization of methyl-coenzyme M reductase in *Methanobacterium thermoautotrophicum, Arch. Microbiol.* 147:190–194.

Allmansberger, R., Bokranz, M., Kröckel, L., Schallenberg, J., and Klein, A., 1989. Conserved gene structures and expression signal in methanogenic archaebacteria, *Can. J. Microbiol.* 35: 52–57.

Ankel-Fuchs, D., and Thauer, R. K., 1986. Methane formation from methyl-coenzyme M in a system containing methyl-coenzyme M reductase, component B and reduced cobalamin, *Eur. J. Biochem.* 156:171–177.

Ankel-Fuchs, D., Jaenchen, R., Gebhardt, N. A., and Thauer, R. K., 1984. Functional relationship between protein-bound and free factor F430 in *Methanobacterium, Arch. Microbiol.* 137: 332–337.

Ankel-Fuchs, D., Hüster, R., Mörschel, E., Albracht, S. P. J., and Thauer, R. K., 1986. Structure and function of methyl-coenzyme M reductase and of factor F430 in methanogenic bacteria, *Syst. Appl. Microbiol.* 7:383–387.

Ankel-Fuchs, D., Böcher, R., Thauer, R. K., Noll, K. M., and Wolfe, R. S., 1987. 7-Mercaptoheptanoylthreonine phosphate functions as component B in ATP-independent methane formation from methyl-CoM with reduced cobalamin as electron donor, *FEBS Lett.* 213:123–127.

Berkessel, A., 1991. Methyl-coenzyme M reductase: Model studies on pentadentate nickel complexes and a hypothetical mechanism, *Bioorgan. Chem.* 19:101–115.

Bobik, T. A., Olson, K. D., Noll, K. M., and Wolfe, R. S., 1987. Evidence that the heterodisulfide of coenzyme M and 7-mercaptoheptanoylthreonine phosphate is a product of the methylreductase reaction in *Methanobacterium, Biochem. Biophys. Res. Commun.* 149:455–460.

Bokranz, M., and Klein, A., 1987. Nucleotide sequence of the methyl coenzyme M reductase gene cluster from *Methanosarcina barkeri, Nucleic Acids Res.* 15:4350–4351.

Bokranz, M., Bäumner, G., Allmansberger, R., Ankel-Fuchs, D., and Klein, A., 1988. Cloning and characterization of the methyl coenzyme M reductase genes from *Methanobacterium thermoautotrophicum, J. Bacteriol.* 170:568–577.

Bonacker, L. G., Bauder, S., and Thauer, R. K., 1992. Differential expression of the two methyl-coenzyme M reductases in *Methanobacterium thermoautotrophicum* as determined immunochemically via isoenzyme-specific antisera, *Eur. J. Biochem.* 206:87–92.

Brandis, A., Thauer, R. K., and Stetter, K. O., 1981. Relatedness of strains ΔH and Marburg of *Methanobacterium thermoautotrophicum, Zentralbl. Bakteriol. Parasitenkd. Infektionskr. Hyg. Abt. 1 Orig. Reihe C* 2:311–317.

Brenner, M. C., Ma, L., Johnson, M. K., and Scott, R. A., 1992. Spectroscopic characterization of the alternate form of S-methylcoenzyme M reductase from *Methanobacterium thermoautotrophicum, Biochim. Biophys. Acta* 1120:160–166.

Cheeseman, M. R., Ankel-Fuchs, D., Thauer, R. K., and Thompson, J., 1989. The magnetic properties of the nickel cofactor F430 in the enzyme methyl-coenzyme M reductase of *Methanobacterium thermoautotrophicum, Biochem. J.* 260:613–616.

Cram, D. S., Sherf, B. A., Libby, R. T., Mattaliano, R. J., Ramachandran, K. L., and Reeve, J. N., 1987. Structure and expression of the genes, *mcrBDCGA*, which encode the subunits of component C of methyl coenzyme M reductase in *Methanococcus vannielli, Proc. Natl. Acad. Sci. USA* 84:3992–3996.

Daniels, L., Sparling, R., and Sprott, G. D., 1984. The bioenergetics of methanogenesis, *Biochim. Biophys. Acta* 768:113–163.

Deppenmeier, U., Blaut, M., Mahlmann, A., and Gottschalk, G., 1990. Reduced coenzyme F_{420}: heterodisulfide oxidoreductase, a proton-translocating redox system in methanogenic bacteria, *Proc. Natl. Acad. Sci. USA* 87:9449–9453.

Diakun, G. P., Piggott, B., Tinton, H. J., Ankel-Fuchs, D., and Thauer, R. K., 1985. An extended-X-ray-absorption-fine-structure (e.x.a.f.s) study of coenzyme F430 from *Methanobacterium thermoautotrophicum, Biochem. J.* 232:281–284.

Diekert, G., Gilles, H.-H., Jaenchen, R., and Thauer, R. K., 1980a. Incorporation of 8 succinate per mol nickel into factors F_{430} by *Methanobacterium thermoautotrophicum, Arch. Microbiol.* 128:256–262.

Diekert, G., Jaenchen, R., and Thauer, R. K., 1980b. Biosynthetic evidence for a nickel tetrapyrrolic structure of factor F_{430} from *Methanobacterium thermoautotrophicum, FEBS Lett.* 119:118–120.

Diekert, G., Klee, B., and Thauer, R. K., 1980c. Nickel, a component of factor F_{430} from *Methanobacterium thermoautotrophicum, Arch. Microbiol.* 124:103–106.

Diekert, G., Weber, B., and Thauer, R. K., 1980d. Nickel dependence of factor F_{430} content in *Methanobacterium thermoautotrophicum, Arch. Microbiol.* 127:273–278.

Diekert, G., Konheiser, U., Piechulla, K., and Thauer, R. K., 1981. Nickel requirement and factor F_{430} content of methanogenic bacteria, *J. Bacteriol.* 148:459–464.

DiMarco, A. A., Bobik, T. A., and Wolfe, R. S., 1990. Unusual coenzymes of methanogenesis, *Annu. Rev. Biochem.* 59:355–394.

Eidsness, M. K., Sullivan, R. J., Schwartz, J. R., Hartzell, P. L., Wolfe, R. S., Flank, A.-M., Cramer, S. P., and Scott, R. A., 1986. Structural diversity of F_{430} from *Methanobacterium thermoautotrophicum.* A nickel X-ray absorption spectroscopic study, *J. Am. Chem. Soc.* 108:3120–3121.

Ellefson, W. L., and Wolfe, R. S., 1980. Role of component C in the methylreductase system of *Methanobacterium, J. Biol. Chem.* 255:8388–8389.

Ellefson, W. L., and Wolfe, R. S., 1981. Component C of the methylreductase system of *Methanobacterium, J. Biol. Chem.* 256:4259–4262.

Ellefson, W. L., Whitman, W. B., and Wolfe, R. S., 1981. Nickel-containing factor F_{430}: Chromophore of the methylreductase of *Methanobacterium, Proc. Natl. Acad. Sci. USA* 79:3707–3710.

Ellermann, J., Kobelt, A., Pfaltz, A., and Thauer, R. K., 1987. On the role of N-7-mercaptoheptanoyl-O-phospho-L-threonine (component B) in the enzymatic reduction of methyl-coenzyme M to methane, *FEBS Lett.* 220:358–362.

Ellermann, J., Hedderich, R., Böcher, R., and Thauer, R. K., 1988. The final step in methane formation. Investigations with highly purified methyl-CoM reductase (component C) from *Methanobacterium thermoautotrophicum* (strain Marburg), *Eur. J. Biochem.* 172:669–677.

Ellermann, J., Rospert, S., Thauer, R. K., Bokranz, M., Klein, A., Voges, M., and Berkessel, A., 1989. Methyl-coenzyme-M reductase from *Methanobacterium thermoautotrophicum* (strain Marburg). Purity, activity, and inhibitors, *Eur. J. Biochem.* 184:63–68.

Eschenmoser, A., 1986. Chemistry of corphinoids, *Ann. N. Y. Acad. Sci.* 471:108–129.

Färber, G., Keller, W., Kratky, C., Jaun, B., Pfaltz, A., Spinner, C., Kobelt, A., and Eschenmoser, A., 1991. Coenzyme F430 from methanogenic bacteria: Complete assignment of configuration based on an X-ray analysis of 12,13-diepi-F430 pentamethylester and on NMR spectroscopy, *Helv. Chim. Acta* 74:697–716.

Fässler, A., Pfaltz, A., Müller, P. M., Farooq, S., Kratky, C., Kräutler, B., and Eschenmoser, A., 1982. Herstellung und eigenschafter einiger hydrocorphinoider nickel(II)-komplexe, *Helv. Chim. Acta* 65:812–827.

Fässler, A., Pfaltz, A., Kräutler, B., and Eschenmoser, A., 1984. Chemistry of corphinoids: Synthesis of a nickel(II) complex containing the chromophore system of coenzyme F430, *J. Chem. Soc., Chem. Commun.* 1984:1365–1367.

Fässler, A., Kobelt, A., Pfaltz, A., Eschenmoser, A., Bladon, C., Battersby, A. R., and Thauer, R. K., 1985. Zur Kenntnis des Faktors F430 aus methanogenen bakterien: Absolute konfiguration, *Helv. Chim. Acta* 68:2287–2298.

Ferry, J. G., 1992. Biochemistry of methanogenesis, *Crit. Rev. Biochem. Mol. Biol.* 27:473–503.

Friedmann, H. C., Klein, A., and Thauer, R. K., 1990. Structure and function of the nickel porphinoid, coenzyme F_{430}, and of its enzyme, methyl coenzyme M reductase, *FEMS Microbiol. Rev.* 87:339–348.

Furenlid, L. R., Renner, M. W., and Fajer, J., 1990a. EXAFS studies of nickel(II) and nickel(I) factor 430 M. Conformational flexibility of the F430 skeleton, *J. Am. Chem. Soc.* 112:8987–8989.

Furenlid, L. R., Renner, M. W., Smith, K. M., and Fajer, J., 1990b. Structural consequences of nickel versus macrocycle reductions in F430 models: EXAFS studies of a Ni(I) anion and Ni(II) π anion radicals, *J. Am. Chem. Soc.* 112:1634–1635.

Furenlid, L. R., Renner, M. W., Szalda, D. J., and Fujita, E., 1991. EXAFS studies of Ni^{II}, Ni^{I}, and Ni^{I}-CO tetraazamacrocycles and the crystal structure of (5,7,7,12,14,14-hexamethyl-1,4,8,11-tetraazacyclotetradeca-4,11-diene)nickel(I) perchlorate, *J. Am. Chem. Soc.* 113:883–892.

Gilles, H., and Thauer, R. K., 1983. Uroporphyrinogen III, an intermediate in the biosynthesis of the nickel-containing factor F_{430} in *Methanobacterium thermoautotrophicum, Eur. J. Biochem.* 135:109–112.

Gunsalus, R. P., and Wolfe, R. S., 1978. Chromophoric factors F_{342} and F_{430} of *Methanobacterium thermoautotrophicum, FEMS Microbiol. Lett.* 3:191–193.

Gunsalus, R. P., and Wolfe, R. S., 1980. Methylcoenzyme M reductase from *Methanobacterium thermoautotrophicum.* Resolution and properties of the components, *J. Biol. Chem.* 255:1891–1895.

Gunsalus, R. P., Romesser, J. A., and Wolfe, R. S., 1978. Preparation of coenzyme M analogues and their activity in the methyl conenzyme M reductase system of *Methanobacterium thermoautotrophicum, Biochemistry* 17:2374–2377.

Hamilton, C. L., Scott, R. A., and Johnson, M. K., 1989. The magnetic and electronic properties of *Methanobacterium thermoautotrophicum* (strain ΔH) methyl coenzyme M reductase and its nickel tetrapyrrole cofactor F_{430}. A low temperature magnetic circular dichroism study, *J. Biol. Chem.* 264:11605–11613.

Hamilton, C. L., Ma, L., Renner, M. W., and Scott, R. A., 1991. Ni(II) and Ni(I) forms of pentaalkylamide derivatives of cofactor F_{430} of *Methanobacterium thermoautotrophicum, Biochim. Biophys. Acta* 1074:312–319.

Hartzell, P. L., and Wolfe, R. S., 1986a. Requirement of the nickel tetrapyrrole F_{430} for *in vitro* methanogenesis: Reconstitution of methylreductase component C from its dissociated subunits, *Proc. Natl. Acad. Sci. USA* 83:6726–6730.

Hartzell, P. L., and Wolfe, R. S., 1986b. Comparative studies of component C from the methylreductase system of different methanogens, *Syst. Appl. Microbiol.* 7:376–382.

Hartzell, P. L., Donnelly, M. I., and Wolfe, R. S., 1987. Incorporation of coenzyme M into component C of methylcoenzyme M methylreductase during *in vitro* methanogenesis, *J. Biol. Chem.* 262:5581–5586.

Hartzell, P. L., Escalante-Semerena, J. C., Bobik, T. A., and Wolfe, R. S., 1988. A simplified methylcoenzyme M methylreductase assay with artificial electron donors and different preparations of component C from *Methanobacterium thermoautotrophicum* ΔH, *J. Bacteriol.* 170:2711–2715.

Hausinger, R. P., Orme-Johnson, W. H., and Walsh, C., 1984. Nickel tetrapyrrole cofactor F_{430}: Comparison of the forms bound to methyl coenzyme M reductase and protein free in cells of *Methanobacterium thermoautotrophicum, Biochemistry* 23:801–804.

Hedderich, R., and Thauer, R. K., 1988. *Methanobacterium thermoautotrophicum* contains a soluble enzyme system that specifically catalyzes the reduction of the heterodisulfide of coenzyme M and 7-mercaptoheptanoylthreonine phosphate with H_2, *FEBS Lett.* 234:223–227.

Hedderich, R., Berkessel, A., and Thauer, R. K., 1990. Purification and properties of heterodisulfide reductase from *Methanobacterium thermoautotrophicum* (strain Marburg), *Eur. J. Biochem.* 193:255–261.

Helveston, M. C., and Castro, C. E., 1992. Nickel(I) octaethylisobacteriochlorin anion. An exceptional nucleophile. Reduction and coupling of alkyl halides by anionic and radical processes. A model for factor F-430, *J. Am. Chem. Soc.* 114:8490–8496.

Hüster, R., Gilles, H.-H., and Thauer, R. K., 1985. Is coenzyme M bound to factor F430 in methanogenic bacteria? Experiments with *Methanobrevibacter ruminantium, Eur. J. Biochem.* 148:107–111.

Jablonski, P. E., and Ferry, J. G., 1991. Purification and properties of methyl coenzyme M methylreductase from acetate-grown *Methanosarcina thermophila, J. Bacteriol.* 173:2481–2487.

Jaenchen, R., Diekert, G., and Thauer, R. K., 1981. Incorporation of methionine-derived methyl groups into factor F_{430} by *Methanobacterium thermoautotrophicum, FEBS Lett.* 130:133–136.

Jaun, B., 1990. Coenzyme F430 from methanogenic bacteria: Oxidation of F430 pentamethyl ester to the nickel(III) form, *Helv. Chim. Acta* 73:2209–2217.

Jaun, B., and Pfaltz, A., 1986. Coenzyme F430 from methanogenic bacteria: Reversible one-electron reduction of F430 pentamethyl ester to the nickel(I) form, *J. Chem. Soc., Chem. Commun.* 1986:1327–1329.

Jaun, B., and Pfaltz, A., 1988. Coenzyme F430 from methanogenic bacteria: Methane formation by reductive carbon–sulphur bond cleavage of methyl sulfonium ions catalyzed by F430 pentamethyl ester, *J. Chem. Soc., Chem. Commun.* 1988:293–294.

Jetten, M. S., Pierik, A. J., and Hagen, W. R., 1991. EPR characterization of a high-spin system in carbon monoxide dehydrogenase from *Methanothrix soehngenii, Eur. J. Biochem.* 202:1291–1297.

Jones, W. J., Nagle, D. P., Jr., and Whitman, W. B., 1987. Methanogens and the diversity of archaebacteria, *Microbiol. Rev.* 51:135–177.

Keltjens, J. T., Whitman, W. B., Caerteling, C. G., van Kooten, A. M., Wolfe, R. S., and Vogels, G. D., 1982. Presence of coenzyme M derivatives in the prosthetic group (coenzyme MF_{430}) of methylcoenzyme M reductase from *Methanobacterium thermoautotrophicum, Biochem. Biophys. Res. Commun.* 108:495–503.

Keltjens, J. T., Caerteling, A. M., van Kooten, A. M., van Dijk, H. F., and Vogels, G. D., 1983a. Chromophoric derivatives of coenzyme MF_{430}, a proposed coenzyme of methanogenesis in *Methanobacterium thermoautotrophicum, Arch. Biochem. Biophys.* 223:235–253.

Keltjens, J. T., Caerteling, C. G., van Kooten, A. M., van Dijk, H. F., and Vogels, G. D., 1983b. 6,7-Dimethyl-8-ribityl-5,6,7,8-tetrahydrolumazine, a proposed constituent of coenzyme MF_{430} from methanogenic bacteria, *Biochim. Biophys. Acta* 743:351–358.

Keltjens, J. T., Hermans, J. M. H., Rijsdijk, G. J. F. A., van der Drift, C., and Vogels, G. D., 1988. Interconversion of F_{430} derivatives of methanogenic bacteria, *Antonie van Leeuwenhoek* 54:207–220.

Klein, A., Allmansberger, R., Bokranz, M., Knaub, S., Müller, B., and Muth, E., 1988. Comparative analysis of genes encoding methyl coenzyme M reductase in methanogenic bacteria, *Mol. Gen. Genet.* 213:409–420.

Kratky, C., Fässler, A., Pfaltz, A., Kräutler, B., Jaun, B., and Eschenmoser, A., 1984. Chemistry of corphinoids: Structural properties of corphinoid nickel(II) complexes related to coenzyme F430, *J. Chem. Soc., Chem. Commun.* 1984:1368–1371.

Kratky, C., Waditschatka, R., Angst, C., Johansen, J. E., Plaquevent, J. C., Schreiber, J., and Eschenmoser, A., 1985. Die sattelkonformation der hydroporphinoiden nickel(II)-komplexe: Struktur, ursprung und stereochemische consequenzen, *Helv. Chim. Acta* 68:1312–1337.

Krone, U. E., Laufer, K., Thauer, R. K., and Hogenkamp, H. P. C., 1989. Coenzyme F_{430} as a possible catalyst for the reductive dehalogenation of chlorinated C_1 hydrocarbons in methanogenic bacteria, *Biochemistry* 28:10061–10065.

Krzycki, J. A., and Prince, R. C., 1990. EPR observation of carbon monoxide dehydrogenase, methylreductase, and corrinoid in intact *Methanosarcina barkeri* during methanogenesis from acetate, *Biochim. Biophys. Acta* 1015:53–60.

Lin, S.-K., and Jaun, B., 1991. Coenzyme F430 from methanogenic bacteria: Detection of a paramagnetic methylnickel(II) derivative of the pentamethyl ester by ^{2}H-NMR spectroscopy, *Helv. Chim. Acta* 74:1725–1735.

Lin, S.-K., and Jaun, B., 1992. Coenzyme F430 from methanogenic bacteria: Mechanistic studies on the reductive cleavage of sulfonium ions catalyzed by F430 pentamethyl ester, *Helv. Chim. Acta* 75:1478–1490.

Livingston, D. A., Pfaltz, A., Schreiber, J., Eschenmoser, A., Ankel-Fuchs, D., Moll, J., and Thauer, R. K., 1984. Zur kenntnis des faktors F430 aus methanogenen bakterien: Struktur des proteinfreien faktors, *Helv. Chim. Acta* 67:334–351.

Mayer, F., Rohde, M., Salzmann, M., Jussofie, A., and Gottschalk, G., 1988. The methanoreductosome: A high-molecular-weight enzyme complex in the methanogenic bacterium strain Göl that contains components of the methylreductase system, *J. Bacteriol.* 170:1438–1444.

Moura, I., Moura, J. J. G., Santos, H., Xavier, A. V., Burch, G., Peck, H. D., Jr., and LeGall, J., 1983. Proteins containing the factor F_{430} from *Methanosarcina barkeri* and *Methanobacterium thermoautotrophicum:* Isolation and properties, *Biochim. Biophys. Acta* 742:84–90.

Mucha, H., Keller, E., Weber, H., Lingens, F., and Trosch, W., 1985. Sirohydrochlorin, a precursor of factor F_{430} biosynthesis in *Methanobacterium thermoautotrophicum, FEBS Lett.* 190:169–173.

Nagle, D. P., Jr., and Wolfe, R. S., 1983. Component A of the methyl coenzyme M methylreductase system of *Methanobacterium thermoautotrophicum:* Resolution into four components, *Proc. Natl. Acad. Sci. USA* 80:2151–2155.

Noll, K. M., and Wolfe, R. S., 1986. Component C of the methylcoenzyme M methylreductase system contains bound 7-mercaptoheptanoylthreonine phosphate (HS-HTP), *Biochem. Biophys. Res. Commun.* 139:889–895.

Noll, K. M., and Wolfe, R. S., 1987. The role of 7-mercaptoheptanoylthreonine phosphate in the methylcoenzyme M methylreductase system from *Methanobacterium thermoautotrophicum, Biochem. Biophys. Res. Commun.* 145:204–210.

Olson, K. D., Won, H., Wolfe, R. S., Hare, D., and Summers, M. F., 1990. Stereochemical studies of coenzyme F430 based on 2D NOESY back-calculations, *J. Am. Chem. Soc.* 112:5884–5886.

Olson, K. D., McMahon, C. W., and Wolfe, R. S., 1991. Photoactivation of the 2-(methylthio)ethanesulfonic acid reductase from *Methanobacterium, Proc. Natl. Acad. Sci. USA* 88:4099–4103.

Olson, K. D., Chmurkowska-Cichowlas, L., McMahon, C. W., and Wolfe, R. S., 1992. Structural modifications and kinetic studies of the substrates involved in the final step of methane formation in *Methanobacterium thermoautotrophicum, J. Bacteriol.* 174:1007–1012.

Ossmer, R., Mund, T., Hartzell, P. L., Konheiser, U., Kohring, G. W., Klein, A., Wolfe, R. S., Gottschalk, G., and Mayer, F., 1986. Immunocytochemical localization of component C of the methylreductase system in *Methanococcus voltae* and *Methanobacterium thermoautotrophicum, Proc. Natl. Acad. Sci. USA* 83:5789–5792.

Pfaltz, A., 1988. Structure and properties of coenzyme F_{430}, in *The Bioinorganic Chemistry of Nickel* (J. R. Lancaster, Jr., ed.), VCH Publishers, New York, pp. 275–298.

Pfaltz, A., Jaun, B., Fässler, A., Eschenmoser, A., Jaenchen, R., Gilles, H. H., Diekert, G., and Thauer, R. K., 1982. Zur kenntnis des faktors F430 aus methanogenen bakterien: Struktur des porphinoiden ligand-systems, *Helv. Chim. Acta* 65:828–865.

Pfaltz, A., Livingston, D. A., Jaun, B., Diekert, G., Thauer, R. K., and Eschenmoser, A., 1985. Zur kenntnis des faktors F430 aus methanogenen bakterien: Über die natur der isolierungs-sartefakte von F430, ein beitrag zur chemie von F430 und zur konformationellen stereochemie der ligandperipherie von hydroporphinoiden nickel(II)-komplexen, *Helv. Chim. Acta* 68: 1338–1358.

Pfaltz, A., Kobelt, A., Hüster, R., and Thauer, R. K., 1987. Biosynthesis of coenzyme F430 in methanogenic bacteria. Identification of $15,17^3$-seco-F430-17^3-acid as an intermediate, *Eur. J. Biochem.* 170:459–467.

Renner, M. W., Furenlid, L. R., Barkigia, K. M., Forman, A., Shim, H.-K., Simpson, D. J., Smith, K. M., and Fajer, J., 1991. Models of factor 430. Structural and spectroscopic studies of Ni(II) and Ni(I) hydroporphyrins, *J. Am. Chem. Soc.* 113:6891–6898.

Rospert, S., Linder, D., Ellermann, J., and Thauer, R. K., 1990. Two genetically distinct methyl-coenzyme M reductases in *Methanobacterium thermoautotrophicum* strain Marburg and ΔH, *Eur. J. Biochem.* 194:871–877.

Rospert, S., Böcher, R., Albracht, S. P. J., and Thauer, R. K., 1991. Methyl-coenzyme M reductase preparations with high specific activity from H_2-preincubated cells of *Methanobacterium thermoautotrophicum, FEBS Lett.* 291:371–375.

Rospert, S., Voges, M., Berkessel, A., Albracht, S. P. J., and Thauer, R. K., 1992. Substrate-analogue-induced changes in the nickel-EPR spectrum of active methyl-coenzyme-M reductase from *Methanobacterium thermoautotrophicum, Eur. J. Biochem.* 210:101–107.

Rouvière, P. E., and Wolfe, R. S., 1987. Use of subunits of the methylreductase protein for taxonomy of methanogenic bacteria, *Arch. Microbiol.* 148:253–259.

Rouvière, P. E., and Wolfe, R. S., 1988. Novel biochemistry of methanogenesis, *J. Biol. Chem.* 263:7913–7916.

Rouvière, P. E., Bobik, T. A., and Wolfe, R. S., 1988. Reductive activation of the methyl coenzyme M methylreductase system of *Methanobacterium thermoautotrophicum* ΔH, *J. Bacteriol.* 170:3946–3952.

Schönheit, P., Moll, J., and Thauer, R. K., 1979. Nickel, cobalt, and molybdenum requirement for growth of *Methanobacterium thermoautotrophicum, Arch. Microbiol.* 123:105–107.

Scott, R. A., Hartzell, P. L., Wolfe, R. S., LeGall, J., and Cramer, S. P., 1986. Nickel X-ray absorption spectroscopy of *Methanobacterium thermoautotrophicum* S-methyl coenzyme-M reductase, in *Frontiers of Bioinorganic Chemistry* (A. V. Xavier, ed.), VCH Publishers, New York, pp. 20–26.

Shelnutt, J. A., 1987. Axial ligation-induced structural changes in nickel hydrocorphinoids related to coenzyme F_{430} detected by Raman difference spectroscopy, *J. Am. Chem. Soc.* 109:4169–4173.

Sherf, B. A., and Reeve, J. N., 1990. Identification of the *mcrD* gene product and its association with component C of methyl coenzyme M reductase in *Methanococcus vannielii, J. Bacteriol.* 172:1828–1833.

Shiemke, A. K., Eirich, L. D., and Loehr, T. M., 1983. Resonance Raman spectroscopic characterization of the nickel cofactor, F_{430}, from methanogenic bacteria, *Biochim. Biophys. Acta* 748:143–147.

Shiemke, A. K., Hamilton, C. L., and Scott, R. A., 1988a. Structural heterogeneity and purification of protein-free F_{430} from the cytoplasm of *Methanobacterium thermoautrophicum, J. Biol. Chem.* 263:5611–5616.

Shiemke, A. K., Scott, R. A., and Shelnutt, J. A., 1988b. Resonance Raman spectroscopic investigation of axial coordination in *M. thermoautotrophicum* methyl reductase and its nickel tetrapyrrole cofactor F_{430}, *J. Am. Chem. Soc.* 110:1645–1646.

Shiemke, A. K., Kaplan, W. A., Hamilton, C. L., Shelnutt, J. A., and Scott, R. A., 1989a. Structural and spectroscopic characterization of exogeneous ligand binding to isolated factor F_{430} and its conformational isomers, *J. Biol. Chem.* 264:7276–7284.

Shiemke, A. K., Shelnutt, J. A., and Scott, R. A., 1989b. Coordination chemistry of F_{430}. Axial ligation equilibrium between square-planar and bis-aquo species in aqueous solution, *J. Biol. Chem.* 264:11236–11245.

Thauer, R. K., 1990. Energy metabolism of methanogenic bacteria, *Biochim. Biophys. Acta* 1018: 256–259.

Wackett, L. P., Hartwieg, E. A., King, J. A., Orme-Johnson, W. H., and Walsh, C. T., 1987a. Electron microscopy of nickel-containing methanogenic enzymes: Methyl reductase and F420-reducing hydrogenase, *J. Bacteriol.* 169:718–727.

Wackett, L. P., Honek, J. F., Begley, T. P., Wallace, V., Orme-Johnson, W. H., and Walsh, C. T., 1987b. Substrate analogues as mechanistic probes of methyl-S-coenzyme M reductase, *Biochemistry* 26:6012–6018.

Wackett, L. P., Honek, J. F., Begley, T. P., Shames, S. L., Niederhoffer, E. C., Hausinger, R. P., Orme-Johnson, W. H., and Walsh, C. T., 1988. Methyl-S-coenzyme-M reductase: A nickel-dependent enzyme catalyzing the terminal redox step in methane biogenesis, in *The Bioinorganic Chemistry of Nickel* (J. R. Lancaster, Jr.), VCH Publishers, New York, pp. 249–274.

Weil, C. F., Cram, D. S., Sherf, B. A., and Reeve, J. N., 1988. Structure and comparative analysis of the genes encoding component C of methyl coenzyme M reductase in the extremely thermophilic archaebacterium *Methanothermus fervidus, J. Bacteriol.* 170:4718–4726.

Weil, C. F., Sherf, B. A., and Reeve, J. N., 1989. A comparison of the methyl reductase genes and gene products, *Can. J. Microbiol.* 35:101–108.

Whitman, W. B., and Wolfe, R. S., 1980. Presence of nickel in the factor F_{430} from *Methanobacterium bryantii, Biochem. Biophys. Res. Commun.* 92:1196–1201.

Whitman, W. B., and Wolfe, R. S., 1985. Activation of the methylreductase system from *Methanobacterium bryantii* by corrins, *J. Bacteriol.* 164:165–172.

Won, H., Olson, K. D., Wolfe, R. S., and Summers, M. F., 1990. Two-dimensional NMR studies of native coenzyme F430, *J. Am. Chem. Soc.* 112:2178–2184.

Won, H., Olson, K. D., Hare, D. R., Wolfe, R. S., Kratky, C., and Summers, M. F., 1992. Structural modelling of small molecules by NMR: Solution-state structure of 12,13-diepimeric coenzyme F430 and comparison with the X-ray structure of the pentamethyl ester derivative, *J. Am. Chem. Soc.* 114:6880–6892.

Zimmer, M., and Crabtree, R. H., 1990. Bending of the reduced porphyrin of factor F430 can accommodate a trigonal-bipyramidal geometry at nickel: A conformational analysis of this nickel-containing tetrapyrrole, in relation to archaebacterial methanogenesis, *J. Am. Chem. Soc.* 112:1062–1066.

Microbial Nickel Metabolism 7

7.1 Introduction

As described in Chapters 3–6, four nickel-dependent enzymes have been isolated and characterized from various microorganisms—urease, hydrogenase, CO dehydrogenase, and methyl coenzyme M reductase. In addition, specific accessory proteins have been identified as being involved in the functional incorporation of nickel ion into urease and hydrogenase. Intracellular nickel processing functions may also be needed for nickel metallocenter assembly in CO dehydrogenase and for the synthesis of the nickel-containing coenzyme F_{430}, a component of methyl coenzyme M reductase. These aspects of microbial nickel metabolism will not be repeated here. Rather, this chapter will focus on nickel ion transport into the microbial cell, nickel ion toxicity and resistance mechanisms in microbes, and other features related to microbial nickel metabolism.

7.2 Microbial Nickel Transport

The demonstration of widespread nickel requirements among microorganisms has sparked interest in the area of cellular nickel transport. In these studies, great care must be taken to distinguish true nickel transport into the cell from simple nickel ion binding to cell walls and extracellular components. For example, short-term nickel ion incorporation studies with *Methanothrix concilii* provided evidence for a nearly nickel-specific uptake system with a transport constant of 91 μM; however, this apparent uptake is actually due to specific adsorption of nickel ion to the cell as shown by reproduction of the saturation kinetics using isolated cell sheath preparations (Baudet *et al.*, 1988). Nickel binding as observed in this case is generally not energy-dependent, whereas true transport of the metal ion generally requires a source of energy. Thus, it is important in transport studies to verify the energy dependence of uptake. The extracellular nickel ion binding found in *M. concilii* may be a prerequisite to actual transport of nickel ion into this nickel-dependent microorganism. Indeed, Kaltwasser and Frings (1980) have suggested that initial adsorption may be the first step in many nickel transport systems.

In addition to problems associated with energy-independent extracellular adsorption of nickel ion, other concerns further complicate detailed nickel transport analyses. For example, the metal ion composition of the medium in which the cells are grown may influence the level of biosynthesis of nickel transport systems. Furthermore, the concentration of free metal ions in the medium will be very different from their total concentrations because of binding by medium components [e.g., see the excellent review by Hughs and Poole (1991)]. Therefore, it is important to use growth and transport assay buffers that do not sequester nickel ion and prevent it from being transported. To examine the specificity for nickel transport, one must assess the degree to which other metal ions compete with nickel ion for uptake. Finally, entry of nickel ion into certain microbes has been shown to occur by multiple mechanisms.

Below, nickel transport is described in detail for two systems, enteric bacteria and *Alcaligenes eutrophus*. Other microbial nickel transport studies are reviewed only briefly. The key features of these transport systems are summarized in Table 7-1.

7.2.1 Nickel Transport into Enteric Bacteria

This section summarizes evidence demonstrating that enteric bacteria such as *Escherichia coli* and *Salmonella typhimurium* can transport nickel ion into the cell by systems designed for magnesium ion entry as well as by a high-affinity nickel-specific transporter.

Abelson and Aldous (1950) noted that low magnesium ion concentrations lead to heightened toxicity of nickel ion and several other cations toward various microorganisms, including *E. coli*. They followed up this observation to show that cellular uptake of radioactive ^{57}Ni (a high-energy β^+ emitter with a 36-h half-life; more recent studies typically use ^{63}Ni, a low-energy β^- emitter with a 92-year half-life) diminishes with increasing magnesium ion concentrations. A similar antagonism between magnesium ion concentration and nickel uptake was noted by Webb (1970b), who additionally showed that nickel uptake is energy-dependent. Furthermore, Jasper and Silver (1977) quoted a personal communication from D. L. Nelson indicating that both magnesium ion and cobalt ion inhibit nickel uptake into *E. coli* and that a *corA* mutant that is defective in magnesium transport also fails to transport

Table 7.1. Properties of Microbial Nickel Transport Systems

Microorganism	Type[a]	K_T[b]	Metal ion inhibition[c]	Reference(s)
Acetogenium kivui	Ni	2.3 μM	Negligible	Yang et al. (1989)
Alcaligenes eutrophus H16	Mg	17 μM	Mg (7.2 μM), Mn (26 μM), Zn (20 μM), Co (14 μM)	Lohmeyer and Friedrich (1987)
Alcaligenes eutrophus H16	Ni (hoxN)	0.34 μM	Negligible	Lohmeyer and Friedrich (1987); Eberz et al. (1989)
Alcaligenes eutrophus AE104	Mg	40 μM	Mg (20 μM), Mn, Zn, Co, Cd	Nies and Silver (1989)
Anabaena cylindrica	Ni	17 nM	Negligible	Campbell and Smith (1986)
Bacillus subtilis	Mg-citrate	ND[d]	Mg, Mn, Co	Willecke et al. (1973)
Bradyrhizobium japonicum JH	Mg	62 μM	Mg (48 μM), Co (22 μM), Mn (12 μM), Zn (8 μM)	Fu and Maier (1991a,b)
Clostridium pasteurianum	Mg	85 μM	Mg, Zn	Bryson and Drake (1988)
Clostridium thermoaceticum	Ni	3.2 μM	Negligible	Lundie et al. (1988)
Escherichia coli	Mg (corA)	ND	Mg, Co	Jasper and Silver (1977)
Escherichia coli	Ni (nik)	ND	ND	Wu et al. (1991)
Methanobacterium bryantii	Ni	3.1 μM	Co	Jarrell and Sprott (1982)
Neurospora crassa	Mg	0.29 mM	Mg (90 μM), Co, Cu, Zn, Mn	Mohan et al. (1984)
Rhodobacter capsulatus[e]	Mg	5.5 μM	Mg, Cu, Mn, Co, Zn	Takakuwa (1987)
Saccharomyces cerevesiae	Mg	0.5 mM	ND	Fuhrmann and Rothstein (1968a)
Salmonella typhimurium	Mg (corA)	237 μM	Mg (20 μM)	Snavely et al. (1991b)
Salmonella typhimurium	Mg (mgtA)	5.1 μM	Mg (5 μM)	Snavely et al. (1991b)
Salmonella typhimurium	Mg (mgtB)	1.7 μM	Mg (5 μM)	Snavely et al. (1991b)

[a] Ni, Nickel-specific transporter; Mg, magnesium-specific transporter; Mg-citrate, citrate-specific transporter. For those systems that have been genetically characterized, the locus name is provided.

[b] K_T, transport constant; concentration of nickel at which the rate of transport is one-half the maximum rate.

[c] Where available, the K_i values for competing metal ions are provided in parentheses.

[d] ND, Not determined.

[e] Formerly Rhodopseudomonas capsulata.

nickel ion effectively. These studies were consistent with the magnesium transport system accounting for a major portion of nickel uptake at high nickel ion concentrations in enteric bacteria.

The most detailed analysis of magnesium transporter-dependent nickel uptake in enterics has been carried out with *S. typhimurium* (Hmiel *et al.,* 1986, 1989; Snavely *et al.,* 1989a,b, 1991b). This microorganism possesses three distinct magnesium transport systems (CorA, MgtA, and MgtB). By constructing mutant strains to study each system in isolation from the other two, each transporter was shown to act on nickel ion. For a triple mutant, uptake of ^{63}Ni was undetectable under the conditions used. The CorA protein (M_r 42,000) is constitutively synthesized at all magnesium levels, whereas the levels of transport by the MgtA and MgtB proteins (M_r 91,000 and 102,000, respectively) are significantly increased at low magnesium ion concentrations due to enhanced protein synthesis. Using *lacZ*-transcriptional fusions, the MgtA system was found to exhibit a greater level of induction at low magnesium ion concentration (1 μM) compared to the MgtB system (800-fold versus 30-fold), and only transcription of the latter system is inhibited by calcium ion. Nickel transport assays were linear for at least 10 min for the CorA system and at least 30 min for the MgtA and MgtB systems. Kinetic analysis of transport at varied nickel ion concentrations provided transport constant (K_T) values of 237, 5.1, and 1.7 μM for the CorA, MgtA, and MgtB systems, respectively. Magnesium ion is a competitive inhibitor of nickel transport for each system with K_i values of 20, 5, and 5 μM, respectively. Sequence analysis revealed that the *mgtB* gene encodes a member of the membrane-bound cation transport P-type ATPases that most resembles eucaryotic calcium-ATPases (Snavely *et al.,* 1991a). The weight of the evidence is clear that there are three distinct *S. typhimurium* transport systems that normally function in magnesium transport but also happen to transport nickel ion. Similarly, Park *et al.* (1976) have provided evidence that multiple genes are involved in magnesium transport in *E. coli*. It is likely that each of the *E. coli* magnesium transporters also can transport nickel ion.

In natural environments, the concentration of magnesium generally exceeds that of nickel by at least a factor of 10^3; thus, one would not expect magnesium transport-dependent nickel uptake to be significant. In addition to this system, at least some enteric bacteria possess a high-affinity nickel-specific transport system. The first clue for such a system in *E. coli* came from Wu and Mandrand-Berthelot (1986), who reported the presence of the *hydC* locus that, when mutated, led to the loss of all hydrogenase activity. They found that hydrogenase activity was completely restored by addition of 500 μM nickel ion to the medium and proposed that the locus is involved in some aspect of nickel transport or processing. Subsequent analysis of the *hydC* mutant (Wu *et al.,* 1989) demonstrated that its intracellular concentration of nickel is less than 1% of that of the parent strain. Both the intracellular nickel content and hydrogenase activity, however, were shown to be restored by growing cells under low-magnesium-ion conditions. Here, nickel uptake was proposed to occur by the magnesium transport systems in the cell. Nickel transport would be enhanced under these conditions both because there is less competition with magnesium ion and because of increased synthesis of the magnesium transport proteins with higher affinity for nickel ion. Attempts to distinguish the proposed high-affinity, low-capacity nickel uptake system from the low-affinity, high-capacity magnesium transport systems in the wild-type microbe by direct transport assays were unsuccessful. Nevertheless, the authors argued that *hydC* encodes a nickel-specific transport system. This locus has now been cloned and shown to encode a multicistronic operon that is clearly distinct from the single *mgtB* gene; rather, the nickel transport gene cluster has features typical of those encoding periplasmic binding protein transport systems as found for molybdenum transport (Wu *et al.*, 1991). The authors proposed that the locus name be changed to *nik* to reflect its role in nickel transport rather than in hydrogenase activity.

7.2.2 Nickel Transport into *Alcaligenes eutrophus*

A nickel requirement for chemolithotrophic growth was first demon-strated in two strains of *Alcaligenes* sp. (formerly *Hydrogenomonas*) by Bartha and Ordal (1965); thus, it is not surprising that the earliest studies on nickel-specific uptake were carried out with this microorganism. Tabillion and Kaltwasser (1977) demonstrated that two strains of autotrophically grown *A. eutrophus* cells accumulate nickel 280-fold over the concentration in the me-dium, with optimal growth occurring at 0.3 μM nickel ion. Moreover, they showed that nickel transport is energy-dependent and highly specific; that is, there is only slight inhibition of nickel accumulation by zinc, cobalt, man-ganese, and copper ions. They did not, however, test the effect of magnesium ion on nickel uptake.

Nickel transport in *A. eutrophus* was further characterized by Lohmeyer and Friedrich (1987), who provided evidence for two distinct uptake systems. Most of the nickel taken up by the cell is associated with a transport system exhibiting a K_T of 17 μM for nickel ion. Consistent with this uptake occurring by a magnesium transporter, nickel uptake by this system is competitively inhibited by magnesium, manganese, cobalt, and zinc ions. In the presence of 0.8 mM magnesium ion, however, a small amount of nickel transport is still observed, and this uptake is not affected by manganese or zinc ions. The second, nickel-specific transport system was shown to possess a K_T of 0.34 μM for nickel ion. Magnesium-transporter-dependent uptake of nickel ion by a different strain of *A. eutrophus* (AE104) has also been described by Nies and Silver (1989). These investigators, however, were unable to detect a high-affinity, nickel-specific transport system in this microbe.

A major advance in characterizing the *A. eutrophus* nickel-specific transport system was the demonstration by Eberz *et al.* (1989) that the genetic locus for nickel uptake is part of the plasmid-encoded hydrogenase gene cluster. They obtained mutants that require high levels of nickel ion (albeit, still only 0.5 μM) for autotrophic growth in medium containing 0.8 mM magnesium ion. They showed the mutants to be deficient in nickel transport, hydrogenase, and urease activities. Finally, the investigators were able to restore the wild-type phenotype by complementation with a cloned DNA sequence (*hoxN*) containing a part of the hydrogenase gene cluster. Eitinger and Friedrich (1991) subsequently demonstrated that the *hoxN* locus encodes two proteins (M_r 30,000 and 28,000) and showed that only the former is essential for nickel transport. They sequenced the gene encoding the transport protein to reveal features of a typical integral membrane protein. Surprisingly, the HoxN protein possesses no sequence homology to other transport proteins or indeed to any other proteins in the sequence data bases. As a first step toward purification and characterization of the HoxN protein, Wolfram *et al.* (1991) constructed a gene fusion containing the *Escherichia coli lacZ* promoter and gene, followed by a DNA segment encoding a protease-cleavable linker, followed by the *hoxN* gene. An *E. coli* host containing this construct synthesizes a fusion protein upon induction, and the linker is capable of being cleaved; however, further characterization of the transporter protein has not been reported.

7.2.3 Nickel Transport into Other Microorganisms

Key features of the reported microbial nickel transport systems are summarized in Table 7-1. In each case I have indicated whether nickel uptake occurs via a magnesium transport system or whether the system appears to be specific for nickel ion. For *Bacillus subtilis,* nickel ion can substitute for magnesium ion in a unique system that is intended to transport citrate into the cell as the magnesium–citrate complex (Willecke *et al.,* 1973). Nickel transport constants also are provided in Table 7-1, and the abilities of other metal ions to inhibit nickel transport are summarized. The following paragraphs highlight aspects of nickel transport in bacteria, members of the archaea (formerly termed the archaebacteria), and eucaryotic microbes.

Nickel transport has been characterized in several bacteria other than those described above. Whereas nickel ion appears to be transported by magnesium uptake systems in *Bradyrhizobium japonicum* (Fu and Maier, 1991a,b), *Clostridium pasteurianum* (Bryson and Drake, 1988), and *Rho-*

dobacter capsulatus (formerly *Rhodopseudomonas capsulata*) (Takakuwa, 1987), highly specific nickel transport systems have been reported for *Acetogenium kivui* (Yang *et al.*, 1989), *Anabaena cylindrica* (Campbell and Smith, 1986), and *Clostridium thermoaceticum* (Lundie *et al.*, 1988). A requirement for specific nickel transport may be expected for *A. kivui* and *C. thermoaceticum*, two acetogenic bacteria that possess high levels of nickel-dependent CO dehydrogenase (Chapter 5). In contrast, it is unclear why the cyanobacterium *A. cylindrica*, which does not require significant amounts of nickel ion for growth, would possess the observed highly specific nickel uptake system. With regard to *B. japonicum*, Stults *et al.* (1987) initially reported the presence of a nickel transport system (K_T = 26 μM) in strain SR that is unaffected by magnesium ion but inhibited by copper and zinc ions. Significantly, nickel uptake appeared to be energy-independent in their 1-min assays. Using a different serotype, *B. japonicum* strain JH, and 5-min assays, Fu and Maier (1991a,b) were unable to reproduce these results and only found evidence for energy-dependent uptake of nickel ion by the magnesium transport system. The prior *B. japonicum* results have not been included in Table 7-1.

Among the archaea, the methanogenic bacteria require tremendous quantities of nickel ion for biosynthesis of several essential nickel enzymes (see Chapters 4, 5, and 6); thus, specific nickel transport systems may be expected in these cells. In the case of *Methanothrix concilii*, true nickel transport is obscured by saturable nickel ion binding to an extracellular sheath (Baudet *et al.*, 1988), as already described. In contrast, Jarrell and Sprott (1982) were able to perform detailed transport experiments with *Methanobacterium bryantii*. They demonstrated the presence of an energy-dependent nickel transport system that is only inhibited by cobalt ion among the metal ions tested. Less detailed studies have demonstrated that a nickel transport system also exists in *Methanospirillum hungatei* (Sprott *et al.*, 1982).

Nickel uptake studies in eucaryotic microbes have included work with two fungi, *Saccharomyces cerevisiae* (Fuhrmann and Rothstein, 1968a) and *Neurospora crassa* (Mohan *et al.*, 1984), and two algae, *Chlamydomonas* sp. (Folsom *et al.*, 1986) and *Phaeodactylum tricornutum* (Skaar *et al.*, 1974). The fungal studies (Table 7-1) clearly demonstrated that nickel transport occurs by a magnesium-ion-specific system. Although neither algal study included true transport assays, uptake of nickel ion in cultures is consistent with *Chlamydomonas* nickel transport occurring via a magnesium uptake system. Interestingly, nickel uptake in *P. tricornutum* appears to be phosphate-dependent, but the details and significance of this dependence were not established. Because few eucaryotic microbes have been shown to have a growth dependence on nickel ion, it is not surprising that no nickel-specific uptake system has been reported.

7.3 Nickel Toxicity and Resistance Mechanisms in Microorganisms

The toxic properties of nickel ion toward microorganisms are well known and have been agriculturally important, as exemplified by the extensive use of nickel salts as fungicidal agents for cereal and other crops (Mishra and Kar, 1974). Although this metal ion is toxic at some level to all microbes, various microorganisms have developed moderate to high degrees of resistance to nickel ion. The following three sections will summarize the mechanisms of nickel toxicity in microorganisms, describe environmental factors that influence this toxicity, and detail the mechanisms of nickel resistance observed in various microbes.

7.3.1 Mechanisms of Nickel Toxicity in Microorganisms

Despite our wealth of knowledge concerning the nickel ion concentration levels that inhibit growth or kill various microorganisms (e.g., Babich and Stotzky, 1983), surprisingly little is known regarding the actual mechanisms by which the metal ion alters cellular function. Three obvious possibilities for the site of toxicity are inhibitory interactions with proteins (including enzymes), polynucleotides (including DNA and RNA), and various small molecules in the cell. Each of these possibilities is commented on below.

Inhibition of essential enzymes and proteins by nickel ion is likely to be a key site of toxicity for this metal ion. Early *in vivo* and *in vitro* studies of nickel inhibition in yeast (Fuhrmann and Rothstein, 1968b) indicated that the predominant effect of this metal ion on fermentation is to reduce alcohol dehydrogenase activity. In this case, nickel ion replaces, but fails to substitute catalytically for, the normal zinc ion at the active site. Alcohol dehydrogenase plays a key role in fermentative growth of yeast; thus, blockage of this enzyme leads to cell growth inhibition. Numerous other zinc enzymes exist in microbial cells, and it is likely that replacement of zinc ion by nickel ion is associated with growth inhibition in various microorganisms. Nickel may also substitute for other metal ions in specific proteins. In the case of redox-active iron proteins, for example, replacement of the iron by nickel would likely eliminate any redox activity and lead to loss of function. Such may be the case in *Aspergillus niger,* where nickel toxicity is ameliorated by providing an elevated iron ion concentration in the medium (Adiga *et al.,* 1961). Although nickel reconstitution of numerous metalloproteins has been carried out *in vitro,* the *in vivo* importance of nickel substitution to toxicity has not been well characterized.

As do all metal ions, nickel binds to RNA and DNA. More specifically, this metal ion coordinates the phosphate groups in the RNA or DNA backbone and (for DNA) the N-7 position of adenine, the O-6 function of guanine, the O-2 group of cytosine, and the O-4 position of thymine. Andronikashvili *et al.* (1988) estimated that the log of the stability constant for nickel binding to DNA at 20°C is 5.56, a rather weak level of interaction. Butzow and Eichhorn (1965) demonstrated that at elevated temperatures the metal ion mediates the hydrolysis of the RNA molecule; however, this reaction is unlikely to be significant at typical growth temperatures and moderate nickel ion concentrations. When bound to adenine and guanine bases in DNA, nickel ion also can catalyze depurination reactions at a rate second only to that observed for copper ion (Andronikashvili *et al.*, 1988). Although nickel-catalyzed modifications of DNA can occur *in vivo*, Biggart and Costa (1986) demonstrated that nickel ion is unable to induce base pair or frameshift mutations in *Salmonella* tester strains that were shown to take up this metal ion. Thus, the predominant toxic effects of nickel on microorganisms are unlikely to involve interactions with polynucleotides.

In addition to binding to proteins and polynucleotides, nickel ion binds to amino acids, nucleotides, glutathione, acetyl coenzyme A, and numerous other small molecules in the cell. The stability constants for many of these interactions have been tabulated (Martin, 1988a,b; National Research Council, 1975), and in many cases the binding is very tight (e.g., with a log of the stability constant of approximately 9 for cysteine and histidine, and approximately 5 for the free nucleotides). Unfortunately, it is unclear how important these interactions are with regard to cellular nickel toxicity in comparison to the polymer interactions discussed above.

7.3.2 Environmental Influences on Nickel Toxicity

In assessing the toxicity of nickel toward a particular microorganism, it is essential to carefully consider several details of the culture conditions. For example, a bacterium may tolerate 1 mM (58.7 ppm) nickel salt added to a rich broth culture but be incapable of growth at 100 μM nickel ion in a defined minimal medium. Similarly, a facultative microbe may tolerate higher nickel ion concentrations when growing anaerobically than aerobically. These and many other environmental aspects related to the toxicity of nickel to microorganisms have been examined in a comprehensive review by Babich and Stotzky (1983). The detailed effects of changes in the culture environment for specific microbes will not be presented here; rather, the following paragraphs highlight key parameters that may affect the toxicity of nickel ion toward microorganisms in general. These factors include the levels and types of organic components or inorganic particulates in the medium, the concentrations of other cations, the amount and identity of available anions, and the pH.

Many organic compounds and inorganic particulates are able to chelate nickel ion and thus reduce its bioavailability. In natural environments, humic acids and clays are very important in sequestering a variety of heavy metals (Gadd and Griffiths, 1978). Soluble organic compounds such as citrate, aspartate, and EDTA are also known to ameliorate nickel toxicity. Extending from these results, it is not surprising that commercial media such as peptone, yeast extract, and tryptic soy broth have a tremendous sparing effect on nickel toxicity. Furthermore, inhibitory effects of nickel ion for growth on plates may vary with the gelling matrix. Whereas nickel ion may be highly toxic to cells growing on medium solidified with agar, reduced toxicity may be observed for the same cells grown in the same medium solidified with Gelrite or gelatin because of the ability of these matrices to bind metal ions. The key point to be learned from these studies is that the nickel ion concentration that is available to the cell may be very different from the total nickel ion concentration added to the medium.

The concentrations of other cations in the medium may have a significant impact on the toxicity of nickel ion for a particular microorganism. High levels of magnesium ion may prevent nickel transport into the cell, as described in Section 7.2. Supplementation of the medium with zinc or ferrous ions may spare the cells from nickel toxicity effects by outcompeting nickel ion either for transport or for metal ion binding sites in zinc- or iron-dependent proteins. Alternatively, elevated concentrations of other metal ions may increase nickel toxicity by substituting for nickel ion that would otherwise be sequestered by organic components in the medium, thus increasing the concentration of free nickel ion. Furthermore, addition of other toxic metal ions can lead to synergistic interactions such that the toxicity of the mixture is greater than the sum of the individual toxicities. Clearly, it is essential for an investigator to carefully note the concentrations of other cations when assessing nickel toxicity.

The anion concentrations to which a culture is exposed will also affect the toxicity of nickel ion. Sulfide, phosphate, and carbonate anions will form insoluble salts with nickel ion and decrease its bioavailability. Sulfide is produced by many microorganisms growing anaerobically, and such cells may have a high resistance to nickel ion. Phosphate, a common buffer in defined laboratory media, is often ignored in terms of its interaction with metal ions, including nickel ion. Finally, carbonate concentrations in natural environments vary with the water hardness. Nickel toxicity toward microbes is generally less of a problem in environments with hard water than locations with soft water. In contrast to these anions, the concentration of chloride has little effect on toxicity of nickel ion.

The pH of an environment greatly influences nickel toxicity. Nickel ion can form the following series of complexes with hydroxide ion;

$$Ni^{2+} \leftrightarrow NiOH^+ \leftrightarrow Ni(OH)_2 \leftrightarrow Ni(OH)_3^- \leftrightarrow Ni(OH)_4^{2-}$$

however, this speciation is not important in controlling the toxicity of nickel ion for most microbes. The uncomplexed cation is the only species found at acidic and neutral pH values, and other forms only dominate at pH values higher than 9.5, where most microorganisms are not able to grow. Rather, the pH dependence of nickel toxicity probably is related to the protonation state of organic components in the medium. As the pH decreases, protons compete more effectively with nickel ion for binding to the chelating components, the concentration of free nickel ion increases, and problems associated with nickel toxicity are elevated.

7.3.3 Mechanisms of Nickel Resistance in Microorganisms

Whereas most microorganisms are inhibited in their growth by ~0.1 mM nickel ion, some have developed mechanisms that allow them to survive in the presence of greatly increased nickel ion concentrations. As extreme examples, strains of *Alcaligenes denitrificans, Alcaligenes xylosoxydans, Alcaligenes eutrophus,* and other uncharacterized bacterial species have been isolated that tolerate at least 20 mM NiCl$_2$ (Kaur *et al.,* 1990; Schmidt and Schlegel, 1989; Schmidt *et al.,* 1991). Furthermore, Schlegel *et al.* (1991) have demonstrated that 1% of the bacterial population under nickel-hyperaccumulating plants (see Chapter 8) is resistant to this concentration of nickel ion. Nickel tolerance is often a plasmid-borne trait, as first reported by Smith (1967), working with clinical isolates of *Escherichia coli* that grow on agar medium containing 3 mM nickel ion. However, in other cases, the resistance is chromosomally encoded [e.g., the *A. denitrificans* strains studied by Kaur *et al.* (1990)]. Several hypothetical mechanisms for how nickel resistance could be achieved will be described here, and, where available, microbial examples of these mechanisms will be provided. To assist the reader, these mechanisms are illustrated in Fig. 7-1.

Those microbes that are resistant to very high levels of nickel ion uniformly appear to possess energy-dependent nickel export systems to maintain low intracellular nickel ion concentrations (Fig. 7-1a). The best studied of these systems is that found in *A. eutrophus* CH34. This strain, isolated from a decantation tank of a zinc factory, was shown to possess two plasmids: pMOL28 [163 kilobase pairs (kbp)] specifies resistance to nickel ion as well as cobalt and mercury ions, and pMOL30 (238 kbp) specifies resistance to

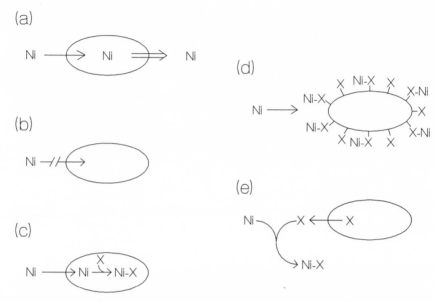

Figure 7-1. Potential mechanisms of nickel ion resistance. A microbial cell (represented by the oval symbol) may develop resistance to nickel ion by (a) exporting nickel ions via a nickel-specific efflux system, (b) decreasing the rate of nickel uptake, (c) sequestering the nickel internally by production of a nickel-binding compound, (d) binding the nickel to nickel-binding substances on the cell surface, or (e) secretion of a chelating agent that prevents nickel transport into the cell.

zinc, cadmium, mercury, and cobalt ions (Mergeay *et al.*, 1985). The minimal inhibitory concentration of nickel ion for the wild-type microbe is 2.5 mM, whereas that of the plasmid-cured strain is only 0.6 mM. Siddiqui and Schlegel (1987) demonstrated that the nickel resistance in this isolate is induced by 30 μM to 3 mM nickel ion, compared to the normal 1 μM concentration that is added as a required trace element for growth. In addition, these investigators showed that the resistant strain accumulates very low levels of nickel compared to the plasmid-cured strain. Siddiqui *et al.* (1988) transferred the pMOL28 plasmid into the better characterized non-nickel-resistant strain *A. eutrophus* N9A and demonstrated that nickel resistance is conferred to the transformant (M220) in an inducible manner. Chemical mutagenesis of M220 led to the isolation of two mutants that constitutively expressed nickel resistance. The mutations were shown to occur on the plasmid, and these plasmids were designated pMOL28.1 and pMOL28.2. As with the wild-type and plasmid-cured CH34 strains, cells containing pMOL28.1 or pMOL28.2 accumulate very little nickel compared to the N9A host. In further studies with N9A and N9A [pMOL28.1], Sensfuss and Schlegel (1988) found (i) that similar levels of ^{63}Ni are accumulated by both strains when anaerobically exposed to 1 μM nickel ion for 24 h and (ii) that upon exposure to oxygen the nickel is instantly eliminated from only the latter strain. These results provide clear evidence that nickel resistance is associated with an energy-dependent nickel efflux system. Inhibitor studies by Varma et al. (1990) were interpreted to suggest that nickel efflux is driven by the proton motive force rather than being ATP-dependent. Such a system contrasts with the well-characterized, ATP-driven efflux systems for arsenate and cadmium ions [reviewed by Silver et al. (1989)]. This distinction was borne out when the pMOL28-derived cobalt and nickel resistance genes were characterized (Liesegang et al., 1993). Of seven open reading frames that were found, none exhibited ATP-binding motifs that one would expect in an ATP-driven efflux system. Further characterization of these gene products is required to assess their functional roles.

Metal ion efflux as a resistance mechanism is not unique to nickel ion in A. eutrophus CH34, as shown by Nies and Silver (1989), who found that the plasmid-localized resistance genes for zinc, cadmium, and cobalt ions each encode inducible energy-dependent extrusion systems. For the latter metal ion, both the Schlegel and Silver laboratories have obtained evidence consistent with the same gene being responsible for the pMOL28-encoded cobalt resistance and the nickel resistance. Whereas Nies et al. (1989) have found that both cobalt and nickel resistance are localized to an 8.5-kbp PstI-EcoRI fragment, Siddiqui et al. (1989) found that a 9.5-kbp KpnI fragment confers dual resistance. It is possible that these fragments encode multiple metal ion efflux genes; however, a single system may be sufficient because nickel and cobalt ions are chemically quite similar. The sequence of the gene encoding the nickel efflux protein has not been reported.

Energy-dependent metal ion efflux is likely to be the resistance mechanism in many other microbes that tolerate high nickel concentrations. Although nickel resistance in A. denitrificans strain 4a-2 is chromosomally encoded and constitutively expressed, the nickel efflux system is probably very similar to that in A. eutrophus CH34 as shown by DNA/DNA hybridization between the A. denitrificans DNA and pMOL28 (Kaur et al., 1990). Hybridization was also observed between pMOL28 and plasmids pTOM8 and pTOM9 from A. xylosoxydans 31A and pGOE2 from A. eutrophus KTO2 (Schmidt et al., 1991). DNA fragments of these plasmids are capable of conferring nickel resistance to several hosts, including E. coli, and appear to be identical as judged by size (14.5 kpb), restriction maps, and pMOL28 hybridization. Interestingly, however, a pTOM9-derived 4.3-kbp subfragment that confers nickel resistance on E. coli does not hybridize to pMOL28. A possibly related feature is that transfer of pTOM9 to A. eutrophus H16 only provides nickel resistance, not resistance to cobalt (Schmidt and Schlegel, 1989). Hence, the similarity in nickel resistance mechanisms between *A. eutrophus* CH34 and these microbes remains an open question.

In the above studies, mutation of nickel-sensitive strains to form highly nickel-resistant strains has never been observed (Siddiqui and Schlegel, 1987; Siddiqui *et al.*, 1988). Generation of a highly active metal ion efflux complex from the inherent genetic repertoire of a microorganism apparently is not easily achieved by simple mutagenesis events. In contrast, development of low levels of nickel ion tolerance has been observed in procaryotes (e.g., Webb, 1970a) and microbial eucaryotes (e.g., Mohan and Sastry, 1983). Below, alternative mechanisms to the metal efflux system are discussed.

A second possible mechanism of nickel resistance is metal ion exclusion by alteration of the uptake system to reduce the specificity or rate of nickel transport into the cell (Fig. 7-1b). For example, Webb (1970a) generated cobalt-tolerant mutants of *Bacillus subtilis, Aerobacter aerogenes,* and *E. coli* and showed that these cells are also resistant to nickel ion. Because the major route of nickel ion entry into these cells is via magnesium transporters, higher concentrations of magnesium ion are required for growth. Analogous results in other procaryotes were reported for *E. coli* by Jasper and Silver (1977) and for *Salmonella typhimurium* by Hmiel *et al.* (1986). Similarly, spontaneously obtained mutants of the eucaryote *Neurospora crassa* were shown to be four-fold more tolerant toward nickel ion than the wild-type microbe (Mohan and Sastry, 1983). One of these strains (Ni^{R3}) was clearly shown to have decreased rates of nickel uptake due to a defective magnesium transport system (Mohan *et al.*, 1984). In contrast, two other mutants possess increased nickel transport rates and accumulate high levels of this metal ion. These results lead us to the third mechanism for nickel resistance, discussed below.

Internal sequestration of nickel ions can serve as another possible resistance mechanism (Fig. 7-1c). Two nickel-tolerant *N. crassa* mutants accumulate high internal levels of nickel, consistent with the presence of an unidentified nickel-binding agent (Mohan and Sastry, 1983; Mohan *et al.*, 1984). Internal sequestration of nickel has also been suggested as an explanation for nickel resistance by *Saccharomyces cerevisiae* strains (Joho *et al.*, 1987). Nickel-resistant *S. cerevisiae* cells were found to possess elevated levels of histidine (Joho *et al.*, 1990), and both the internalized nickel ion and the histidine were shown to be localized to vacuoles (Joho *et al.*, 1992). Although not clearly associated with nickel resistance, binding of nickel to polyphosphate bodies has been described in the cyanobacterium *Nostoc muscorum* (Singh *et al.*, 1992) and this polyanion may function to sequester the metal ion in this and/or other microorganisms. As an alternative to polyphosphate sequestration or binding by an amino acid or other small molecule, it is possible that nickel ion could bind to metallothionein-like peptides. Metal-binding peptides have been identified in both procaryotes and eucaryotic microbes [reviewed by Nies (1992)]; however, the known metal-detoxification peptides are generally associated with non-nickel metal ions such as cadmium.

Microorganisms can potentially use two additional mechanisms of nickel resistance that involve extracellular binding of the metal ion. In the first case (Fig. 7-1d), the metal is bound to extracellular components of the cell such as the capsule, lipopolysaccharide, or sheath. For example, the cell walls of many lichens, fungi, and algae possess high binding capacities for metal ions, including nickel ion (Richardson *et al.*, 1980). Similarly, metal ions are known to bind to acidic polysaccharides in the slime produced by zoogleal bacteria, which can tolerate high levels of heavy metal ions (Gadd and Griffiths, 1978). Alternatively, microbes may secrete factors that bind metal ions to reduce their availability to the cell (Fig. 7-1e). For example, microbes that produce sulfide, citrate, or oxalate often exhibit higher tolerance to heavy metal ions (Gadd and Griffiths, 1978). Furthermore, Wildung *et al.* (1979) reported that 165 of 239 metal-resistant bacteria synthesized and secreted nickel-complexing metabolites belonging to 13 chemical classes. The same authors reported that 59 metal-resistant fungi all produced nickel-complexing factors, and the structures were distinct for every strain.

7.4 Other Aspects of Microbial Nickel Metabolism

Some microorganisms, such as *Chlorella emersonii* (Soeder and Engelmann, 1984), *Chlorella vulgaris* (Bertrand and de Wolf, 1967), and an *Oscillatoria* species (Van Baalen and O'Donnell, 1978) exhibit a growth requirement for nickel ion where the role of the metal ion has not been identified. In addition to such uncharacterized nickel requirements, however, two other examples of microbial nickel metabolism can be discussed. These include an unusual role for nickel ion in cyanobacterial cyanophycin metabolism and a possible nickel storage protein in *Bradyrhizobium japonicum*.

The cyanobacteria are phototrophs that possess photopigments bound to phycobiliproteins, including phycocyanin. In addition to its role in photosynthesis, phycocyanin functions as a nitrogen reserve in the cell. A second reserve molecule that is formed in many cyanobacteria is cyanophycin, a copolymer of arginine and aspartic acid. Some species are capable of growth in nitrogen-free medium by developing specialized cells, termed heterocysts, that carry out the reduction of nitrogen gas using the enzyme nitrogenase. Because nitrogenase catalysis is highly energy dependent, this enzyme is tightly regulated in the cell. Daday *et al.* (1988) noted that when nitrogenase-repressed, nickel-depleted cells of *Anabaena cylindrica* were inoculated into fresh nitrogen-free medium, the cells exhibited a pronounced lag before growth reinitiated. In contrast, when nickel ion was present at 0.68 μM, no growth lag was observed. During the lag period for nickel-depleted cells, phycocyanin levels decreased, cyanophycin transiently accumulated, and the concentration of urea became elevated. Heterocyst differentiation, nitrogenase synthesis, and phycocyanin synthesis were delayed during this ∼10-h period. The authors proposed that nickel ion depletion somehow diverts nitrogen flux from protein synthesis to cyanophycin formation. The cyanophycin is subsequently degraded, with the arginine component forming urea among other products. Because the cells lack nickel ion, they are unable to synthesize an active urease and the urea levels increase. This process is schematically illustrated in Fig. 7-2. The precise role for nickel ion in regulating the flux of nitrogen has not been identified.

Suggestive evidence for a nickel storage protein in *Bradyrhizobium japonicum* has been obtained by studying the pathway for uptake of ^{63}Ni radioactive isotope into hydrogenase (Maier *et al.*, 1990). Heterotrophically grown cultures were shown to accumulate considerably more nickel ion than the amount required for synthesis of their nickel-containing hydrogenase.

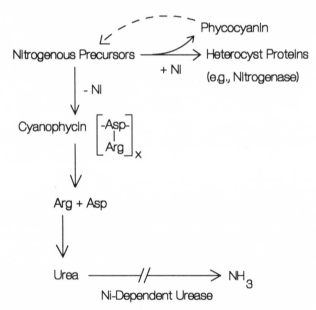

Figure 7-2. Possible role for nickel ion in cyanobacterial cyanophycin metabolism. During a shift from nitrogen-replete to nitrogen-deficient conditions, *Anabaena cylindrica* continues to synthesize the photopigment phycocyanin and initiates new synthesis of heterocyst proteins including nitrogenase. However, in the absence of nickel ion, heterocyst development and nitrogenase synthesis are delayed for approximately 10 h. During this lag phase, phycocyanin levels decrease, cyanophycin concentrations transiently become greatly elevated, and urea increases in concentration (Daday *et al.*, 1988). The role for nickel ion in regulation of the cyanobacterial nitrogen flux is not clear.

The accumulated nickel ion appears to be bound to a cytoplasmic protein in a form that is not able to be exchanged with exogenous nickel ion or removed by dialysis in the presence of EDTA. Importantly, the bound nickel is capable of being mobilized in the cell for use in subsequent hydrogenase synthesis. Partial purification of the nickel-binding protein was achieved by gel filtration and DEAE chromatography; however, loss of nickel during purification prevented further analysis. The possible nickel storage protein is not antigenically related to either hydrogenase subunit. Further studies are needed to verify that the protein does indeed serve as a storage protein for nickel and to characterize its binding properties.

7.5 Perspective

The essential transition metals are typically required in only trace amounts by microorganisms, and elevated metal ion concentrations often are toxic to cells. This situation also exists for nickel, although the requirements for and toxicity of nickel ion can vary greatly among microbes. In the environment, microorganisms encounter a wide range of nickel ion concentrations; thus, it is not surprising that high-affinity nickel-specific transport systems have been identified in several nickel-dependent microbes and that various nickel resistance mechanisms have evolved in others. Indeed, some microbes have been shown to possess both nickel-specific uptake systems and nickel-specific efflux systems that are used to balance the internal cellular concentrations of this metal ion. It appears from limited *Alcaligenes eutrophus* and *Escherichia coli* sequence data that different microbes use distinct mechanisms to specifically transport nickel ion into the cell, and much remains to be characterized about these nickel-specific uptake pathways as well as the magnesium transporter-dependent uptake of nickel ion. If too much nickel enters the cell, the metal ion can be toxic; however, surprisingly little is known about how nickel ion alters cellular function to lead to toxicity. Several resistance mechanisms (efflux, exclusion, internal sequestration, surface binding, and secretion of a chelator) may be used to prevent the internal concentration of free nickel ion from becoming toxic to cells, but many questions remain about the regulation and properties of these systems. At the steady-state concentration found in the cell, nickel ion must be specifically incorporated into nickel-containing enzymes, it may be transiently stored, or it may function in some other metabolic role. Additional nickel-dependent enzymes may yet be identified, and other metabolic functions for nickel ion in microbes are almost certain to be discovered.

References

Abelson, P. H., and Aldous, E., 1950. Ion antagonisms in microorganisms: Interference of normal magnesium metabolism by nickel, cobalt, cadmium, zinc, and manganese, *J. Bacteriol.* 60: 401–413.

Adiga, P. R., Sastry, K. S., Venkatasubramanyam, V., and Sarma, P. S., 1961. Interrelationships in trace-element metabolism in *Aspergillus niger, Biochem. J.* 81:545–550.

Andronikashvili, E. L., Bregadze, V. G., and Monaselidze, J. R., 1988. Interactions between nickel and DNA: Considerations about the role of nickel in carcinogenesis, in *Nickel and Its Role in Biology* (H. Sigel and A. Sigel, eds.), *Metal Ions in Biological Systems,* Vol. 23, Marcel Dekker, New York, pp. 331–367.

Babich, H., and Stotzky, G., 1983. Toxicity of nickel to microbes: Environmental aspects, *Adv. Appl. Microbiol.* 29:195–265.

Bartha, R., and Ordal, E. J., 1965. Nickel-dependent chemolithotrophic growth of two *Hydrogenomonas* strains, *J. Bacteriol.* 89:1015–1019.

Baudet, C., Sprott, G. D., and Patel, G. B., 1988. Adsorption and uptake of nickel in *Methanothrix concilii, Arch. Microbiol.* 150:338–342.

Bertrand, D., and de Wolf, A., 1967. Le nickel, oligoélément dynamique pour les végétaux supérieurs, *C. R. Acad. Sci.* 265:1053–1055.

Biggart, N. W., and Costa, M., 1986. Assessment of the uptake and mutagenicity of nickel chloride in *Salmonella* tester strains, *Mutat. Res.* 175:209–215.

Bryson, M. F., and Drake, H. L., 1988. Energy-dependent transport of nickel by *Clostridium pasteurianum, J. Bacteriol.* 170:234–238.

Butzow, J. J., and Eichhorn, G. L., 1965. Interactions of metal ions with polynucleotides and related compounds. IV. Degradation of polyribonucleotides by zinc and other divalent metal ions, *Biopolymers* 3:95–107.

Campbell, P. M., and Smith, G. D., 1986. Transport and accumulation of nickel ions in the cyanobacterium *Anabaena cylindrica, Arch. Biochem. Biophys.* 244:470–477.

Daday, A., Mackerras, A., H., and Smith, G. D., 1988. A role for nickel in cyanobacterial nitrogen fixation and growth via cyanophycin metabolism, *J. Gen. Microbiol.* 134:2659–2663.

Eberz, G., Eitinger, T., and Friedrich, B., 1989. Genetic determinants of a nickel-specific transport system are part of a plasmid-encoded hydrogenase gene cluster in *Alcaligenes eutrophus, J. Bacteriol.* 171:1340–1345.

Eitinger, T., and Friedrich, B., 1991. Cloning, nucleotide sequence, and heterologous expression of a high-affinity nickel transport gene from *Alcaligenes eutrophus, J. Biol. Chem.* 166:3222–3227.

Folsom, B. R., Popescu, A., Kingsley-Hickman, P. W., and Wood, J. M., 1986. A comparative study of nickel and aluminum transport and toxicity in freshwater green algae, in *Frontiers in Bioinorganic Chemistry* (A. V. Xavier, ed.), VCH Publishers, New York, pp. 391–398.

Fu, C., and Maier, R. J., 1991a. Identification of a locus within the hydrogenase gene cluster involved in intracellular nickel metabolism in *Bradyrhizobium japonicum, Appl. Environ. Microbiol.* 57:3502–3510.

Fu, C., and Maier, R. J., 1991b. Competitive inhibition of an energy-dependent nickel transport system by divalent cations in *Bradyrhizobium japonicum* JH, *Appl. Environ. Microbiol.* 57: 3511–3516.

Fuhrmann, G. F., and Rothstein, A., 1968a. The transport of Zn^{+2}, Co^{+2}, and Ni^{+2} into yeast cells, *Biochim. Biophys. Acta* 163:325–330.

Fuhrmann, G. F., and Rothstein, A., 1968b. The mechanism of the partial inhibition of fermentation in yeast by nickel ions, *Biochim. Biophys. Acta* 163:331–338.

Gadd, G. M., and Griffiths, A. J., 1978. Microorganisms and heavy metal toxicity, *Microb. Ecol.* 4:303–317.

Hmiel, S. P., Snavely, M. D., Miller, C. G., and Maguire, M. E., 1986. Magnesium transport in *Salmonella typhimurium:* Characterization of magnesium influx and cloning of a transport gene, *J. Bacteriol.* 168:1444–1450.

Hmiel, S. P., Snavely, M. D., Florer, J. B., Maguire, M. E., and Miller, C. G., 1989. Magnesium transport in *Salmonella typhimurium:* Genetic characterization and cloning of three magnesium transport loci, *J. Bacteriol.* 171:4742–4751.

Hughs, M. N., and Poole, R. K., 1991. Metal speciation and microbial growth—the hard (and soft) facts, *J. Gen. Microbiol.* 137:725–734.

Jarrell, K. F., and Sprott, G. D., 1982. Nickel transport in *Methanobacterium bryantii, J. Bacteriol.* 151:1195–1203.

Jasper, P., and Silver, S., 1977. Magnesium transport in microorganisms, in *Microorganisms and Minerals* (E. D. Weinberg, ed.), Marcel Dekker, New York, pp. 7–47.

Joho, M., Imada, Y., and Murayama, T., 1987. The isolation and characterization of Ni^{+2} resistant mutants of *Saccharomyces cerevisiae, Microbios* 51:183–190.

Joho, M., Inouhe, M., Tohoyama, H., and Murayama, T., 1990. A possible role of histidine in a nickel resistant mechanism of *Saccharomyces cerevisiae, FEMS Microbiol. Lett.* 66:333–338.

Joho, M., Ishikawa, Y., Kunikane, M., Inouhe, M., Tohoyama, H., and Murayama, T., 1992. The subcellular distribution of nickel in Ni-sensitive and Ni-resistant strains of *Saccharomyces cerevisiae, Microbios* 71:149–159.

Kaltwasser, H., and Frings, W., 1980. Transport and metabolism of nickel in microorganisms, in *Nickel in the Environment* (J. O. Nriagu, ed.), John Wiley & Sons, New York, pp. 463–491.

Kaur, P., Roß, K., Siddiqui, R. A., and Schlegel, H. G., 1990. Nickel resistance of *Alcaligenes denitrificans* strain 4a-2 is chromosomally coded, *Arch. Microbiol.* 154:133–138.

Liesegang, H., Lemke, K., Siddiqui, R. A., and Schlegel, H.-G., 1993. Characterization of the inducible nickel and cobalt resistance determinant *cnr* from pMOL28 of *Alcaligenes eutrophus* CH34, *J. Bacteriol.* 175:767–778.

Lohmeyer, M., and Friedrich, C. G., 1987. Nickel transport in *Alcaligenes eutrophus, Arch. Microbiol.* 149:130–135.

Lundie, L. L., Jr., Yang, H., Heinonen, J. K., Dean, S. I., and Drake, H. L., 1988. Energy-dependent, high-affinity transport of nickel by the acetogen *Clostridium thermoaceticum, J. Bacteriol.* 170:5705–5708.

Maier, R. J., Pihl, T. D., Stults, L., and Sray, W., 1990. Nickel accumulation and storage in *Bradyrhizobium japonicum, J. Bacteriol.* 56:1905–1911.

Martin, R. B., 1988a. Nickel ion binding to amino acids and peptides, in *Nickel and Its Role in Biology* (H. Sigel and A. Sigel, eds.), *Metal Ions in Biological Systems,* Vol. 23, Marcel Dekker, New York, pp. 124–164.

Martin, R. B., 1988b. Nickel ion binding to nucleosides and nucleotides, in *Nickel and Its Role in Biology* (H. Sigel and A. Sigel, eds.), *Metal Ions in Biological Systems,* Vol. 23, Marcel Dekker, New York, pp. 315–330.

Mergeay, M., Nies, D., Schlegel, H. G., Gerits, J., Charles, P., and van Gijsegem, F., 1985. *Alcaligenes eutrophus* CH34 is a facultative chemolithotroph with plasmid-bound resistance to heavy metals, *J. Bacteriol.* 162:328–334.

Mishra, D., and Kar, M., 1974. Nickel in plant growth and metabolism, *Bot. Rev.* 40:395–452.

Mohan, P. M., and Sastry, K. S., 1983. Interrelationships in trace-element metabolism in metal toxicities in nickel-resistant strains of *Neurospora crassa, Biochem. J.* 212:205–210.

Mohan, P. M., Rudra, M. P. P., and Sastry, K. S., 1984. Nickel transport in nickel-resistant strains of *Neurospora crassa, Curr. Microbiol.* 10:125–128.

National Research Council, 1975. *Nickel,* National Academy of Sciences, Washington, D.C.

Nies, D. H., 1992. Resistance to cadmium, cobalt, zinc, and nickel in microbes, *Plasmid* 27:17–28.

Nies, D. H., and Silver, S., 1989. Plasmid-determined inducible efflux is responsible for resistance to cadmium, zinc, and cobalt in *Alcaligenes eutrophus, J. Bacteriol.* 171:896–900.

Nies, A., Nies, D. H., and Silver, S., 1989. Cloning and expression of plasmid genes encoding resistances to chromate and cobalt in *Alcaligenes eutrophus, J. Bacteriol.* 171:5065–5070.

Park, M. H., Wong, B. B., and Lusk, J. E., 1976. Mutants in three genes affecting transport of magnesium in *Escherichia coli:* Genetics and physiology, *J. Bacteriol.* 126:1096–1103.

Richardson, D. H. S., Beckett, P. J., and Nieboer, E., 1980. Nickel in lichens, bryophytes, fungi and algae, in *Nickel in the Environment* (J. O. Nriagu, ed.), John Wiley & Sons, New York, pp. 367–406.

Schlegel, H. G., Cosson, J.-P., and Baker, A. J. M., 1991. Nickel-hyperaccumulating plants provide a niche for nickel-resistant bacteria, *Bot. Acta* 104:18–25.

Schmidt, T., and Schlegel, H. G., 1989. Nickel and cobalt resistance of various bacteria isolated from soil and highly polluted domestic and industrial wastes, *FEMS Microbiol. Lett.* 62:315–328.

Schmidt, T., Stoppel, R. D., and Schlegel, H. G., 1991. High-level nickel resistance in *Alcaligenes xylosoxydans* 31A and *Alcaligenes eutrophus* KTO2, *Appl. Environ. Microbiol.* 57:3301–3309.

Sensfuss, C., and Schlegel, H. G., 1988. Plasmid pMOL28-encoded resistance to nickel is due to specific efflux, *FEMS Microbiol. Lett.* 55:295–298.

Siddiqui, R. A., and Schlegel, H. G., 1987. Plasmid pMOL28-mediated inducible nickel resistance in *Alcaligenes eutrophus* strain CH34, *FEMS Microbiol. Lett.* 43:9–13.

Siddiqui, R. A., Schlegel, H. G., and Meyer, M., 1988. Inducible and constitutive expression of pMOL28-encoded nickel resistance in *Alcaligenes eutrophus* N9A, *J. Bacteriol.* 170:4188–4193.

Siddiqui, R. A., Benthin, K., and Schlegel, H. G., 1989. Cloning of pMOL28-encoded nickel resistance genes and expression of the genes in *Alcaligenes eutrophus* and *Pseudomonas* spp., *J. Bacteriol.* 171:5071–5078.

Silver, S., Nuciforma, G., Chu, L., and Misra, T. K., 1989. Bacterial resistance ATPases: Primary pumps for exporting cations and anions, *Trends Biochem. Sci.* 14:76–80.

Singh, A. L., Asthana, R. K., Srivastava, S. C., and Singh, S. P., 1992. Nickel uptake and its localization in a cyanobacterium, *FEMS Microbiol. Lett.* 99:165–168.

Skaar, H., Rystad, B., and Jensen, A., 1974. The uptake of ^{63}Ni by the diatom *Phaeodactylum tricornutum, Physiol. Plant* 32:353–358.

Smith, D. H., 1967. R factors mediate resistance to mercury, nickel, and cobalt, *Science* 156:114–116.

Snavely, M. D., Florer, J. B., Miller, C. G., and Maguire, M. E., 1989a. Magnesium transport in *Salmonella typhimurium:* Expression of cloned genes for three distinct Mg^{2+} transport systems, *J. Bacteriol.* 171:4752–4760.

Snavely, M. D., Florer, J. B., Miller, C. G., and Maguire, M. E., 1989b. Magnesium transport in *Salmonella typhimurium:* $^{28}Mg^{2+}$ transport by the CorA, MgtA, and MgtB systems, *J. Bacteriol.* 171:4761–4766.

Snavely, M. D., Miller, C. G., and Maguire, M. E., 1991a. The *mgtB* Mg^{2+} transport locus of *Salmonella typhimurium* encodes a P-type ATPase, *J. Biol. Chem.* 266:815–823.

Snavely, M. D., Gravina, S. A., Cheung, T. T., Miller, C. G., and Maguire, M. E., 1991b. Magnesium transport in *Salmonella typhimurium*. Regulation of *mgtA* and *mgtB* expression, *J. Biol. Chem.* 266:824–829.

Soeder, C. J., and Engelmann, G., 1984. Nickel requirement in *Chlorella emersonii*, *Arch. Microbiol.* 137:85–87.

Sprott, G. D., Jarrell, K. F., Shaw, K. M., and Knowles, R., 1982. Acetylene as an inhibitor of methanogenic bacteria, *J. Gen. Microbiol.* 128:2453–2462.

Stults, L. W., Mallick, S., and Maier, R. J., 1987. Nickel uptake in *Bradyrhizobium japonicum*, *J. Bacteriol.* 169:1398–1402.

Tabillion, R., and Kaltwasser, H., 1977. Energieabhangige ^{63}Ni-aufnahme bei *Alcaligenes eutrophus* stamm H1 and H16, *Arch. Microbiol.* 113:145–151.

Takakuwa, S., 1987. Nickel uptake in *Rhodopseudomonas capsulata*, *Arch. Microbiol.* 149:57–61.

Van Baalen, C., and O'Donnell, R., 1978. Isolation of a nickel-dependent blue-green alga, *J. Gen. Microbiol.* 105:351–353.

Varma, A. K., Sensfuß, C., and Schlegel, H. G., 1990. Inhibitor effects on the accumulation and efflux of nickel ions in plasmid pMOL28-harboring strains of *Alcaligenes eutrophus*, *Arch. Microbiol.* 154:42–49.

Webb, M., 1970a. The mechanism of acquired resistance to Co^{+2} and Ni^{+2} in Gram-positive and Gram-negative bacteria, *Biochim. Biophys. Acta* 222:440–445.

Webb, M., 1970b. Interrelationships between the utilization of magnesium and the uptake of other bivalent cations by bacteria, *Biochim. Biophys. Acta* 222:428–439.

Wildung, R. E., Garland, T. R., and Drucker, H., 1979. Nickel complexes with soil microbial metabolites—mobility and speciation in soils, in *Chemical Modeling in Aqueous Systems* (A. Jenne, ed.), ACS Symposium Series No. 93, American Chemical Society, Washington, D.C., pp. 181–200.

Willecke, K., Gries, E.-M., and Oehr, P., 1973. Coupled transport of citrate and magnesium in *Bacillus subtilis*, *J. Biol. Chem.* 248:807–814.

Wolfram, L., Eitinger, T., and Friedrich, B., 1991. Construction and properties of a triprotein containing the high-affinity nickel transporter of *Alcaligenes eutrophus*, *FEBS Lett.* 283:109–112.

Wu, L. F., and Mandrand-Berthelot, M.-A., 1986. Genetic and physiological characterization of new *Escherichia coli* mutants impaired in hydrogenase activity, *Biochimie* 68:167–179.

Wu, L.-F., Mandrand-Berthelot, M.-A., Waugh, R., Edmonds, C. J., and Boxer, D. H., 1989. Nickel deficiency gives rise to the defective hydrogenase phenotype of *hydC* and *fnr* mutants in *Escherichia coli*, *Mol. Microbiol.* 3:1709–1718.

Wu, L.-F., Navarro, C., and Mandrand-Berthelot, M.-A., 1991. The *hydC* region contains a multicistronic operon (*nik*) involved in nickel transport in *Escherichia coli*, *Gene* 107:37–42.

Yang, H., Daniel, S. L., Hsu, T., and Drake, H. L., 1989. Nickel transport by the thermophilic acetogen *Acetogenium kivui*, *Appl. Environ. Microbiol.* 55:1078–1081.

Plant Nickel Metabolism

8.1 Introduction

Interactions between nickel ions and plants are diverse and agriculturally significant (Mishra and Kar, 1974). At low metal ion concentrations, nickel is an essential nutrient for many plants and enhances several physiological functions. In contrast, excess levels of nickel can lead to growth inhibition, necrosis, or even plant death. Some plants classified as hyperaccumulators, however, can tolerate and may even prefer elevated levels of nickel (i.e., over 1000 μg of nickel per gram dry weight of tissue).

The following sections describe the functions of and requirements for nickel in plants, the mechanisms for nickel uptake from the environment and systemic transport within plants, the toxic effects of nickel on plants, and the unique properties of nickel-hyperaccumulating plants.

8.2 Nickel Functions and Requirements in Plants

A physiological requirement for nickel ion in certain plants has only been ascertained in the last few years [reviewed by Dalton *et al.* (1988)]. In part, the recent identification of nickel as a plant micronutrient was delayed because of the ready availability of nickel ion as a component of soils and the very low levels of this metal ion that are typically required for growth. For example, Eskew *et al.* (1984) demonstrated that the total amount of nickel ion required by a soybean (*Glycine max*) plant to complete its life cycle can be obtained from the initial nickel content of the seed. In many cases, the requirement for nickel ion has been shown to be related to metabolic flux of plant nitrogen through urea and the corresponding need for functional urease, a nickel-dependent enzyme described in Chapter 3. The following paragraphs summarize evidence that nickel is a required trace metal ion in selected plants and detail the physiological functions of this metal ion.

Since 1975, when nickel was shown to be a component of jack bean (*Canavalia ensiformis*) urease (Dixon *et al.,* 1975), evidence has accumulated that urease plays an important role in the nitrogen metabolism of many plants. For example, Eskew *et al.* (1983) grew soybean plants in nickel-depleted nutrient solutions and showed that urea accumulates to 2.5% dry weight in the leaf tips, leading to necrotic lesions. Similarly, cowpeas (*Vigna unguiculata*) were shown to develop comparable lesions when grown in the absence of nickel ion (Eskew *et al.,* 1984; Walker *et al.,* 1985). Consistent with these affects arising directly from the absence of a functional urease, analogous increases in leaf-tip necrosis and elevated urea concentrations were generated by growing soybean, wheat (*Triticum aestivum*), or sorghum (*Sorghum bicolor*) plants in the presence of urea and a urease inhibitor (Krogmeier *et al.,* 1989a,b). The obvious interpretation of these results is that nitrogen metabolism in many plants includes a major flux through urea, probably arising from the breakdown of arginine and other nitrogen-rich compounds. The absence of urease activity alters the flow of nitrogen and leads to the buildup of inhibitory concentrations of urea in the plant.

The lack of a functional urease may account for other effects noted in nickel-depleted plants. For example, nitrogen turnover is likely to be greatest during the catabolism of seed protein to produce new tissues in a growing seedling; thus, it would be reasonable to expect that seed germination would exhibit a nickel dependence. In this regard, Mishra and Kar (1974) and Welch (1981) reviewed a large body of early literature that describes the ability of nickel to stimulate germination rates in numerous plants. Similarly, Brown *et al.* (1987a) reported that barley (*Hordeum vulgare*) grain exhibits a reduced germination rate, seedlings show depressed vigor, and plants fail to produce viable grain when provided with nickel-depleted nutrient solution. Interestingly, nickel deficiency leads to significant changes in levels of amino acids, malate, inorganic anions, and total nitrogen in barley tissues (Brown *et al.,* 1990). Elevated urea levels as well as alterations in growth behavior were noted for nickel-depleted wheat, barley, and oat (*Avena sativa*) plants (Brown *et al.,* 1987b). Decreased root growth in seedlings and a 23% reduction in crop yield were reported for nickel-deprived versus nickel-replete pea (*Pisum sativum*) plants (Horak, 1985b). Finally, the stimulatory effects of nickel on growth of black locust (*Robinia pseudoacacia*) correlates with the urease activities in this tree (Benchemsi-Bekkari and Pizelle, 1992). These alterations in growth properties may all arise from deficiencies in functional urease leading to a cessation of nitrogen flux through urea; however, it is also possible that additional, as yet uncharacterized nickel-dependent functions may be impaired in these plants.

In contrast to reports that describe specific plant requirements for nickel ion as detailed above, numerous other studies have noted an enhancement of growth yield by nickel ion that may not represent a true stimulation of the plant. Mishra and Kar (1974) and Welch (1981) have summarized reports of apparent growth enhancement in various crops where the plant stimulation may be due to the well-known fungicidal effects of nickel ion. By reducing the negative effects of pathogenic fungi, nickel appears to stimulate plant growth. These results highlight the caution that is necessary in interpreting the effects of nickel ion on plants in field studies.

8.3 Nickel Transport in Plants

Nickel is found in all plants and in all plant tissues, typically at concentrations of 0.05–5 $\mu g/g$ dry weight (Vanselow, 1966). However, the nickel content varies with the soil composition, the type of plant, and the type of tissue. In order for the nickel to move from the environment to the various plant tissues, the metal ion must be transported into the plant roots and then throughout the plant. Nickel uptake from the environment and systemic transport of nickel ion are each discussed below.

8.3.1 Uptake of Nickel from the Environment

Plants appear to obtain nickel from the environment primarily as the divalent cation (Mishra and Kar, 1974). As described for nickel transport in microorganisms (Chapter 7), many physical properties of the environment affect nickel transport. The most obvious parameter determining the rate and amount of nickel ion uptake into plants is the nickel ion concentration in the environment. However, organic components and inorganic particulates in the root environment can chelate nickel ion, leading to lower effective concentrations for the plant. Similarly, the presence of certain anions can lead to precipitation of nickel as an insoluble salt, thus decreasing metal ion availability to the plant. Other metal ions can compete with nickel ion for transport proteins, leading to inhibition of transport. Alternatively, other cations can displace nickel ion that is bound to various external chelators, thus increasing the free nickel ion concentrations and enhancing the rate of nickel uptake. Finally, the pH of the environment determines the protonation status of external nickel chelators. At low pH values, the nickel-binding components are protonated, the free nickel ion concentration increases, and the amount of nickel ion taken up by the plant is likely to increase.

The most detailed analysis of plant nickel ion uptake that has been reported made use of 21-day-old soybean plants (Cataldo *et al.*, 1978a). The amount of nickel ion absorbed by the whole plant over 2 h (monitored by using trace levels of ^{63}Ni) was assessed for samples that were treated with nickel ion concentrations ranging from 0.01 to 200 μM. The nickel uptake kinetics are very complicated, as is characteristic of the kinetics for uptake of other ions in higher plants. The authors interpreted their results in terms of the presence of three absorption phases. The high-affinity (low-concentration) phase was stated to possess a K_T (transport constant) of 0.51 μM; however the data used to obtain this value apparently only included concentrations up to 0.25 μM. This phase is associated with a very low rate of nickel uptake (12.9 μg nickel/g dry weight of root per h). The second phase is associated with a faster rate of transport (175 $\mu g/g$ dry weight per h) and possesses a K_T of 8.6 μM. Finally, a low-affinity phase was noted with a K_T of 379 μM and a high rate of transport (1870 $\mu g/g$ dry weight per h), although again the values must be judged with caution since the data only included nickel concentrations up to 200 μM. The ability for 5 μM concentrations of other metal ions to compete with nickel transport was determined over a range of nickel ion concentrations from 1 to 5 μM. Copper and zinc ions were found to be competitive inhibitors of nickel ion uptake. Ferrous ion and cobalt also inhibit, but not competitively. The effects of magnesium ion were not determined. The authors suggested that these divalent cations may all be taken up by a common absorption process.

8.3.2 Systemic Transport of Nickel

After entering the plant root, nickel ion is distributed to all plant tissues. Rather than being uniformly distributed in the plant, the nickel ion concentration varies with the tissue, the age of the plant, and the species. For example, in oat plants the leaves contain more nickel than the stems, the blades more than the leaf sheaths, the flowers more than the peduncles, and grain more than the straw (Mishra and Kar, 1974). In many plants, especially among the Fabaceae, nickel is accumulated at the highest concentration in the seed (Horak, 1985a). In the case of soybean, the nickel ion is initially distributed in vegetative tissues and is remobilized at maturation during the seed-fill process (Cataldo et al., 1978b). How nickel distribution in plants is governed is completely unknown, but the disbursement may simply be regulated by the affinity and levels of nickel-binding sinks, such as the urease protein, in each of the tissues. In contrast to the sparsity of information regarding factors that control the relative nickel concentrations in various tissues, the mechanisms by which nickel is transported have been studied in several plants. The results of these experiments are described in the following paragraphs.

Nickel is translocated from roots to other plant tissues via the xylem sap. Tiffin (1971) examined the xylem exudate and root sap from tomato (*Lyco-*

persicon esculentum), cucumber (*Cucumis sativus*), corn (*Zea mays*), carrot (*Daucus carota*), and peanut (*Arachis hypogaea*) plants that were treated with various concentrations of nickel ion containing ^{63}Ni. Whereas the nickel ion in the nutrient solution behaves as the dication during electrophoretic analysis, the migration of nickel in the exudates was toward the anode as if it was associated with an anionic carrier. In tomato, only the carrier-bound nickel was observed at low nickel ion concentrations, but the carrier could be saturated at >2 μM nickel ion concentrations, leading to the presence of free nickel ion. In root sap of the other plants, the levels of nickel carrier appeared to be much higher, requiring nearly 100 μM nickel ion concentrations to saturate. The carriers were electrophoretically indistinguishable among the five plants, but the chemical structure of the organic anion in the sap was not determined.

Following up on the early studies of Tiffin (1971), several authors have examined the speciation of nickel in tissues of various plants. Cataldo *et al.* (1978b) showed that xylem exudate of soybean contains a major anionic species, a major cationic species, and a minor cationic species of nickel, whereas no free nickel ion was observed. In contrast, for soybean plants provided with 10 μM NiCl$_2$, ~8% of the nickel in roots and ~2% of the metal ion in leaves is thought to exist as the dication. Analysis of soybean root and leaf extracts provided evidence for changes in the chemical form of some of the bound nickel in the different tissues. However, further chemical characterization of these different nickel forms in soybean tissues was limited to crude size estimates to show a M_r of between 500 and 100,000. Similar observations were reported for expressed fluids derived from alfalfa (*Medicago sativa*) and potato (*Solanum tuberosum*), where both cationic and anionic species of nickel were shown to possess approximate sizes of 2000 daltons (Theisen and Blincoe, 1984, 1988). It is unclear, however, whether certain of the multiple species observed in homogenized tissues from these various plants may have been generated during the tissue disruption process, for example, binding of nickel by released amino acids or other components. Furthermore, none of the nickel species were structurally characterized. It is likely, however, that several of the observed species are organic acid complexes of nickel ion. Many plants are known to possess organic acids such as citrate and malate in their xylem fluids.

In addition to the initial translocation from roots to the various plant tissues, nickel can also be remobilized from one tissue to another. For example, Cataldo *et al.* (1978b) demonstrated the movement of nickel from leaves to seeds during senescence of soybean plants. Metal ion movement of this type is likely to occur via the phloem. Using both pea and geranium (*Pelargonium zonale*), Neumann and Chamel (1986) directly demonstrated that nickel was transported from the leaves to other plant tissue through the phloem. The chemical form of nickel that was transported in these experiments was not analyzed; however, Wiersma and Van Goor (1979) showed that the phloem exudate of caster bean (*Ricinus communis*) was anionic and possessed an M_r of 1000–5000.

8.4 Nickel Toxicity and Resistance in Plants

Soils typically contain between 5 and 500 ppm nickel with an average of approximately 100 ppm (Swain, 1955); yet when provided to plants in nutrient solutions, nickel ion is toxic to most species at levels in the 10 μM (~0.6 ppm) range (Mishra and Kar, 1974). The actual concentration of nickel that is toxic depends on the plant and on the growth conditions. Much of the nickel ion in soils and in nutrient solutions is not readily available to the plant owing to the same factors as already described with regard to nickel uptake involving soil nickel-binding components (Section 8.3.1).

The symptoms of nickel toxicity are manifested as induced iron-deficiency chlorosis (yellowing of the leaves), foliar necrosis, stunted growth, plant deformation, and unusual spottings on the leaves or stems (Hutchinson, 1981; Mishra and Kar, 1974). Unfortunately, the biochemical details of nickel toxicity have not been elucidated. Inhibition of enzyme action and interference with energy metabolism are two reasonable sites of action for nickel ion. For example, Morgutti *et al.* (1984) showed that nickel ion inhibited phosphoenolpyruvate carboxylase activity in maize roots and found nickel-induced alterations in proton pumping in this tissue. Pandolfini *et al.* (1992) reported that nickel treatment of wheat led to enhanced extracellular peroxidase activity, enhanced lipid peroxidation (as monitored by increased malondialdehyde levels), and increased leakage of potassium ion from roots. Furthermore, the overall growth inhibition and reduced water content of nickel-treated wheat plants was only found when the nickel ion concentration exceeded a threshold value of 20 μM. Numerous other enzymes and processes can be affected by nickel in various plants.

Whereas many plants are very susceptible to low concentrations of nickel ion, some plants are more tolerant and others are highly resistant to this metal ion. Verkleij and Schat (1990) reviewed the general mechanisms of metal resistance used by plants. They divided the mechanisms into two categories: avoidance and tolerance. Avoidance mechanisms involve the organism's ability to prevent the metal ion from crossing the plasmalemma of a root cell by reducing the rate of metal transport, by increasing the metal binding capacity of the cell wall, and by secreting metal-binding compounds. These avoidance mechanisms used by plants are identical to three of the mechanisms used by microorganisms for nickel resistance (see Fig. 7-1b, d, and e, respectively).

Tolerance mechanisms used by plants include the production of intracellular metal-binding compounds, compartmentation of the metal ion, changes in plant metabolism away from metal-sensitive pathways and toward the use of metal-tolerant enzymes, and alterations in membrane structures. The first of these mechanisms is especially important in many plants and is also utilized by microorganisms (Fig. 7-1c). The resistance mechanisms used by plants are best understood in the hyperaccumulating plants that thrive in soils with very high levels of this cation (described in Section 8.5). Much less is known about the resistance mechanisms in the more common plants that exhibit only moderate levels of nickel resistance. The following paragraph focuses on one aspect of resistance that may be operative in such plants.

Phytochelatins, found in both monocotyledonous and dicotyledonous plants, are peptides consisting of 2 to 11 γ-glutamylcysteine units with a carboxyl-terminal glycine (Rauser, 1990). Rather than being synthesized on a ribosome, these compounds are made by a γ-glutamylcysteine dipeptidyl transpeptidase (phytochelatin synthase) from glutathione (Grill *et al.*, 1989). The thiolates participate in the binding of several heavy metal ions, and phytochelatins are thought to be functionally analogous to animal metallothioneins; that is, they appear to function in the accumulation, detoxification, and metabolism of metal ions (Grill *et al.*, 1987). [Interestingly, at least one plant, *Mimulus guttatus*, also has been shown to possess a metallothionein gene (de Miranda *et al.*, 1990).] Using suspension cultures of *Rauvolfia serpentina*, Grill *et al.* (1987) demonstrated that nickel ions are capable of inducing the synthesis of phytochelatins, although cadmium, lead, zinc, antimony, and silver ions lead to higher levels of induction. Despite the apparent regulation of phytochelatin synthesis by nickel ion and the ability of phytochelatins to bind this metal ion, no evidence has been presented that these peptides function in providing nickel resistance in plants. In contrast to this lack of knowledge regarding nickel resistance mechanisms in plants that exhibit moderate tolerance levels, the following section describes the well-characterized mechanisms found in plants that are resistant to very high levels of this metal ion.

8.5 Hyperaccumulators of Nickel

Whereas the leaves of most plants possess concentrations of nickel that are less than 10 μg/g dry weight (0.001%), the dried leaves of certain species [termed nickel hyperaccumulators by Brooks *et al.* (1977)] contain more than 1000 μg of nickel per gram (0.1% nickel). Indeed, the dry weight of foliage from many of these plants consists of over 1% nickel, with the record currently held by *Psychotria douarrei* at 4.75% nickel (Baker and Brooks, 1989; Baker and Walker, 1990; Brooks, 1987). Not unexpectedly, these plants are found in nickel-rich soils; however, it may be surprising that the plants further concentrate the metal ion severalfold over the soil levels. For example, in an early study of hyperaccumulators, Severne and Brooks (1972) demonstrated the presence of 23% nickel in leaf ash of *Hybanthus floribundus* that was grown on soil containing 670 ppm nickel (0.067%), or a leaf ash:soil ratio of 343. The first section below describes the phylogenetic and geographic distributions of plants that hyperaccumulate nickel. A second section then details the biochemistry of nickel uptake, transport, and storage in these plants. Finally, the function of nickel hyperaccumulation is discussed in terms of potential advantages to the plant.

8.5.1 Distribution of Nickel Hyperaccumulators

More than 150 hyperaccumulators of nickel have been identified [reviewed by Baker and Brooks (1989), Baker and Walker (1990), and Brooks (1987)]. The genus names for these plants and the number of species within each genus are shown in Table 8-1. The 43 genera that have been identified include representatives of 26 families that are located in 18 orders in 6 subclasses and in 2 classes. Except for *Luzula,* belonging to the class Liliopsida, all of the nickel hyperaccumulators belong to the class Magnoliopsida. By far, the largest genus of nickel hyperaccumulators is *Alyssum,* containing 48 species. It is important to note, however, that not all *Alyssum* species accumulate nickel and that nonhyperaccumulating species of the other genera are also known. This list of nickel hyperaccumulators is unlikely to grow much larger because identification has included both thorough field sampling in nickel-rich environments and extensive analysis of samples from numerous herbaria [e.g., Brooks *et al.* (1977) sampled nearly 2000 herbarium specimens to show that 128 of 240 *Hybanthus* samples and 104 of 150 *Homalium* samples possess high levels of nickel]. Although many are tropical evergreens, the nickel hyperaccumulators include representatives of herbs, shrubs, and trees from both temperate and tropical zones (Baker and Walker, 1990).

All of the nickel hyperaccumulators were found in one or more of the following regions: New Caledonia, western Australia, southern Europe and Asia Minor, the Malay Archipelago, Cuba, western United States, and Zimbabwe. The ultramafic or serpentine soils in these areas are enriched in nickel and other metals but have only low nutrient concentrations (Brooks, 1987). The nickel resistance exhibited by hyperaccumulators provides these plants a competitive edge in these environments because nonhyperaccumulators have difficulty in tolerating such high nickel levels. Other parts of the globe, including many sites in Asia and North America, possess similar metal-rich

Table 8.1. Plants That
Hyperaccumulate Nickel[a]

Genus	Number of species
Agatea	1
Alyssum	48
Argophyllum	2
Berkheya	1
Blepharis	1
Bornmuellera	6
Breckenridgea	1
Buxus	1
Cardamine	1
Casearia	1
Chrysanthemum	1
Cleidion	1
Cochlearia	2
Dichapetalum	1
Dicoma	1
Diospyros	1
Dipterocarpus	1
Geissois	7
Homalium	7
Hybanthus	5
Lasiochlamys	1
Leucocroton	1
Linaria	1
Luzula	1
Merremia	1
Minuartia	1
Myristica	1
Oncotheca	1
Pearsonia	1
Peltaria	2
Phyllanthus	11
Planchonella	1
Psychotria	2
Rhus	1
Rinorea	2
Saxifraga	2
Sebertia	1
Stachys	1
Strepanthus	1
Thlaspi	17
Trichospermum	1
Trifolium	1
Xylosma	11

[a] From Baker and Walker (1990).

soils, but no hyperaccumulators have been found in these areas. It has been suggested that prior glaciation of these regions may have prevented the evolutionary development of the unique physiology that is found in these plants (Baker and Brooks, 1989).

8.5.2 Biochemistry of Nickel Metabolism in Hyperaccumulators

A discussion of nickel biochemistry in hyperaccumulators must include consideration of metal uptake into the plant, systemic transport, and the tissue distribution. As described in the next few paragraphs, some of these aspects are much better understood than others. Speculations about the function of nickel in hyperaccumulators are reserved for Section 8.5.3.

Nickel uptake from soil by hyperaccumulators has been studied by Morrison et al. (1980). The nickel concentration in dry leaf matter was examined as a function of the provided nickel concentration for several nickel hyperaccumulating Alyssum species and closely related, nonhyperaccumulating species of the same genera. The leaf nickel content of the nonhyperaccumulators varies in a roughly linear fashion as a function of nickel concentration up to toxic levels ($\sim$400 μg/g) and is always below the metal concentration provided in the nutrient solution. In contrast, the leaf nickel content of the hyperaccumulators is greatly elevated over the concentration that is provided at relatively low nickel levels and then saturates at high nickel levels. For example, at a nickel concentration of 100 μg/g in the nutrient solution, hyperaccumulator dry leaf matter possesses approximately 1000 μg/g (0.1% nickel). Increasing the level of nickel in the medium up to 10,000 μg/g leads to an increase in the leaf content to similar values. At a concentration of 7000 μg/g, the rate of nickel uptake by 5-week-old A. euboeum is linear for 5 days, at which time the nickel content in leaves is saturated. It appears that the plant attempts to maintain high nickel concentrations in its foliage at all concentrations of nickel and thus has devised an efficient uptake system to extract nickel from the soil at lower nickel concentrations. The metal ion uptake system in nickel hyperaccumulators is likely to possess a high degree of specificity toward nickel. Cobalt, a metal ion that shares several physical characteristics with nickel, is present at a nickel:cobalt ratio of approximately 7:1 in many of the nickel-rich soils where these plants are found, yet the plants take up only trace amounts of cobalt compared to the enormous levels of nickel that are accumulated. Based on a comparison of stability constants for binding of nickel and cobalt by a range of ligands, Still and Williams (1980) proposed that selection for nickel could be achieved by using a ligand that binds the metal ion in octahedral coordination with two or more nitrogen donor atoms and at least one oxygen ligand. Brooks (1987) has described the hypothetical nickel-specific uptake system in the root membrane as a selector (see Fig. 8-1).

After nickel enters the hyperaccumulating plant, it is distributed to various tissues via the xylem. The root uptake system may release nickel ion into the xylem fluid where it is rapidly bound by organic components, or it is possible that the nickel is directly transferred from the uptake selector to a systemic transporter molecule without the presence of a free nickel ion intermediate. In either case the form of the metal in the xylem is not free nickel ion, but rather it is bound to one or more organic carrier molecules. Incredibly, the bluish green sap of *Sebertia acuminata* contains 11.2% nickel on a fresh weight basis or 25.7% nickel on a dry weight basis (Jaffré *et al.*, 1976)! Infrared spectroscopic and gas/liquid chromatographic–mass spectrometric methods were used to demonstrate that the nickel complex in the sap of this plant

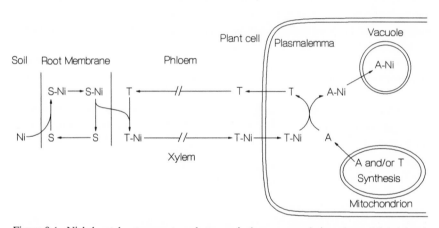

Figure 8-1. Nickel uptake, transport, and storage in hyperaccumulating plants. Nickel ion is thought to be taken up from the soil by a membrane-bound nickel-specific transport system that has been termed the selector (S) (Brooks, 1987). The properties of this putative selector have not been determined; however, Still and Williams (1980) hypothesized that, if present, the selector probably binds the metal ion in an octahedral complex with at least two nitrogen atom donors in order to achieve the observed metal ion specificity. The nickel is then complexed by one or more organic acids, termed transport ligands (T), and distributed throughout the plant via the xylem. As one example, the sap of *Sebertia acuminata* contains over 25% of its dry weight as nickel, where the metal is in a complex with two citrate molecules (Jaffré *et al.*, 1976). Within the various plant tissues, nickel is found in vacuoles as water-soluble, polar complexes (Brooks *et al.*, 1981). The complexes again involve organic acids (citrate, malate, malonate, etc.), the so-called acceptor ligands (A), which may be identical to the organic acids used in xylem for nickel transport to the tissues. The mechanisms for uptake of the nickel complexes into the plant cell or into the vacuole have not been characterized. Transport ligands and acceptor ligands are shown as being synthesized in mitochondria (e.g., using enzymes of the tricarboxylic acid cycle); however, this speculation has not been verified. Newly synthesized or recycled transport ligands are provided to the roots via the phloem. Modified from Fig. 8.9 of Brooks (1987).

consists primarily of two citrate molecules per nickel ion (Lee *et al.*, 1977). Numerous other plants, including hyperaccumulators and nonhyperaccumulators, possess citrate and other organic acids in the xylem fluid that could be used to transport nickel and other metal ions. Simply the presence of these compounds, however, is not sufficient for a plant to be resistant to metal ions. In Fig. 8-1 the organic molecules that bind nickel are referred to as the transport ligands.

The nickel is distributed from the xylem to the various tissues of hyperaccumulating plants in a nonuniform manner. As one example, Table 8-2 provides the tissue distribution of nickel in *Alyssum heldreichii*, where the largest sink for the metal is in the leaves (Brooks, 1987). Although in this case the growing tissues possess higher nickel concentrations than tissues that are less active, in *S. acuminata* the trunk bark had higher levels than the leaves (2.45% vs. 1.17%, respectively) (Jaffré *et al.*, 1976). No studies have addressed the stability of nickel once it reaches the various tissues. For example, it is reasonable to suppose that the nickel may be remobilized during maturation of the plant.

Within tissues, nickel exists as water-soluble, polar complexes that appear to be located in vacuoles (Brooks *et al.*, 1981). In the case of *S. acuminata*, as well as for *Geissois pruniosa*, 12 *Homalium* strains, *Hybanthus austrocaledonicus*, and *Hybanthus caledonicus*, chemical analysis demonstrated that the major form of this nickel is a citrato complex. By contrast, other plants

Table 8.2. Distribution of Nickel in *Alyssum heldreichii*[a]

Tissue	Percent of total weight	Ni concentration ($\mu g/g$)	Percent of total Ni
Roots			
Lower	11.9	4330	4.5
Upper	9.4	9150	7.6
Stems			
Lower	10.0	7190	6.4
Middle	5.4	9660	4.6
Upper	3.4	16740	5.0
Lower lateral	10.0	17060	15.1
Leaves			
Lower stem	27.2	12150	29.2
Midstem	12.9	11890	13.5
Upper stem	7.6	14070	9.5
Apical buds	2.2	23400	4.6
Seeds	—	1880	—

[a] From Brooks (1987)

such as *Alyssum bertolonii, Alyssum serpyllifolium,* and *Pearsonia metallifera* possess nickel that is complexed with noncitrate organic acids (Lee *et al.,* 1978). For example, Pelosi *et al.* (1976) and Brooks *et al.* (1981) provided evidence consistent with nickel complexes of malic acid and malonic acid in *A. bertolonii* and *A. serpyllifolium.* Furthermore, *Psychotria douarrei* uses primarily malic acid in the anionic complex but also had some of the citrato species (Kersten *et al.,* 1980). Other examples where multiple organic acids are used to bind nickel include *Phyllanthus serpentinus,* where nickel is bound by approximately equal amounts of citrate and malate (Kersten *et al.,* 1980), and *Dichapetalum gelonioides,* where citrate and malate are present in a nearly 1:2 ratio (Homer *et al.,* 1991). In these cases, however, it must be pointed out that the log of the stability constant for the monocitrate complex of nickel ($\log K = 5.47$) far exceeds that of the malate ($\log K = 3.30$) or that of the dicitrate complex ($\log K = 2.33$) so that the monocitrate compound has the greatest physiological importance. In addition to the citrate, malate, and malonate complexes, other organic acids have been observed (e.g., tartaric acid and homocitrate), and the nickel dication is usually present to some extent. In Fig. 8-1, the organic acids that bind nickel in the plant vacuoles are termed the acceptor ligands, as in the terminology of Brooks (1987). The acceptor ligands may be the same as the transport ligands.

A few further comments about the organic components of the nickel complexes are worth noting. Organic acids exhibit poor discrimination between nickel and other metal ions, such as cobalt, consistent with the plant's selectivity for nickel residing in the root uptake system rather than being due to a preference for nickel by the complexing agents (Still and Williams, 1980). Lee *et al.* (1978) demonstrated a strong correlation between the level of nickel and the level of citrate in a range of plant hyperaccumulators. Furthermore, levels of the organic acids in plant leaf tissue have been shown to vary with the soil nickel content for two plants (Pancaro *et al.,* 1978). Using *Alyssum bertolonii* and *Alyssum pintodasilvae,* these investigators found an order of magnitude increase in organic acids for samples grown in serpentine soils compared to the levels in samples grown in garden soil containing less than 40 μg of nickel per gram. Brooks *et al.* (1981) hypothesized a mechanism for regulation of the organic acid level. They speculated that inhibition of mitochondrial malate dehydrogenase activity by nickel ion could result in the buildup of malate, and perhaps other organic acids, which would subsequently chelate nickel ion and release the inhibition. It is not clear how hyperaccumulators are able to withdraw such large amounts of organic acids from the tricarboxylic acid cycle without growth inhibition, and this requirement may be the major constraint to more widespread occurrence of plants that hyperaccumulate nickel.

8.5.3 Functional Advantage of Nickel Hyperaccumulation

Although Severne and Brooks (1972) speculated that nickel may be an essential metal ion for growth of one hyperaccumulating species (*Hybanthus floribundus*), most, if not all, such species are facultative rather than obligate accumulators of this metal. The function of nickel in hyperaccumulators remains unclear; however, several possible relationships can be hypothesized. First, accumulation of nickel in leaves followed by annual leaf loss may be a detoxification mechanism that reduces the nickel concentrations in the rhizosphere (see Brooks, 1987). In such a manner, the nickel may be removed from the undersoil near the root tips, harmlessly stored in vacuoles, and deposited in the topsoil, where it may be washed away more easily. Such plants have been proposed to drive a nickel cycle that results in the continuous percolation of the topsoil by nickel ions, providing a niche from which nickel-resistant bacteria are readily isolated (Schlegel *et al.*, 1991). This scenario is clearly not important in all hyperaccumulators since many are perennials. A second possibility has been considered for hyperaccumulators that grow in arid environments. Severne (1974) suggested that epidermal accumulation of nickel in *Hybanthus floribundus* could reduce the rate of cuticular transpiration. Alternatively, nickel complexes may be produced as osmotolerant solutes in order to reduce intracellular osmotic stress. To date, no direct evidence has been provided to support either of these hypotheses, and many hyperaccumulators are not located in arid environments. A third suggestion, which has not been supported by field observation (Baker and Walker, 1990), is that nickel accumulation may protect the plant from consumption by herbivores. A fourth hypothesis for a functional role for nickel in hyperaccumulators involves enhanced resistance to pathogens. When found in soils containing low concentrations of nickel, the hyperaccumulators appear to be more susceptible to fungal pathogens than other plants and are poorly competitive with nonhyperaccumulators, whereas they are resistant to pathogens when grown in nickel-rich soils. Finally, a fifth hypothesis suggests that nickel hyperaccumulation is not an advantage in itself, but that this happens to be an effective mechanism of detoxification that allows the plant to thrive in environments where it is free of competition by other plants that cannot tolerate the high nickel levels. Investigation of the functional roles for hyperaccumulation remains an important area for future studies.

In addition to possible advantages to the plant, one can consider possible biotechnological advantages to humans of plant nickel hyperaccumulation. For example, Baker and Brooks (1989) suggested that potentially arable land may be detoxified of nickel by continuous cropping of hyperaccumulators. In addition, biological metal mining may be possible; however, the current low world price for nickel precludes this option from being economically viable.

8.6 Perspective

Several plants have exhibited a nutritional requirement for nickel. In many cases, the role of the metal ion has been shown to be related to urease involvement in a poorly characterized metabolic flux of nitrogen through urea. In other plants, the role of nickel is even less clear, and understanding the metabolic requirements for nickel ion remains an important target for future studies. The biochemical properties of nickel uptake systems have not been elucidated, but specific nickel transporters are unlikely to be present in plants except in the case of hyperaccumulators. Once in a plant, nickel ion is highly mobile as chelated complexes in the xylem. Upon reaching the various plant tissues, nickel is also found as low-molecular-weight chelated complexes, and, at least in the hyperaccumulators, these complexes appear to be located in vacuoles. For some plants, nickel transport between various tissues can also occur by the phloem system. Although this metal ion is toxic to plants, the mechanisms of toxicity remain poorly characterized, and the mechanisms of resistance are not well understood. These issues and the phenomenal ability of nickel hyperaccumulators to tolerate and further concentrate already elevated nickel concentrations will continue to inspire further experiments on plant–nickel interactions.

References

Baker, A. J. M., and Brooks, R. R., 1989. Terrestrial higher plants which hyperaccumulate metallic elements—a review of their distribution, ecology and phytochemistry, *Biorecovery* 1:81–126.

Baker, A. J. M., and Walker, P. L., 1990. Ecophysiology of metal uptake by tolerant plants, in *Heavy Metal Tolerance in Plants: Evolutionary Aspects* (A. J. Shaw, ed.), CRC Press, Boca Raton, Florida, pp. 155–177.

Benchemsi-Bekkari, N., and Pizelle, G., 1992. *In vivo* urease activity in *Robinia pseudoacacia, Plant Physiol.* 30:187–192.

Brooks, R. R., 1987. The distribution and phytochemistry of plants which hyperaccumulate nickel, *Serpentine and Its Vegetation: A Multidisciplinary Approach,* Croom Helm, London, pp. 85–108.

Brooks, R. R., Lee, J., Reeves, R. D., and Jaffré, T., 1977. Detection of nickeliferous rocks by analysis of herbarium specimens of indicator plants, *J. Geochem. Explor.* 7:49–77.

Brooks, R. R., Shaw, S., and Marfil, A. A., 1981. The chemical form and physiological function of nickel in some Iberian *Alyssum* species, *Physiol. Plant.* 51:167–170.

Brown, P. H., Welch, R. M., and Cary, E. E., 1987a. Nickel: A micronutrient essential for higher plants, *Plant Physiol.* 85:801–803.

Brown, P. H., Welch, R. M., Cary, E. E., and Checkai, R. T., 1987b. Beneficial effects of nickel on plant growth, *J. Plant Nutr.* 10:2125–2135.

Brown, P. H., Welch, R. M., and Madison, J. T., 1990. Effect of nickel deficiency on soluble anion, amino acid, and nitrogen levels in barley, *Plant Soil* 125:19–27.

Cataldo, D. A., Garland, T. R., and Wildung, R. E., 1978a. Nickel in plants. I. Uptake kinetics using intact soybean seedlings, *Plant Physiol.* 62:563–565.

Cataldo, D. A., Garland, T. R., Wildung, R. E., and Drucker, H., 1978b. Nickel in plants. II. Distribution and chemical form in soybean plants, *Plant Physiol.* 62:566–570.

Dalton, D. A., Russell, S. A., and Evans, H. J., 1988. Nickel as a micronutrient element for plants, *Biofactors* 1:11–16.

de Miranda, J. R., Thomas, M. A., Thurman, D. A., and Thomsett, A. B., 1990. Metallothionein genes from the flowering plant *Mimulus guttatus*, *FEBS Lett.* 260:277–280.

Dixon, N. E., Gazzola, C., Blakeley, R. L., and Zerner, B., 1975. Jack bean urease (EC 3.5.1.5). A metalloenzyme. A simple biological role for nickel?, *J. Am. Chem. Soc.* 97:4131–4133.

Eskew, D. L., Welch, R. M., and Cary, E. E., 1983. Nickel: An essential micronutrient for legumes and possibly all higher plants, *Science* 222:621–623.

Eskew, D. L., Welch, R. M., and Norvell, W. A., 1984. Nickel in higher plants. Further evidence for an essential role, *Plant Physiol.* 76:691–693.

Grill, E., Winnacker, E.-L., and Zenk, M. H., 1987. Phytochelatins, a class of heavy-metal-binding peptides from plants, are functionally analogous to metallothioneins, *Proc. Natl. Acad. Sci. USA* 84:439–443.

Grill, E., Löffler, S., Winnacker, E.-L., and Zenk, M. H., 1989. Phytochelatins, the heavy-metal-binding peptides of plants, are synthesized from glutathione by a specific γ-glutamylcysteine dipeptidyl transpeptidase (phytochelatin synthase), *Proc. Natl. Acad. Sci. USA* 86:6838–6842.

Homer, F. A., Reeves, R. D., Brooks, R. R., and Baker, A. J. M., 1991. Characterization of the nickel-rich extract from the nickel hyperaccumulator *Dichapetalum gelonioides*, *Phytochemistry* 30:2141–2145.

Horak, O., 1985a. Zur bedeutung des nickels für Fabaceae. I. Vergleichende untersuchungen über den gehalt teile und samen an nickel und anderen elementen, *Phyton* 25:135–146.

Horak, O., 1985b. Zur bedeutung des nickels für Fabaceae. II. Nickelaufnahme und nickelbedarf von *Pisum sativum* L., *Phyton* 25:310–307.

Hutchinson, T. C., 1981. Nickel, in *Effect of Heavy Metal Pollution on Plants* (N. W. Lepp, ed.), Applied Science Publishers, London, pp. 171–211.

Jaffré, T., Brooks, R. R., Lee, J., and Reeves, R. D., 1976. *Sebertia acuminata:* A hyperaccumulator of nickel from New Caledonia, *Science* 193:579–580.

Kersten, W. J., Brooks, R. R., Reeves, R. D., and Jaffré, T., 1980. Nature of nickel complexes in *Psychotria douarrei* and other nickel-accumulating plants, *Phytochemistry* 19:1963–1965.

Krogmeier, M. J., McCarty, G. W., and Bremner, J. M., 1989a. Phytotoxicity of foliar-applied urea, *Proc. Natl. Acad. Sci. USA* 86:8189–8191.

Krogmeier, M. J., McCarty, G. W., and Bremner, J. M., 1989b. Potential phytotoxicity associated with the use of soil urease inhibitors, *Proc. Natl. Acad. Sci. USA* 86:1110–1112.

Lee, J., Reeves, R. D., Brooks, R. R., and Jaffré, T., 1977. Isolation and identification of a citrato-complex of nickel from nickel-accumulating plants, *Phytochemistry* 16:1503–1505.

Lee, J., Reeves, R. D., Brooks, R. R., and Jaffré, T., 1978. The relationship between nickel and citric acid in some nickel-accumulating plants, *Phytochemistry* 17:1033–1035.

Mishra, D., and Kar, M., 1974. Nickel in plant growth and metabolism, *Bot. Rev.* 40:395–452.

Morgutti, S., Sacchi, G. A., and Cocucci, S. M., 1984. Effects of Ni^{+2} on proton extrusion, dark CO_2 fixation and malate synthesis in maize roots, *Physiol. Plant.* 60:70–74.

Morrison, R. S., Brooks, R. R., and Reeves, R. D., 1980. Nickel uptake by *Alyssum* species, *Plant Sci. Lett.* 17:451–457.

Neumann, P. M., and Chamel, A., 1986. Comparative phloem mobility of nickel in nonsenescent plants, *Plant Physiol.* 81:689–691.

Pancaro, L., Pelosi, P., Gambi, O. V., and Galloppini, C., 1978. Further contribution on the relationship between nickel and malic and malonic acids in *Alyssum, G. Bot. Ital.* 112:282–283.

Pandolfini, T., Gabbrielli, R., and Comparini, C., 1992. Nickel toxicity and peroxidase activity in seedlings of *Triticum aestivum* L., *Plant Cell Environ.* 15:719–725.

Pelosi, P., Fiorentini, R., and Galoppini, C., 1976. On the nature of nickel compounds in *Alyssum bertolonii* Desv.-II, *Agric. Biol. Chem.* 40:1641–1642.

Rauser, W. E., 1990. Phytochelatins, *Annu. Rev. Biochem.* 59:61–86.

Schlegel, H. G., Cosson, J.-P., and Baker, A. J. M., 1991. Nickel-hyperaccumulating plants provide a niche for nickel-resistant bacteria, *Bot. Acta* 104:18–25.

Severne, B. C., 1974. Nickel accumulation by *Hybanthus floribundus, Nature (London)* 248:807–808.

Severne, B. C., and Brooks, R. R., 1972. A nickel-accumulating plant from Western Australia, *Planta* 103:91–94.

Still, E. B., and Williams, R. J. P., 1980. Potential methods for selective accumulation of nickel(II) by plants, *J. Inorg. Biochem.* 13:35–40.

Swain, D. J., 1955. The trace element content of soils, Commonwealth Bureau of Soil Science, Technical Communication No. 48 HMSO, London.

Theisen, M. O., and Blincoe, C., 1984. Biochemical form of nickel in alfalfa, *J. Inorg. Biochem.* 21:137–146.

Theisen, M. O., and Blincoe, C., 1988. Isolation and partial characterization of nickel complexes in higher plants, *Biol. Trace Elem. Res.* 16:239–251.

Tiffin, L. O., 1971. Translocation of nickel in xylem exudate of plants, *Plant Physiol.* 48:273–277.

Vanselow, A. P., 1966. Nickel, in *Diagnostic Criteria for Plants and Soils* (H. D. Chapman, ed.), University of California Division of Agricultural Sciences, Davis, California, pp. 302–309.

Verkleij, J. A. C., and Schat, H., 1990. Mechanisms of metal tolerance in higher plants, in *Heavy Metal Tolerance in Plants: Evolutionary Aspects* (A. J. Shaw, ed.), CRC Press, Boca Raton, Florida, pp. 179–194.

Walker, C. D., Graham, R. D., Madison, J. T., Cary, E. E., and Welch, R. M., 1985. Effects of Ni deficiency on some nitrogen metabolites in cowpeas (*Vigna unguiculata* L. Walp), *Plant Physiol.* 79:474–479.

Welch, R. M., 1981. The biological significance of nickel, *J. Plant Nutr.* 3:345–356.

Wiersma, D., and Van Goor, B. J., 1979. Chemical forms of nickel and cobalt in phloem of *Ricinus communis, Physiol. Plant.* 45:440–442.

Animal Nickel Metabolism 9

9.1 Introduction

All animals, including humans, are constantly exposed to nickel ion and particulate nickel compounds through the food we eat and the air we breath. A small portion of the ingested nickel is absorbed by cells lining the small intestine, and additional small amounts of nickel are assimilated by pulmonary cells after inhalation. The absorbed nickel ion is systemically transported to all tissues by proteinaceous and low-molecular-weight nickel-binding components in the serum. Although steady-state levels of nickel are fairly uniform in various tissues of the body, when radiolabeled nickel ion is administered to an animal, the metal ion is rapidly accumulated in the kidney. The kidney and urinary tract serve as the major route of elimination for absorbed nickel ion, while nonabsorbed nickel compounds are eliminated in the feces. Cellular internalization of nickel ion can occur by action of metal ion transport proteins, whereas lipophilic nickel complexes appear capable of diffusion into cells, and certain cells can phagocytize nickel particles. Although it remains obscure whether nickel is an essential or even beneficial trace metal ion in humans, low concentrations of nickel do appear to facilitate optimal growth of several animals. The functional roles for nickel in animals, however, are only poorly understood. In contrast, the toxic, carcinogenic, and other harmful effects of certain nickel species have been well documented in various systems. This chapter will describe the metabolic flux of nickel ion in animals, examine the evidence that nickel is essential for animal growth, and detail the harmful effects of nickel compounds on animal cells.

9.2 Flux of Nickel in Animals

Nickel is present in all animals; for example, there is a total of approximately 10 mg of nickel in the human body (Schroeder *et al.,* 1962). The steady-state level of nickel is governed by various components of the nickel flux in the body. Considerations of nickel flux must include the following topics: mechanisms by which nickel enters the body, systemic transport of nickel, tissue distribution of nickel, and mechanisms by which nickel is eliminated from the body. In addition, an important topic related to nickel flux involves cellular internalization processes. An attempt to schematically illustrate the flux of nickel in animals is provided in Fig. 9-1, and the topics listed above are described in detail in the following sections.

9.2.1 Uptake of Nickel from the Environment

Uptake of nickel from the environment includes two distinct steps: entry into the body and subsequent absorption by body tissue. These steps are considered separately below.

For humans and other animals, the major routes of nickel intake include ingestion and inhalation; however, minor amounts of nickel can enter dermally, during hemodialysis, via metal prostheses, and by other routes. Estimates for the total dietary exposure levels of humans include 300–600 μg/day (Sunderman, 1977) or 160 μg/day (Nieboer *et al.,* 1988a), but the level can vary widely depending on the diet. Compared to the ingested nickel levels, the amount of nickel that enters the body by inhalation is usually negligible [estimated at 0.16 μg/day (Nieboer *et al.,* 1988b)]; however, in certain workplace or environmental conditions inhalation of nickel-containing dust or gaseous nickel compounds can be significant. Depending on the particle size, much of the nickel may be cleared from the respiratory tract by mucociliary action followed by ingestion. Alternatively, certain nickel-containing particles tend to be trapped in the alveolar region of the lung or in the tracheobronchial airways. By contrast, nickel carbonyl, a gaseous form of nickel, can freely disperse throughout the entire respiratory system.

After entering the body, only a portion of the ingested or inhaled nickel is absorbed. With regard to ingested nickel, 1–4% is absorbed by humans (Nieboer *et al.,* 1988a), and the remainder is simply eliminated in the feces. The mechanism of absorption of ingested nickel is poorly understood, but it clearly involves transport across the intestinal barrier. Using *in situ* perfusion studies of rat jejunum, Foulkes and McMullen (1986) demonstrated that nickel absorption is biphasic. The first phase involves a saturable step associated with crossing the brush border membrane, whereas the second step involves transfer into the body and is not saturable. Respiratory absorption of insoluble nickel particles such as nickel oxide does not occur significantly over a short time range. These recalcitrant compounds can remain in lung tissues for years and can cause subsequent damage (see Section 9.4). In contrast, soluble nickel

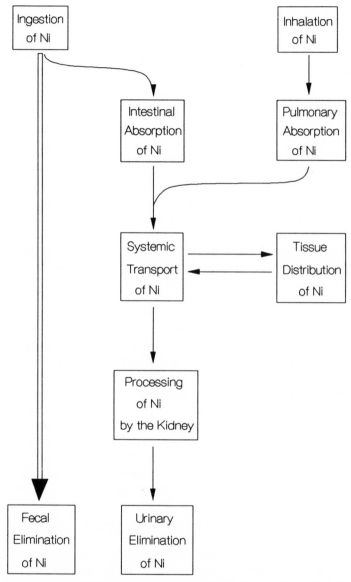

Figure 9-1. Overview of nickel flux in animals. Nickel enters an animal primarily by ingestion or inhalation. A small portion of the ingested nickel is subsequently absorbed in the intestines, whereas most is eliminated in the feces. Gaseous nickel compounds, including the highly toxic substance nickel carbonyl, or aerosols containing soluble nickel salts can be absorbed directly by the lungs, whereas phagocytic processes are required for uptake of particulate nickel compounds in the respiratory tract. The absorbed nickel is systemically transported throughout an animal while bound to proteins and small molecules in the bloodstream. Although all tissues possess a steady-state nickel concentration, the absorbed metal ion is concentrated in kidney tissue and eliminated from the body via the urinary tract.

compounds and the gaseous substance nickel carbonyl are absorbed very rapidly, although the detailed mechanisms for absorption are not completely understood. Whether ingested or inhaled, absorbed nickel compounds are distributed to all body tissues by the bloodstream.

9.2.2 Systemic Transport of Nickel

Upon entry into the bloodstream, nickel ion is bound to specific serum components and rapidly distributed throughout an animal [reviewed by Sarkar (1984)]. The steady-state levels of nickel in the bloodstream will vary with the animal species, the diet, and the environmental conditions, but, as one example, the concentration of nickel in whole blood and serum of humans is 0.34 ± 0.28 and 0.28 ± 0.24 μg/liter, respectively (Nieboer et al., 1992). [These values recently were revised downward from previous, less reliable estimates that included nickel derived from the use of nickel-contaminated heparin during sample collection. For example, a large body of literature quotes human serum levels of 2.6 μg/liter.]

One of the earliest studies to characterize substances that are involved in systemic transport of nickel utilized rabbit serum because it possesses a higher nickel concentration than serum of other species. Nomoto et al. (1971) demonstrated the presence of three forms of nickel in rabbit serum: 16% is present as ultrafiltrable material, 40% is associated with albumin, and 44% is associated with a large protein they termed "nickeloplasmin." These ratios are not uniform across species and are likely to vary with the status of the animal and the level of nickel exposure. For example, Nomoto and Sunderman (1988) found that the same three fractions were present in human serum in relative proportions of 40%, 34%, and 26%, whereas Lucassen and Sarkar (1979) added radiolabeled nickel ion to human serum and found 4.2% associated with the low-molecular-weight fraction, 95.7% associated with albumin, and less than 0.1% associated with the large protein fraction. Additional nickel-binding components may also be present in some species. The following paragraphs describe the properties of the nickel-binding fractions in more detail.

The predominant low-molecular-weight form of nickel in serum is a complex of nickel with the free amino acid L-histidine (Lucassen and Sarkar, 1979). Histidine has a great affinity for nickel at the pH of serum and is thought to form the Ni(His)$_2$ complex shown in Fig. 9-2. Although nickel ion can complex weakly with many other free amino acids, cysteine is probably the only other amino acid that contributes significantly to nickel transport in serum. Jones et al. (1980) have used computer simulations to characterize the predominant low-molecular-weight species in serum and have suggested

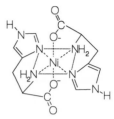

Figure 9-2. Structure of the Ni(His)$_2$ complex, a major form of nickel in the serum.

that over 40% of the nickel may be present as the Ni(His)(Cys) complex. No experimental data have been published to support this assertion.

The most important nickel-binding protein in serum is albumin. The metal ion binds in a 1:1 complex to the amino terminus of this highly abundant blood protein, which is also known to bind copper ion (Tsangaris *et al.*, 1969). In the case of human albumin, nickel binds with an association constant of $3.7 \times 10^9 \ M^{-1}$ and generates an absorbance with a maximum at 420 nm and a shoulder at 450–480 nm, consistent with square-planar or square-pyramidal coordination geometry (Glennon and Sarkar, 1982a). Laurie and Pratt (1986) suggested on the basis of UV-visible and circular dichroism spectroscopic studies of bovine serum albumin that the nickel–protein complex exists as an equilibrium mixture of square-planar and octahedral species, in a ratio of 70:30 at pH 7.4. They suggested that low pH increases the concentration of the square-planar species. In contrast, ^{13}C- and ^{1}H-nuclear magnetic resonance (NMR) studies of the binding of nickel to the 24-residue peptide from the amino terminus of the human protein demonstrated that solely the square-pyramidal structure shown in Fig. 9-3 is formed (Laussac and Sarkar, 1984). Ligands to nickel include the amino-terminal aspartyl α-NH$_2$, the deprotonated peptide nitrogens from the second and third residues (alanine and histidine), the imidazole $N(1)$ nitrogen of the third residue, and the carboxyl group from the amino-terminal aspartyl residue. The four nitrogens are thought to coordinate in a planar arrangement with the carboxyl group coordinating axially. The complex is only stable at neutral to high pH, presumably because of the requirement for deprotonated amide nitrogens.

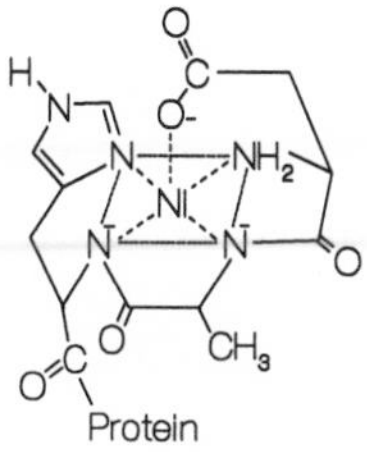

Figure 9-3. Structure of the nickel site in human serum albumin. Albumin serves as the major nickel-binding protein in the serum. The amino-terminal α-NH$_2$, an imidazole nitrogen, and two deprotonated peptide nitrogens bind nickel ion in a plane. The aspartyl carboxyl group supplies the axial ligand in the square-pyramidal structure (Laussac and Sarkar, 1984).

Nickel can exchange between albumin and free histidine via a ternary complex that binds the metal ion very tightly (association constant of 1.7 × 10^{16} M^{-1}; Glennon and Sarkar, 1982a). Using L-histidine and the tripeptide L-aspartyl-L-alanyl-L-histidine-N-methylamide, Tabata and Sarkar (1992) studied the kinetics and mechanism of nickel transfer between these two ligands. They demonstrated that the rate-limiting step in nickel transfer from peptide to free amino acid is the protonation of a peptide nitrogen (Fig. 9-4). In contrast, the rate-determining step in transfer from the amino acid to peptide is the formation of the ternary complex, requiring deprotonation of a peptide nitrogen (Fig. 9-4). The overall equilibrium constant favors the histidine complex over the peptide complex by a factor of nearly 2000 (Tabata and Sarkar, 1992). Although copper also forms complexes with both histidine and serum albumin, the equilibrium constant only favors the amino acid complex by sixfold. Differences in the relative affinities for free amino acid versus protein of the two metal ions may relate to their distinct tissue distributions and clearance rates. These processes for nickel exchange may be relevant to nickel exchange reactions associated with cellular uptake processes.

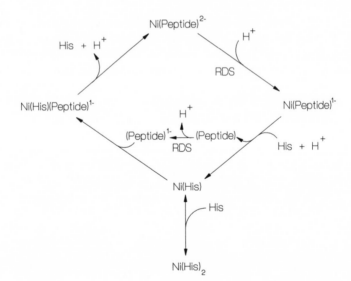

Figure 9-4. Mechanisms of nickel exchange between albumin and L-histidine. Using L-histidine and a tripeptide that corresponds to the amino terminus of human serum albumin, Tabata and Sarkar (1992) characterized the kinetics and mechanism of transfer between these species. The rate-determining step (RDS) for both exchange reactions involves a proton transfer. These reactions may have significance with regard to other nickel exchange reactions in an animal, for example, the exchange from the bound form in the serum to a cellular nickel transport protein.

In contrast to the serum albumin from human, bovine, rabbit, rat, and several other sources, the proteins isolated from canine, porcine, and chicken sources fail to exhibit tight binding of nickel ion. Glennon and Sarkar (1982b) demonstrated that the substitution of tyrosine for histidine at residue 3 is responsible for the low affinity of dog albumin for nickel ion. Decock-Le Reverend et al. (1987) used ^{1}H-NMR spectroscopy to characterize the weak nickel binding site present in a 24-residue amino-terminal peptide isolated from dog albumin. They showed that nickel is coordinated by the amino-terminal $-NH_2$ and deprotonated amide nitrogens from residues 2, 3, and 4 (alanine, tyrosine, and lysine). There was no evidence for coordination by the phenolate oxygen of tyrosine. The porcine protein also has a tyrosine substitution for histidine and is likely to bind nickel in a fashion similar to that for dog albumin. In contrast, the avian protein possesses a glutamic acid residue inserted prior to the histidine. The shift in position of histidine to residue 4 apparently precludes the chicken albumin from binding nickel (Predki et al., 1992b).

The high-molecular-weight nickeloplasmin serum protein has been identified as α_2-macroglobulin (Nomoto et al., 1971; Nomoto and Sunderman, 1988). This glycoprotein, reported to bind 0.8 nickel ion per tetrameric protein molecule (M_r 720,000), is probably equivalent to the serum fraction that Himmelhoch et al. (1966) earlier found to tightly bind nickel. α_2-Macroglobulin is the major zinc-binding protein in the serum, and it is possible that the nickel is bound at the zinc binding site. The inability to remove nickel from holoprotein by dialysis and the failure of apoprotein to bind nickel demonstrates that this metal ion is not in a simple equilibrium with the protein. Rather, these data are consistent with the binding site being buried in the protein. The coordination geometry of nickel bound to α_2-macroglobulin has not been reported.

Less well characterized proteins also may contribute to nickel transport in serum. For example, Morgan (1981) has demonstrated that a histidine-rich glycoprotein can reversibly bind nickel. This protein (M_r 58,000) contains 10% histidine, an excellent chelator of metal ions. In vitro binding studies indicated that each protein molecule can bind 9.5 nickel atoms with a dissociation constant of 1.3 μM. This protein also binds heme, copper, zinc, and other metals and may serve as a general carrier of metal ions. Its contribution to nickel transport in serum is unknown.

9.2.3 Tissue Distribution of Nickel

Nickel that is present in serum is rapidly distributed throughout the body. The concentrations of nickel in various animal tissues have been assessed by numerous investigators [e.g., the results of several studies were summarized by Nielsen (1986)]; however, appropriate precautions to avoid contamination have not always been taken, and optimal quantitative methods were not always used. As an example of a well carried out analysis, Rezuke *et al.* (1987) measured nickel concentrations in adult human postmortem tissues using electrothermal atomic absorption spectrophotometric methods (Table 9-1). Average nickel concentrations in various organs are rather uniform, with some elevation observed in lung, thyroid, and adrenal tissues. In general, the concentrations found in individual tissues across other species fall within an order of magnitude of each other (Nielsen, 1986).

The tissue distribution of nickel may change depending on the status of the animal. An especially intriguing finding related to this is the dramatic increase in human serum levels of nickel following a myocardial infarction. D'Alonzo and Pell (1963) were the first to observe this phenomenon, with 19 of 20 patients possessing elevated serum nickel concentrations. In a more extensive study, Sunderman *et al.* (1970) verified the presence of increased serum nickel concentrations in 73 patients who had suffered a myocardial infarction and found elevated levels (defined in that study as more than 4.2 μg/liter) of the metal ion up to five days later in some cases. Because the serum nickel concentrations do not correlate with levels of creatine kinase or cardiac lactate dehydrogenase and because heart tissue does not possess sufficient quantities of nickel to account for the elevations observed in the serum, Sunderman (1977) suggested that the hypernickelemia arises from leukocytosis and leukocytolysis, rather than from direct cardiac damage. The results of McNeeley *et al.* (1971) showing nickel elevation in patients who had suffered

Table 9.1. Human Tissue Distribution of Nickel

Tissue	Nickel concentration (μg/kg dry weight)[a]	
	Mean ± SD	Median (range)
Lung	173 ± 94	130 (71–371)
Thyroid	141 ± 83	126 (41–240)
Adrenal	132 ± 84	126 (53–341)
Kidney	62 ± 43	54 (19–171)
Heart	54 ± 40	51 (10–110)
Liver	50 ± 31	38 (11–102)
Brain	44 ± 16	51 (20–65)
Spleen	37 ± 31	21 (9–95)
Pancreas	34 ± 25	37 (7–71)

[a] From Rezuke *et al.* (1987).

a severe stroke or extensive external burns appear to be consistent with this view regarding the source of nickel. Whether the nickel that is endogenous to or released from leukocytes has any physiological role is unknown.

In addition to assessing the steady-state concentrations of nickel in various organs, it is important to characterize the flux of the metal ion through these tissues. To elucidate the spatial and temporal distribution of exogenously added nickel in animals, researchers have typically administered trace levels of ^{63}Ni-labeled nickel salts and monitored the tissue levels of radioactivity over time. Nickel ion is rapidly excreted from the body, but it does exhibit transient accumulations in specific tissues. The results from several early studies in mouse, rat, rabbit, and guinea pig (Wase *et al.*, 1954; Smith and Hackley, 1968; Sarkar, 1980; Parker and Sunderman, 1974) are consistent with numerous more recent studies [summarized by Coogan *et al.* (1989)] that all show a rapid localization of nickel ion to the kidney. In addition, substantial concentrations of nickel are found often in the lung and less often in other tissues. Importantly, nickel ion appears to be readily transported across the placenta, and significant concentrations of nickel have been observed in the developing fetus [reviewed by Coogan *et al.* (1989)]. The tissue distribution in animals provided with ^{63}Ni does not appear to depend significantly on the method or level of nickel administration.

9.2.4 Elimination of Nickel

Most ingested nickel is not absorbed by the body, but rather it is eliminated in the feces. Absorbed nickel, in contrast, is eliminated predominantly in urine with some loss of the metal ion in sweat, bile, saliva, and hair. Sunderman (1977) summarized the results from several human studies that demonstrated a mean nickel elimination in feces of 258 µg/day compared to the value for urine of 2.6 µg/day. Profuse sweating may at times account for a major route of nickel elimination as shown by nickel concentrations of 52 µg/liter for sweat versus 2.2 µg/liter for urine. Onkelinx and Sunderman (1980) summarized a series of animal studies that monitored the kinetics of nickel disappearance from plasma. They demonstrated that the disappearance curve could be fit to a two-compartment model where nickel is eliminated from compartment 1, which includes all serum nickel-binding species, and this pool is in equilibrium with a second compartment, comprised of the remaining tissues (Fig. 9-5). More recently, this two-compartment model has been shown to hold in humans (Sunderman, 1992). Thus, elimination of nickel from the body is determined by the first-order rate constant for elimination and the serum nickel concentration.

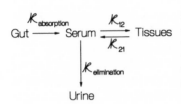

Figure 9-5. Two-compartment model governing the kinetics of nickel disappearance from plasma. The rate of nickel elimination from the body is governed by k_{elim}, the first-order rate constant for nickel elimination, and the serum nickel concentration. The latter concentration depends on the rate of absorption, $k_{absorption}$, and the rates for exchange with tissues, k_{12} and k_{21}.

In the kidney, low-molecular-weight complexes of nickel are filtered through the glomerulus, and much of the nickel is subsequently reabsorbed in the tubules [reviewed by Templeton (1992)]. Ligand exchange is thought to occur concomitantly with glomerular filtration and/or tubule reabsorption. The discussion below focuses on the mechanisms of nickel elimination via urine with particular emphasis on the nickel-binding components in kidney.

Sunderman *et al.* (1981, 1983) demonstrated that most of the nickel localized in the rat kidney is associated with a low-molecular-weight fraction, although significant levels of radioactivity are associated with at least five soluble, presumably cytoplasmic, proteins. To date, there is no evidence that the proteins play a role in urinary excretion of nickel; however, it is appealing to consider that these proteins may transfer the metal ion to, or be precursors of, the small complexes. Characterization of the kidney-derived nickel-binding proteins has been limited to estimation of their apparent sizes (M_rs of 168,000, 84,000, 51,000, 24,000, and 10,000). In contrast, the low-molecular-weight fraction has been extensively studied, as described below, and appears to include nickel complexes that are very similar in structure to those found in urine (Predki *et al.*, 1992a).

The structural properties of the major, low-molecular-weight nickel-binding fraction in kidney have been described from several species. Although Sunderman *et al.* (1981, 1983) estimated that the predominant nickel-binding fraction in rat kidney possesses an M_r of <2000, Abdulwajid and Sarkar (1983) claimed that the major form of bound nickel was associated with a much larger glycopeptide (M_r 15,000–16,000). This glycine- and proline-rich protein was purified, and the carbohydrate was shown to be of the high-mannose type. The authors speculated that the protein may be derived from the renal basement membrane. The presence of this larger peptide in rat or other kidney tissues has not been substantiated by others. The same laboratory later showed that the low-molecular-weight species from several sources was comprised of two components, a peptide of approximately 35 residues and a sulfated oligosaccharide fraction containing uronic acid and neutral sugars (Templeton and Sarkar, 1985). The human peptide was purified and shown to bind one nickel ion per M_r 3500 with a dissociation constant of 11 μM (Templeton and Sarkar, 1986). Importantly, Predki *et al.* (1992a) demonstrated that the peptide was present both intracellularly and extracellularly, using a porcine proximal tubule-like cell line. This result, and the demonstration of acidic nickel-binding peptides in the urine, is consistent with a role for this component in urinary excretion. By contrast, the oligosaccharides only weakly bind nickel and are present only intracellularly in the renal cell line.

9.2.5 Cellular Internalization of Nickel

Animal cells appear to be capable of internalizing nickel by three distinct general mechanisms: uptake via metal ion transport systems, diffusion of lipophilic nickel compounds through the membrane, and phagocytosis. Each of these processes, illustrated in Fig. 9-6, is described in more detail in the following paragraphs.

A major route for cellular uptake of nickel by microbial cells involves the action of metal ion transport systems, especially those that function in magnesium transport (see Chapter 7). Animal cells also are known to possess magnesium transport systems [reviewed by Romani and Scarpa (1992)], and nickel ion may be able to substitute for the magnesium ion in these systems. For example, the findings of Luo *et al.* (1993) that magnesium deprivation enhances and magnesium supplementation diminishes nickel-induced embryotoxicity and teratogenicity in frog embryos are consistent with competition

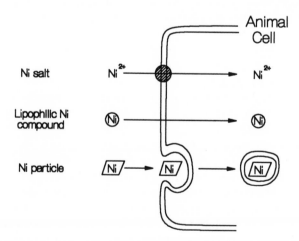

Figure 9-6. Mechanisms of nickel internalization by animal cells. Nickel can enter cells by three processes: divalent cation uptake via membrane-associated metal ion transport systems, simple diffusion of lipophilic nickel compounds, and engulfment processes termed phagocytosis (for large particles) or pinocytosis (for small particles).

of these metal ions for the same transporter system. Unlike the microbial case, however, magnesium ion influx is coupled to sodium ion efflux in animal cells, and nickel ion competition has not been carefully examined. Subcellular nickel ion transport also is likely to occur by metal ion transport systems in animal cells. As one example, nickel appears to substitute for sodium ion in a sodium–calcium exchange process in sarcolemmal vesicles of the bullfrog ventricle (Brommundt and Kavaler, 1987). No studies have directly examined intracellular nickel ion transport processes. Returning to the question of whole-cell internalization of nickel, Nieboer *et al.* (1984a) observed divalent nickel ion uptake by human lymphocytes, B lymphoblasts, and macrophages and demonstrated that the uptake process can be inhibited by complexing agents (EDTA, L-histidine, serum albumin, D-penicillamine) that sequester the metal ion. These data are consistent with uptake by a metal ion transport system. Kinetic characterization of these transport processes and examination of competition by other metal ions are important experiments that remain to be done.

In the case of lipophilic nickel compounds, the metal ion can enter the cell by simple diffusion. This system appears to be operative in the case of nickel carbonyl, a gaseous form of nickel. In addition, Menon and Nieboer (1986) have shown that nickel chelates of lipophilic ligands such as diethyl-dithiocarbamate and pyrrolidinedithiocarbamate can diffuse into human peripheral mononuclear leukocytes. In contrast, the nonlipophilic ligand dithiocarbamate does not promote nickel uptake into the cells.

Particulate nickel compounds can be taken up by certain cells via engulfment processes termed phagocytosis (for large particles) or pinocytosis (for tiny particles). These processes are of special importance in the respiratory tract, where cells may encounter particulate nickel compounds from inhaled dust. A phagocytic mechanism was first noted for nickel subsulfide particles by Costa and Mollenhauer (1980a,b), using cultured Syrian hamster embryo cells and Chinese hamster ovary cells. Numerous investigations in a range of cell types (e.g., Costa *et al.*, 1981; Abbracchio *et al.*, 1982; Kuehn *et al.*, 1982; Heck and Costa, 1983; Nieboer *et al.*, 1984b) have subsequently demonstrated that uptake of various types of nickel-containing particles is dependent on the degree of crystallinity, the size, and the charge of the particle. Recently, uptake studies of two nickel sulfides (Ni_3S_2 and NiS) by guinea pig alveolar macrophages (Shirali *et al.*, 1991) and rat lymphocytes (Hildebrand *et al.*, 1991) have revealed that the internalized particles have lost sulfur and replaced it, at least in part, with phosphorus. The internalized particles are bound to numerous cellular organelles including mitochondria, endoplasmic reticulum, Golgi vesicles, nuclear membranes, and the euchromatin. The importance of internalized nickel particles will become clear in Section 9.4.5, concerned with the carcinogenicity of certain nickel compounds.

The ability of animal cells to internalize nickel ion and nickel compounds naturally leads to questions concerning the beneficial and harmful effects of nickel after entering a cell. These topics are discussed in Sections 9.3 and 9.4.

9.3 Essentiality and Functional Roles of Nickel in Animals

Although nickel is found in all animal tissues, the importance and possible roles of this metal ion remain unclear for most species. Because nickel is a ubiquitous environmental contaminant, it is difficult to achieve true nickel deficiency in animal studies. Interpretation of the effects of nickel-depleted growth conditions can be clouded by secondary effects; for example, efforts to remove nickel from nutrient sources may alter the concentrations of other metal ions. Furthermore, low nickel concentrations may lead to subtle ancillary influences in an animal; for example, nickel deficiency has been suggested to decrease iron uptake in rats (Kirchgessner and Schnegg, 1980). Nielsen *et al.* (1984) have suggested that test animals in many nickel-deficiency experiments may have been iron-deficient, whereas the control animals in these studies may have been partially spared from iron deficiency by the supplemental nickel that was provided. Thus, apparent requirements for nickel may be artifactual in some cases. Despite these caveats, evidence has accumulated from several laboratories that strongly implicate nickel as a beneficial trace element in animals. Furthermore, in some species the function of nickel has been at least partially identified.

Below, I discuss evidence that nickel probably plays an important role in chickens, monogastric mammals, ruminants, and invertebrates. Unfortunately, information about the importance of nickel in other animals has not been reported. At the end of this section, I provide speculations about possible functions for nickel in these species. Additional details concerning the essentiality of nickel in animals can be found in reviews by Anke *et al.* (1983, 1984), Nielsen (1984, 1986), and Spears (1984).

9.3.1 Evidence That Nickel Is Required in Chickens

The first studies to provide evidence of beneficial effects of nickel for animals were carried out in chickens. Nielsen and Sauberlich (1970) provided chicks with a diet containing less than 80 ppb nickel and noted pigmentation changes and unusual swelling in the legs. Subsequent work by the same laboratory (Nielsen, 1971) reported additional manifestations of nickel deficiency such as a dermatitis on the shank skin, a change in texture and color of the liver, and a reduction in ether-extractable lipids. Following up these initial studies, Sunderman *et al.* (1972) provided chicks with diets containing 44 ppb nickel. These investigators were unable to reproduce any of the gross effects previously described in chicks; however, they did observe an ultrastructural change in the liver involving dilation of the perimitochondrial rough endoplasmic reticulum.

In further studies by Nielsen and Ollerich (1974), the nickel concentration in the diet was reduced to 3–4 ppb. This work allowed the researchers to conclude that several of the gross changes observed earlier were artifactual; however, they did continue to note changes in shank skin pigmentation. The ultrastructural changes in liver mitochondria noted by Sunderman *et al.* (1972) were confirmed, and intracellular structural changes were extended to include dilation of the perinuclear space and condensation of chromatin and pyknotic nuclei. More importantly, several biochemical changes (oxygen uptake, lipid content, and phospholipid content) in the livers were noted. Although the observed changes were very small and could perhaps be partially attributed to changes in iron status, they were reproducible in other experiments described in that paper and in subsequent studies (Nielsen *et al.,* 1975a). The authors concluded that the results were significant and consistent with an essential role for nickel in chickens.

9.3.2 Evidence That Nickel Is Required in Monogastric Mammals

The monogastric mammal of most interest to the reader is the human animal; however, no information is available regarding the requirements or roles for nickel in this species. Studies have been reported for rats and miniature pigs. One important feature of these mammalian studies is that some experiments have included raising multiple generations under nickel-deficient conditions, whereas in the previous studies with chicks the animals were reared on low-nickel rations for only a few weeks. It is possible that information obtained from the rat and pig studies concerning the essentiality of nickel may be applicable to the human case.

The first report demonstrating an effect of nickel deficiency in rats was that of Nielsen and Ollerich (1974). Several other laboratories previously had searched in vain for changes in rats arising from nickel depletion, but the latter investigators used plastic cages and diets containing very low nickel concentrations to succeed in this quest. They noted a reduced oxygen uptake by liver homogenates provided with α-glycerolphosphate (4.17 μl O_2 h^{-1} mg^{-1} for nickel-supplemented vs. 3.20 μl O_2 h^{-1} mg^{-1} for nickel-depleted animals) and a shift from polysomes to monosomes in liver cells. In addition, female rats raised under nickel-deficient conditions were found to lose 15% of their young around the time of birth, compared to no losses in the control group.

Extending their previous work, Nielsen *et al.* (1975b) raised several generations of rats under nickel-depleted conditions and observed a series of changes in the animals. These effects included increased mortality at birth, decreased weight, alterations in hair development, altered appearance of the liver, increased rates of oxidation by liver extracts (note that this is opposite to the finding of the earlier report), decreased cholesterol levels, and ultrastructural changes in the liver rough endoplasmic reticulum. The opposite effects that were seen in the two studies for oxidative rates of liver homogenates illustrate the complications that arise in identifying effects due to nickel deficiency. Other subtle environmental effects, especially relative to iron nutritional status, may confound interpretations related to nickel requirements. Nevertheless, the authors suggested that nickel deficiency does exhibit reproducible effects in rats consistent with a functional role for this metal ion.

The proposal that nickel has a functional role in rats has been buttressed by work in another laboratory. Nickel depletion reproducibly led to negative effects on animal weight in several studies by Kirchgessner and Schnegg, as summarized by these authors in 1980. At a biochemical level, Kirchgessner and Schnegg (1976) demonstrated that the malate dehydrogenase level in nickel-deficient rats was only two-thirds of that in nickel-replete animals. Even more significantly, they found that glucose-6-phosphate dehydrogenase levels were markedly reduced in the male F_1 generation (38 mU/mg protein in nickel-deficient versus 209 mU/mg protein in control animals). Later studies by these same authors demonstrated significant changes in activity levels for a wide variety of enzymes [reviewed by Kirchgessner and Schnegg (1980)]. A key enzyme that exhibited reduced levels was α-amylase, leading the investigators to propose that the global effects may arise from an overall decrease in energy efficiency due to inadequate levels of starch hydrolysis. An additional important contribution by these authors was the observation that nickel deficiency led to changes in iron, copper, and zinc levels in several tissues. The iron concentration appeared to be affected the most, consistent with a reduction in iron absorption under low-nickel-status conditions. In sum, the data are reasonably compelling that nickel has an important role in rats.

Anke *et al.* (1984) summarized the results of several studies showing an effect of nickel deficiency in miniature pigs. In these studies, the nickel-deficient mothers were provided with food containing nickel at a dry weight concentration of 100 ppb. Differences in the growth rates of nickel-deprived and control pigs could only be observed in the second-generation animals. For example, the live weights at days 1, 14, and 28 for nickel-depleted piglets (0.358, 1.070, and 1.986 kg) were significantly reduced compared to those of the controls (0.375, 1.277, and 2.279 kg). Furthermore, the rates of survival of nickel-deficient piglets were significantly reduced compared to those of control animals (3.1 vs. 6.3 piglets per sow at 56 days). With regard to effects on metabolism of other metal ions, decreases in the concentrations of calcium and zinc were observed in the low-nickel animals.

9.3.3 Evidence That Nickel Is Required in Ruminants

Ruminants (e.g., sheep, goats, and cows) possess a forestomach containing a rich population of microorganisms that are important in fermentative metabolism of feedstock. The following paragraphs review the evidence that nickel is important to each of these mammals. This work identifies at least one potentially important site of action for this metal ion in certain animals.

Several studies have examined the effects of nickel depletion on lamb development. In early work, Spears et al. (1978) were unable to demonstrate an effect on weight gain for animals provided with a nickel-depleted diet (containing 65 ppb nickel) compared to a control group; however, serum protein and alanine transaminase levels were found to be depressed. Furthermore, when lambs were provided with diet that was low in protein, Spears et al. (1979) were able to observe clear differences in the rate of development (0.135 vs. 0.179 kg/day) for unsupplemented animals versus animals supplemented with 5 ppm nickel. The growth effects appeared to correlate with activity levels of ruminal urease, an enzyme that had been shown to require nickel (Spears et al., 1977; Spears and Hatfield, 1978; Chapter 3).

Analogous to the case of lambs, several effects of nickel deficiency have been demonstrated in goats provided with food containing nickel at a dry weight concentration of 100 ppb. Anke et al. (1980, 1984) summarized the results of long-term goat studies that found several small, but statistically significant, nickel-dependent changes: compared to control animals, the second-generation goats provided with a low-nickel diet exhibited growth retardation (2.8 kg compared to 3.2 kg at birth, and 15.8 kg compared to 19.9 kg at 56 days), increased mortality (0.5 kid versus 0.9 kid per mother), and other effects (e.g., depressed levels of blood hemoglobin and hematocrit in lactating animals). Furthermore, decreased ruminal ammonia concentrations in the low-nickel animals are consistent with reduced urease levels. Surprisingly, the nickel concentrations in the test and control kids were not significantly different in the various organs tested. In contrast, zinc levels were significantly less in the skeletal ash of nickel-deficient goats. Furthermore, iron absorption appeared to be decreased in goats provided with limited nickel concentrations. Thus, it appears that nickel deficiency in goats may interfere with normal rumen metabolism and with utilization of other metal ions.

The benefits of nickel supplementation to ruminants may be a general phenomenon, especially for animals provided with a low-protein diet. For example, Spears et al. (1979) noted that nickel supplementation increased weight gain along with increasing urease activity levels in steers. How nickel enhancement of rumen urease activity may benefit the animal is described in Section 9.3.5.

9.3.4 Evidence That Nickel Is Required in Invertebrates

By careful control of experimental growth conditions, several laboratories have obtained evidence consistent with a nickel requirement in chicks, rats, pigs, lambs, goats, and steers; however, similar nickel-deficiency experiments have not been carried out with any of the other many types of animals. Nevertheless, a reasonable case can be made that nickel plays a role in at least two invertebrates.

Urease is known to be present in selected invertebrates (Barnes and Crossland, 1976; Cooley *et al.,* 1976; McDonald *et al.,* 1980; Simmons, 1961). In the case of the land snail *Otala lactea,* McDonald *et al.* (1980) partially purified the urease from hepatopancreas tissue and showed that it, like all other ureases (Chapter 3), contains nickel. The role of urease in these organisms is not clear, but it may play an important role in nitrogen metabolism if a major flux of nitrogen goes through urea. Nickel-deficiency experiments need to be carried out to demonstrate whether nickel has an essential role in these invertebrates.

Tunicates, marine organisms that sometimes are termed sea slugs, are known to concentrate a wide variety of metals. Rayner-Canham *et al.* (1985) demonstrated that at least one of these species, *Halocynthia pyriformis,* possesses rather high levels of nickel and speculated that the metal may be essential for its growth. The form of nickel in this organism is unknown; however, a nickel-containing chlorin was shown by Bible *et al.* (1988) to be synthesized in *Trididemnum solidum.* Purification and structural characterization of the compound, designated tunichlorin, revealed a 2-devinyl-2-hydroxymethyl-pyropheophorbide *a* composition as illustrated in Fig. 9-7. This compound is present at very low concentrations, estimated to be $10^{-5}\%$ of the animal. Because the tunicate is associated with symbiotic algae, it is unclear whether the nickel chlorin is synthesized by the animal or the algal partner. Furthermore, the functional role of the compound is unknown.

Turning away from the sparse information available regarding nickel in invertebrates, the following section will return to birds and mammals to consider further the possible roles of nickel in these animals.

9.3.5 Possible Roles for Nickel in Animals

Nickel deficiency clearly leads to a variety of effects in several animals consistent with an important, and perhaps essential, role for this metal ion.

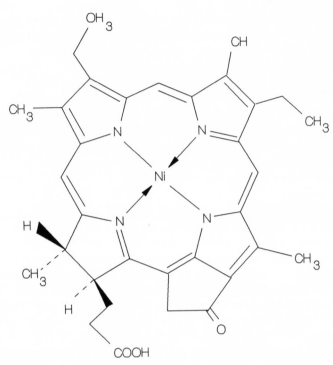

Figure 9-7. Structure of tunichlorin isolated from the tunicate *Trididemnum solidum* (Bible *et al.*, 1988).

Anke *et al.* (1983, 1984) suggested that the nickel requirement for chicks and rats must be more than 50 μg/kg of dry food, whereas for pigs and ruminants more than 100 μg/kg is required. In all cases, including humans, the requirements are probably less than 500 μg/kg of dry feed. Ruminants consistently exhibit a higher nickel requirement than monogastric mammals, leading to the following suggestion for a urease-related role for nickel in these animals.

A major site of action for nickel in ruminants appears to be the nickel-containing enzyme urease (see Chapter 3). Urease is thought to play a key role in nitrogen cycling in ruminants by allowing the microbes in the fore-stomach to utilize a waste product of the animal as a nitrogen source for growth. Enhanced metabolism in the rumen directly benefits the animal by increasing the production of volatile fatty acids, the substrates used by the ruminants for respiration. Although monogastric animals depend to a lesser extent on nitrogen cycling than the ruminants, ureolytic microbes are present in our intestinal tracts (Suzuki *et al.*, 1979; Wozney *et al.*, 1977), and nickel may exhibit some of its benefits by stimulating urease activity. However, depressed rates of ureolysis are unlikely to explain all of the effects seen during nickel deficiency in animals.

Additional roles for nickel in animals may be related to some of the other nickel-containing enzymes that have already been described: hydrogenase (Chapter 4), carbon monoxide dehydrogenase (Chapter 5), and methyl coenzyme M reductase (Chapter 6). Although animal tissues do not possess these enzymes, they are important to many of the microorganisms inhabiting animal intestinal tracts. Complex food webs are present in the rumen and intestinal tracts of nonruminants. For example, the fermentative production of hydrogen gas by some microbes is intimately linked to the consumption of hydrogen by methanogenic and acetogenic bacteria (Fig. 9-8). Thus, it is conceivable that interruption of the key nickel-dependent enzymes in any of these types of microorganisms could lead to greatly altered digestive processes in nickel-depleted animals. No studies have described the effects of nickel deficiency on levels of hydrogenase, CO dehydrogenase, and methyl coenzyme M reductase in whole animals. Furthermore, the changes in metabolism arising from changes in intestinal microbial populations due to nickel depletion have not been examined.

Nickel may additionally play a role in animals by interacting with enzymes that are normally not regarded as nickel-dependent. For example, at low concentrations nickel ion can replace magnesium ion in the formation of C3 convertase of human complement, resulting in a much more stable enzyme complex (Fishelson and Müller-Eberhard, 1982; Fishelson et al., 1982). Similarly, nickel ion interacts in an intriguing manner with calcineurin: this protein is normally a calcium ion- and calmodulin-dependent phosphoprotein phosphatase, but it can be activated by nickel ion alone at very low concentrations (King and Huang, 1983). In addition, nickel-specific activation of phosphatidylserine biosynthesis has been described in rat brain microsomes (Pullarkat et al., 1981). Whereas phosphatidylserine normally is formed by an energy-independent and calcium-requiring reaction, the nickel-dependent synthesis is thought to additionally require ATP and is suggested to proceed through a pyrophosphatidic acid intermediate. Activation or inhibition of a large range of other enzymes by nickel ion has been documented (Nieboer et al., 1984b). On the basis of recent studies that suggest an interaction of nickel with propionate metabolism, Nielsen (1991) hypothesized that nickel deficiency may affect propionyl-CoA carboxylase. The in vivo importance of these nickel–enzyme interactions is unknown.

Finally, it is important to consider that nickel may play an essential role in animals by a mechanism other than direct enzyme interaction. For example, nickel could stabilize subcellular organelles, certain important membrane components, or required structural proteins. Alternatively, it could facilitate transport of an essential nutrient across the intestinal epithelium or into a

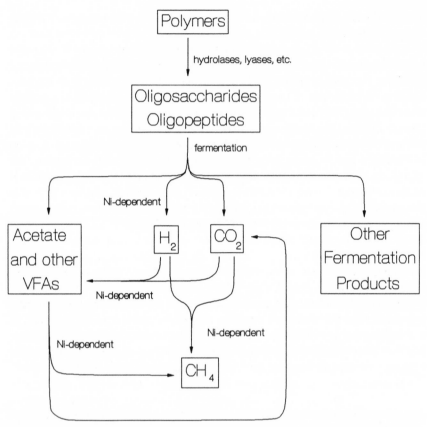

Figure 9-8. Nickel-dependent processes involved in rumen and intestinal metabolism of nutrients. Symbiotic associations between bacterial populations in the rumen or gut rely on interspecies hydrogen transfer involving the generation of H_2 in certain species and consumption of H_2 in others. For example, the uptake of hydrogen can occur in methanogenic or acetogenic microorganisms. The metabolism in these microbes requires the functioning of hydrogenases, methyl coenzyme M reductase, and CO dehydrogenase; all of these proteins are nickel-dependent enzymes. (VFA is the abbreviation for volatile fatty acids.)

cell. In addition, it could interact with polynucleotide components or transcriptional regulatory proteins in the nucleus to act at the genetic level. Clearly, elucidation of the site of action for nickel in animals is an important area of future investigation.

9.4 Harmful Effects of Nickel in Animals

The harmful effects of nickel toward humans and other animals have been the subject of entire books [e.g., *Nickel Toxicology* (Brown and Sunderman, 1980), *Nickel in the Human Environment* (Sunderman *et al.*, 1984), *Progress in Nickel Toxicology* (Brown and Sunderman, 1985), and *Nickel and Human Health: Current Perspectives* (Nieboer and Nriagu, 1992)] and several excellent recent reviews (e.g., Nieboer *et al.*, 1988b; Coogan *et al.*, 1989; Sunderman, 1989a; Christie and Katsifis, 1990; Costa, 1991; Snow, 1992). The following sections cannot adequately cover all aspects of such an important topic, and the reader is directed toward the above references and other authoritative sources (provided below) for more detailed coverage. I will provide only an overview of this area with a focus from a biochemical perspective.

To assist the reader, I have divided discussion of this huge field into six sections. First, I describe the current understanding of the molecular events giving rise to nickel-associated contact dermatitis and other immunological effects. Second, I consider toxic effects of various soluble nickel salts and of the volatile species, nickel carbonyl, on renal, hepatic, and pulmonary tissues. Third, I summarize the ability of nickel compounds to interfere with proper development of the embryo and fetus. Fourth, I provide a synopsis of the carcinogenicity of nickel including the impact on humans, whole-animal studies, and molecular analyses of the nickel-induced tumors. Fifth, I describe tissue culture work that provides an improved understanding of the cellular events associated with the carcinogenic effects of nickel compounds. Finally, I explore the potential chemical and biochemical mechanisms by which nickel compounds may interfere with normal cellular functions.

9.4.1 Immunological Effects of Nickel

Nickel is one of the leading causes of allergic contact dermatitis, with a prevalence of 7–10% among woman and 1–2% among men (Menné, 1992). The source of nickel may be occupational, as in metalworking, electroplating, welding, and related fields, or nonoccupational exposure can occur through contact with nickel-containing earrings, other jewelry, wristwatches, clothing fasteners, and similar items. Key features involved in the development of nickel allergy include prolonged contact with the nickel-containing object and occlusion of the skin contact site (sweat buildup can increase local exposure levels of solubilized nickel); often, a secondary irritation can enhance sensitization by exposing lower layers of the skin. The greater incidence of nickel allergy in women than in men may be associated with their greater use of jewelry.

Nickel-associated contact dermatitis is a typical delayed-type hypersensitivity reaction to a nickel-conjugate antigen [reviewed by Nicklin and Nielsen (1992)]. Delayed-type hypersensitivity responses depend on the proliferation of clonal T cells that are involved in cell-mediated immune reactions upon being stimulated by the antigen. Sinigaglia *et al.* (1985) first described the isolation of T-cell clones that proliferate in response to nickel, a process that involves human leukocyte antigen group A (HLA) class II determinants and additionally requires histocompatible antigen-presenting cells (e.g., Langerhans cells found in the skin). The nickel-specific T-cell clones that these investigators characterized from patients with nickel-contact dermatitis produce high levels of interferon-γ in the presence of nickel. Similar nickel-specific clones have been obtained and more fully characterized by others (e.g., Silvennoinen-Kassinen *et al.*, 1991; Kapsenberg *et al.*, 1992), but the details will not be presented here. Once a person is sensitized to nickel, reexposure to minute levels can elicit contact hypersensitivity (inflammation and eczema) within hours. Indeed, the severity of dermatitis in sensitized individuals can be affected by the levels of nickel taken orally. In addition to the hypersensitivity reaction at the contact site, secondary lesions can occur at more distant sites. Fortunately, the symptoms of nickel-associated contact hypersensitivity can usually be prevented by simply separating the affected individual from the source of nickel.

A separate type of immune reaction related to nickel involves respiratory sensitivity that is referred to as an asthmatic response. This response is antibody-mediated. Nickel (conjugated to albumin or some other protein; Dolovich *et al.*, 1984) binds to a nickel-specific IgE molecule that is subsequently bound to a mast cell [reviewed by Nicklin and Nielsen (1992)]. The stimulated mast cell releases various mediators that result in immediate and acute bronchospasm and/or a late-response reaction that appears several hours after exposure. Nickel-associated asthmatic responses are generally only seen in workers who are occupationally exposed to nickel. A recent example of a study involving nickel-associated asthma has been described by Shirakawa *et al.* (1992). These investigators examined 21 workers with hard-metal asthma (hard metal is an alloy containing 80–95% tungsten carbide, 5–20% cobalt, and 0–5% nickel) and found that sera from six subjects appeared to possess antibodies that recognized both nickel and cobalt bound to albumin.

An additional effect of nickel on immune cells that is not related to the above allergic responses involves the immunotoxicity of nickel. For example, Smialowicz *et al.* (1984) showed that nickel chloride injection into mice (18.3 μg/kg) decreased the isolated splenocyte lymphoproliferative response to T-cell mitogens, but not to a B-cell mitogen, consistent with T-cell inhibition by nickel ion. In addition, the primary antibody response to T-cell-dependent antigen was suppressed in these mice, again consistent with nickel inhibition of T cells. Similarly, Schiffer *et al.* (1991) noted that isolated splenocytes from nickel-fed mice that were immunized with keyhole limpet hemocyanin exhibited decreased antigenic responsiveness compared to splenocytes from control immunized mice that were not supplemented with nickel. Further-more, B cells from the nickel-fed mice exhibited a reduced proliferative re-sponse to lipopolysaccharide mitogen. Moreover, Smialowicz *et al.* (1984, 1987) found that natural killer cell activity was significantly, though only transiently, suppressed by injection of nickel chloride into rats. Nicklin and Nielsen (1992) have summarized other immunotoxic effects of nickel, in-cluding reduction of host resistance to viral or bacterial pathogens and de-creased phagocytic capacity of macrophages. These types of nickel-induced impairments of immune function are examples of the general toxicity of nickel.

9.4.2 General Toxicity of Nickel

Although nickel compounds can damage a wide range of tissues in the body when provided at high levels, it is important to initiate this discussion by highlighting the relatively nontoxic nature of this metal ion (Sunderman, 1988). Unlike many heavy metals (cobalt, chromium, mercury, etc.), nickel ion is relatively innocuous to most tissues in the body unless exposure is at very high levels. For example, dogs were unaffected by administration of a single oral dose of nickel ion at 1–3 g/kg or by a regimen involving 200 days of nickel ion provided at 4–12 mg/kg [these results and additional early data on nickel tolerance are quoted in a book by the National Research Council (1975)]. This is not to say, however, that systemic toxicity from nickel does not occur. For example, a number of studies conducted in the 1800s dem-onstrated that nickel salts do exhibit toxic and even lethal effects when ad-ministered at high levels by stomach tube or by injection [summarized by National Research Council (1975)]. Furthermore, we have already learned in the previous section about immunotoxic effects attributed to nickel ion. The following paragraphs detail additional examples of general toxicity: renal damage resulting from exposure to nickel salts, nickel-induced hepatic toxicity, and lung damage arising from exposure to nickel carbonyl or nickel-containing dust or aerosols.

The action of nickel as a mild nephrotoxin is beginning to be character-ized. Using rats that were injected with nickel chloride at 2–5 mg/kg body weight, Gitlitz *et al.* (1975) noted an acute nickel-induced nephropathy. The resulting proteinuria was short-term, with a return to normal after a few days. More recently, Templeton and co-workers have been characterizing the mechanism of renal damage by nickel ion [reviewed by Templeton (1992)]. Binding of nickel to the glomerular basement membrane (with an association constant of $4.5 \times 10^6 \ M^{-1}$ when expressed in terms of uronic acid content) led to a reduction of net charge by ~44%. These changes may relate to the observed alterations in renal function such as the excessive loss of protein

(Templeton, 1987). Moreover, biosynthesis of glomerular basement membrane, requiring the incorporation of a heparin sulfate proteoglycan, was shown to be inhibited by 10 μM levels of nickel chloride. Thus, many of the toxic effects of nickel in the kidney appear to relate to surface phenomena. Nonetheless, nickel injection can induce other changes in the kidney including DNA–protein and DNA–DNA cross-links as well as DNA strand breaks (Ciccarelli et al., 1981; Ciccarelli and Wetterhahn, 1982). The latter features are of most consequence to nickel-induced carcinogenesis (Section 9.4.4).

Nickel salts also can lead to toxic effects in the liver. For example, Sunderman et al. (1985) injected nickel chloride into rats and observed both dose- and time-dependent increases in liver homogenate malondialdehyde levels as measured by a thiobarbituric acid assay. These investigators attributed the observed chemical changes to nickel-stimulated lipid peroxidation, involving destruction of the normal membrane lipids by free-radical reactions. (A detailed discussion of lipid peroxidation is provided in Section 9.4.6.) Following up on this work, Donskoy et al. (1986) demonstrated that 4-fold elevated levels of thiobarbituric acid-reactive material were accompanied by a 13-fold increase in the amount of conjugated dienes in rat microsomal lipids. Consistent with lipid damage, levels of aspartate aminotransferase (a typical cytoplasmic enzyme) were increased in the serum of nickel-treated rats, and visible and electron microscopic examination of hepatic tissue revealed microvesicular aberrations. As further evidence of free-radical-mediated lipid damage in nickel-treated animals, ethane and ethylene were found in exhaled breath, lipoperoxide concentrations were elevated in serum, and erthyrocytes were deformed in appearance [summarized by Sunderman (1987)]. Athar et al. (1987) noted that glutathione and iron levels increased concomitantly with lipid peroxidation and suggested that peroxidative damage occurs by ferrous ion-catalyzed hydroxyl radical formation; however, several other mechanisms have also been considered (Sunderman, 1987), and the actual mechanism(s) has not been elucidated. In addition to nickel-induced peroxidative damage to liver membranes, similar types of damage caused by nickel have been noted in the kidney and lungs (Sunderman et al., 1985) and the brain (Hasan and Ali, 1981). Peroxidative damage to membrane lipids may interfere with various transport systems, affect membrane-associated enzyme activities, and compromise the membrane integrity to result in profound negative effects on the cell. In contrast, temporal studies of livers from nickel-treated rats (Stinson et al., 1992) indicated that nickel chloride-induced lipid peroxidation does not directly cause the genetic damage observed in nickel-treated cells, as discussed in Sections 9.4.4–9.4.6.

Nickel carbonyl, an industrially important intermediate in nickel refining (see Chapter 1), is primarily toxic toward lung tissue. Acute exposure to high levels of this volatile, colorless liquid has occurred in several industrial accidents, and numerous animal studies have been carried out to better charac- terize the toxic effects of this compound [reviewed by Sunderman (1989b)]. For example, the concentration of nickel carbonyl resulting in 50% lethality (LD_{50}) after 30-min exposure by inhalation was 0.067 mM for mice and 0.24 mM for rats (10 and 33 mmol/liter of air, respectively). In humans, the clinical symptoms of exposure often resemble viral influenza pneumonia, and, in severe untreated exposures, death results from respiratory failure. The most significant pathological changes observed in humans and other animals involve the lung. How nickel carbonyl exerts its toxic effects on pulmonary tissue remains unclear. Nickel, however, is known to be important in the toxicity of $Ni(CO)_4$; that is, this compound does not simply act as a source of carbon monoxide. Because nickel carbonyl does not immediately decompose in the body, as shown by prolonged exhalation of the gas following exposure, the intact compound may have some inherent toxic effects; however, some of the nickel is deposited as the metal in tissues, and most is rapidly converted to nickel ion. The ability for chelating agents to greatly reduce the toxicity of nickel carbonyl is most consistent with the active species being the dication. As reviewed by Sunderman (1977), the immediate toxic effects of the active species may involve inhibition of ATPase activity, and longer term effects may relate to inhibition of RNA polymerase (nickel ion is known to inhibit these processes). It may well be that the lipid-soluble nature of nickel carbonyl is important in providing high intracellular concentrations of the metal ion. In summary, nickel carbonyl appears to primarily affect lung functioning due to its ability to deliver nickel ion into pulmonary cells, where enzymatic ac- tivities are inhibited.

Pulmonary toxicity has also been observed after inhalation of dust and of aerosols that contain soluble nickel and even after parenteral administration of nickel ion. The carcinogenic effects of such compounds will be considered in Section 9.4.4, whereas other toxic manifestations are considered here. A large number of pulmonary toxicity studies have been conducted in various animals in order to assess the relative toxicity of various nickel compounds, the time dependence of toxic changes, the types of pathological changes ob- served, and the changes in enzyme levels detected for several methods of nickel administration (Coogan *et al.*, 1989). As one example of a study to evaluate the pulmonary toxicity of inhaled nickel compounds, Benson *et al.* (1989) subjected rats and mice to aerosols containing either insoluble (Ni_3S_2 or NiO) or soluble ($NiSO_4$) nickel compounds and monitored a series of biochemical parameters in bronchoalveolar lavage fluid. All compounds gen- erated increases in lactate dehydrogenase and β-glucuronidase activities, el- evated the total protein levels, and led to infiltration by neutrophils, consistent with cytotoxic and inflammatory responses. The relative toxicities of the com- pounds were ranked in the order $NiSO_4 > Ni_3S_2 > NiO$. In addition to pulmonary toxicity by inhalation, repeated intramuscular administration of NiSO$_4$ to rats has also been shown to induce pathological reactions in the lungs as well as the liver, thymus, and spleen (Knight et al., 1991). The biochemical processes responsible for toxic effects in the lung have not been characterized but were suggested to be related to peroxidative processes (Knight et al., 1991). Nevertheless, it is clear that potentially toxic effects of nickel (in addition to the carcinogenic effects) should be a concern for individuals exposed to high levels of nickel-containing dust or aerosols.

9.4.3 Embryotoxicity of Nickel

Nickel compounds can have profound influences on the development of an embryo (Leonard and Jacquet, 1984; Coogan et al., 1989). In rodents, for example, maternal exposure to nickel compounds can lead to a decline in the frequency of implantation of fertilized eggs, enhanced resorption of the early developmental stages, elevation in the frequency of stillbirths, and growth abnormalities in the liveborn young (e.g., Schroeder and Mitchner, 1971; Sunderman et al., 1978; Lu et al., 1979). These effects are likely to arise from a combination of direct embryotoxic effects and maternal toxicity. This section will focus on the last consequence of maternal exposure, that is, growth aberrations or teratogenic effects. Pregnant rats are more susceptible to nickel ion toxicity than control animals (Mas et al., 1985), and the dams can develop clear alterations in glucose metabolism when challenged with nickel ion. Sadler (1980) has demonstrated that diabetic serum has teratogenic potential; hence, maternal hyperglycemia resulting from elevated nickel concentrations may give rise to some of the observed teratogenic effects. Although a portion of the embryonic consequences may be secondary to maternal toxicity, other effects involve direct action of nickel on the embryo. Nickel salts are known to readily cross the placenta, and significant concentrations of nickel may transiently accumulate in the embryo and alter its development. However, it is difficult to sort out the effects of maternal toxicity from direct embryonic toxicity of nickel in animals that possess a placenta. An approach to overcome this problem is to monitor the effects in embryos that are cultured in serum outside the uterus, as described for rat embryos by Saillenfait et al. (1991). However, the observed large changes in embryotoxicity between cultured embryos from placental animals and those treated maternally can raise questions regarding the significance of these findings. Alternatively, the effects of nickel on embryonic development in nonplacental animals have been studied, as described below.

The best defined biochemical system for characterizing nickel-induced teratogenic effects is probably that utilizing Xenopus laevis embryos. Hopfer et al. (1991) have shown that nickel is a potent teratogen for Xenopus embryos, with the severity of the malformations dependent on the nickel concentration and the timing of the exposure. By using a protein blotting method that detects nickel-binding proteins [developed by Lin *et al.* (1989)], Beck *et al.* (1992) identified a *Xenopus* nickel-binding protein (*pNiXa*) whose presence coincided with teratogenic susceptibility. The protein was found to be identical to Ep45, a hepatic protein that is thought to function as a serine protease inhibitor. The protein possesses a histidine-rich sequence (*His*-Arg-*His*-Arg-*His*-Glu-Gln-Gln-Gly-*His*-*His*-Asp-Ser-Ala-Lys-*His*-Gly-*His*) that could potentially function as a metal binding site. The authors speculated that the *pNiXa*(Ep45) protein may be the target of nickel teratogenesis. The first step in the chain of events leading to teratogenesis may involve nickel binding to the histidine-rich region of the protein to alter its conformation. It is reasonable, but not yet proven, that the metal-bound form of the protein possesses a distinct capacity for proteinase inhibition compared to metal-free protein (e.g., it may inhibit a different subset of proteases). Because proteases are so important to embryonic development, the changes in protease activity may then lead to the observed teratogenic effects.

9.4.4 Carcinogenicity of Nickel

An association between the nickel industry and cancer of tissues in the respiratory tract has a long and tragic history. For example, Doll (1958) reported that among 293 nickel workers who died between 1938 and 1956 in two areas of South Wales, 75 (25.6%) died of lung cancer and 29 (9.9%) died of nasal cancer. A report of the National Research Council (1975) discussed in detail the epidemiological evidence from nickel workers in Wales, Canada, Norway, and Russia that implicates nickel as a carcinogenic agent in humans. It is important to note, however, that many of the nickel workers included in these statistics were exposed to other carcinogenic metals (chromium, cobalt, arsenic) and organic compounds. Nevertheless, there is general agreement that insoluble nickel dust was a principal carcinogen for the workers in these early studies (Christie and Katsifis, 1990). Fortunately, present-day occupational exposure to nickel dust for workers involved in mining, smelting, and refining nickel has been greatly reduced by careful monitoring of dust levels and the wearing of protective filters. These approaches combined with medical monitoring should lead to a continued decline in the incidence of cancer in nickel workers. Langård and Stern (1984) have warned, however, that occupational exposure to nickel fumes may be an unrecognized cancer hazard in another group of workers—welders who work with stainless steel or high-alloy steels. These authors have stressed that an international epidemiological effort may be required to assess the danger of nickel fumes to such workers.

Animals have a long history in studies of nickel carcinogenesis, including the demonstration in 1943 that mice exposed to nickel dust developed tumors (Campbell, 1943). The results from numerous animal studies [e.g., individual contributions in the volumes edited by Brown and Sunderman (1980, 1985), Sunderman et al. (1984) and Nieboer and Nriagu (1992), as well as references in the reviews by the National Research Council (1975), Sunderman (1977), Coogan et al. (1989), Christie and Katsifis (1990), Costa (1991), and Snow (1992)] have verified that nickel compounds can be highly carcinogenic. The carcinogenic potential of a nickel compound depends on the animal species examined (e.g., hamsters are relatively more tolerant of nickel compounds than rats or mice), the chemical composition of the nickel compound (solubility appears to play a large role, with higher carcinogenicity found for insoluble compounds), the treatment of the compound (there may be large differences based on the particle size and the age of the substance), and the method of administration. A variety of cancer types have been generated in a wide range of animals by injection, implantation, and inhalation of nickel compounds. In general, crystalline nickel subsulfide (Ni_3S_2), crystalline nickel sulfide (NiS), and nickel oxide (NiO) are the most carcinogenic forms of nickel. For example, in contrast to all other known carcinogenic compounds, Ni_3S_2 consistently formed tumors in the Japanese common newt when injected intraocularly (Okamoto, 1987). Insoluble nickel compounds generally are highly effective in inducing tumor formation at the site of injection or implantation. In contrast, soluble nickel salts are generally not carcinogenic when administered by injection (Kasprzak et al., 1983; Knight et al., 1991), although it has been reported that repeated injections of $NiCl_2$ may induce tumor development in the lung [quoted by Coogan et al. (1989)] and the administration of a soluble nickel salt followed by a tumor promoter (sodium barbital) can lead to renal tumors (Kasprzak et al., 1990). In addition, the combination of nickel acetate and sodium barbital was shown to have transplacental effects, causing tumor formation in fetal rat kidney and pituitary (Diwan et al., 1992). Particulate compounds are highly effective in causing tumor formation when inhaled, but soluble compounds and the gaseous substance nickel carbonyl are also carcinogenic when administered through the respiratory system. For many of the carcinogenic compounds, it is important to note that a long latent period may be observed prior to tumor formation.

Carcinogenesis is a multistage process involving a series of biochemical changes that are referred to as initiation, promotion, progression, and metastasis [reviewed by Snow (1992)]. Whole-animals studies are essential for establishing the degree of carcinogenicity for an individual compound in a particular species administered by a specific method; however, they provide little information about the biochemical events that are associated with these steps of carcinogenesis. Examination of biopsy tissues taken at various times after nickel administration by visible and electron microscopic methods (e.g., Lumb and Sunderman, 1988) can provide the temporal sequence for morphological phenomena associated with carcinogenesis, but this approach similarly fails to identify the underlying biochemical changes that take place. In contrast, molecular analysis of tumor tissue can begin to reveal some of the fundamental biochemical alterations associated with carcinogenesis. For example, Christie *et al.* (1988) showed that cell lines derived from mouse rhabdomyosarcomas that were induced with nickel sulfide possess minichromosomes and other chromosome rearrangements. At a higher level of resolution, Higinbotham *et al.* (1992) have demonstrated by sequence analysis that rat renal sarcomas induced by Ni_3S_2 plus iron possess a G-to-T transversion at codon 12 of the K-*ras* oncogene. Such a mutagenic event is consistent with an oxidative reaction associated with nickel. Perhaps related to the carcinogenic process is the observation by Kasprzak *et al.* (1990) that a single parenteral administration of nickel ion is associated with an increased concentration of 8-hydroxyl-2'-deoxyguanosine in rat kidney tissue. The same laboratory demonstrated that nickel injection into pregnant rats led to enhanced levels of 11 DNA base derivatives, where these products are consistent with hydroxyl radical damage (Kasprzak *et al.*, 1992a). In addition, Ciccarelli *et al.* (1981) and Ciccarelli and Wetterhahn (1982) used the alkaline elution assay method to demonstrate the presence of DNA strand breaks (kidney and lung), intrastrand cross-links (kidney), and DNA–protein cross-links (kidney) in rat tissues after intraperitoneal administration of nickel carbonate. A gentle isolation procedure was used to demonstrate the existence of nickel–nucleic acid and nickel–DNA–protein complexes in kidney and liver of nickel-treated rats (Ciccarelli and Wetterhahn, 1984). In the kidney, approximately half of the nickel associated with DNA was in a DNA–protein complex, whereas RNA-associated nickel was nearly exclusively in an RNA–protein complex. Curiously, very different results were found in the liver: DNA-associated nickel was free of protein, and approximately half of RNA-associated nickel was in an RNA–protein complex. These differences may relate to the slow-to-be repaired lesions found in kidney tissue versus the lack of damage in liver tissue. Nickel(II)-induced DNA–protein cross-linking and DNA base oxidation in rat kidney have been shown to be elevated by administration of L-histidine to the animal (Misra *et al.*, 1993). Although tremendous progress has been made in understanding the individual biochemical steps leading to tumor development in a whole animal or in excised tumor tissue, many of the detailed molecular characterization studies of nickel carcinogenicity have focused on tissue culture studies, as detailed below.

9.4.5 Tissue Culture Studies Related to Nickel Carcinogenesis

The pioneering studies of Basrur and Gilman (1967) first demonstrated that a nickel compound can cause a morphological transformation of tissue culture cells. These investigators treated rat embryo muscle cells with Ni_3S_2 and observed a shift toward elongated, multinucleate cells with greatly increased levels of DNA synthesis as detected by increased incorporation of [^{3}H]thymidine. In contrast, cultures derived from an Ni_3S_2-induced rhabdomyosarcoma were unaffected by analogous treatment. Since their work, the phenomenon of nickel-induced transformation has been observed for a variety of nickel compounds in a range of cell types. Transformed cells often can lead to tumor formation when transferred into an animal; thus, it was not surprising when Costa *et al.* (1979) demonstrated that Ni_3S_2-transformed Syrian hamster fetal cells led to the development of undifferentiated sarcomas when implanted into nude mice. The following paragraphs briefly describe some of the findings from culture studies that relate to nickel-induced carcinogenesis.

In order to initiate the process of carcinogenesis, nickel first must enter the animal cell. As described in Section 9.2.5, highly carcinogenic particulate compounds such as Ni_3S_2 are taken up by certain cells via phagocytic processes. The uptake effectiveness of tissue culture cells is enhanced by the crystallinity of the compound and the surface charge, and the uptake rate is highly dependent on particle size (e.g., Costa and Mollenhauer, 1980a,b; Costa *et al.*, 1981; Abbracchio *et al.*, 1982; Kuehn *et al.*, 1982; Heck and Costa, 1983; Nieboer *et al.*, 1984b). Curiously, after internalization of nickel sulfide crystals the sulfur is replaced by phosphate according to energy-dispersive spectrometry measurements (Hildebrand *et al.*, 1991; Shirali *et al.*, 1991). The internalized particles initially appear to reside within lysosomes, but subsequently the particles seem to be bound to several cellular organelles including the nuclear membrane. Costa (1991) summarized evidence that indicates that the dissolution of the particles is not accompanied by loss of nickel from the cell, but rather the nickel appears to enter the nucleus. The high efficiency of cell transformation by insoluble nickel compounds may be related to their ability to maintain an elevated concentration of nickel ions in the nucleus over a prolonged period. Although soluble nickel compounds also can induce transformation of cells in culture (e.g., Pienta *et al.*, 1977; DiPaolo and Castro, 1979; Hansen and Stern, 1983; Lechner *et al.*, 1984), carcinogenesis from soluble nickel compounds is not important *in vivo*. The differences between the *in vivo* results and cell culture studies may occur because (i) elevated levels of nickel ion are not maintained in an animal owing to rapid elimination of the cation whereas high levels can be maintained in culture and (ii) nickel is not continually released into the nucleus from soluble compounds, in contrast to the case of dissolution of attached particles.

Insights into the types of chromosomal alterations that may be associated with carcinogenicity can be revealed by tissue culture studies. For example, Swierenga and Basrur (1968) showed that NiS induced a variety of abnormal mitotic chromosomal forms and reduced the mitotic index of rat embryo muscle cultures. On the basis of histochemical staining results, they proposed that nickel may interfere with sulfhydryl groups in the spindle complex. Similarly, Nishimura and Umeda (1979; Umeda and Nishimura, 1979) observed metaphase chromosomes from mouse mammary carcinoma cells and found numerous changes (gaps, breaks, exchanges, fragmentation, and formation of rings) after treatment with soluble nickel salts. Robison *et al.* (1984) extended these findings to demonstrate the presence of significant levels of single-strand breakage in isolated nucleoids that were treated with nickel chloride as well as in the intact cells. Furthermore, these investigators showed that addition of nickel ions at concentrations that were below that needed to cause detectable DNA lesions induced high levels of DNA repair activity. In general, nickel treatment of cells in culture leads to aberrations in the heterochromatic regions of chromosomes (Sen and Costa, 1985). Sen and Costa (1985) proposed that the specificity may relate to the location of heterochromatin in the nucleus; that is, because heterochromatin lines the inside surface of the nucleus, it is the first site of contact for entering nickel ions. Sen and Costa (1985) and Conway and Costa (1989) found that treatment of Chinese hamster embryo cells with either soluble or particulate nickel compounds led to a high recovery of transformants containing a deletion in the long arm of the X chromosome. The latter authors proposed that transformation was associated with the loss of a senescence gene that was located on the X chromosome. Not surprisingly, the frequency of transformation was greater in cells derived from male embryos (containing a single X chromosome copy) than in those derived from female embryos. Further studies from the same laboratory (Klein *et al.,* 1991a) have provided support for this hypothesis. When a normal X chromosome was transferred into the immortal cell line, the senescence phenotype returned. As another example, heterochromatic abnormalities that involve fusions at the centromeres of mouse cells have led to the development of nickel-resistant cell lines (Wang *et al.,* 1988). These cells are resistant to 6- to 11-fold higher concentrations of nickel ion than the initial cells, yet the nickel uptake and metallothionein levels are unchanged. Interestingly, the nickel-resistant cells exhibited a depressed overall protein phosphorylation activity and were incapable of catalyzing the nickel-dependent phosphorylation of specific proteins as seen in wild-type cells (Wang and Costa, 1991). Yet another type of chromosomal aberration induced by nickel is the formation of DNA–protein cross-links. Patierno and Costa (1985) demonstrated that both particulate (crystalline NiS) and soluble ($NiCl_2$) nickel compounds induced DNA–protein cross-links, as well as single-strand breaks, in Chinese hamster ovary cells. Furthermore, they found that the cross-links were slow to be repaired and the concentration of nickel associated with cross-linkage correlated with the metal ion concentration that inhibited cellular replication.

Tissue culture studies similarly have been used to reveal details about single-base-change mutational events that may relate to possible early steps in the carcinogenic process. Tveito *et al.* (1989) treated human kidney epithelial cells with nickel ion and found that the cells acquired an indefinite life span in culture and formed colonies in soft agar but were incapable of forming a tumor when implanted into an animal. When these cells were transfected with the v-Ha-*ras* oncogene, they became capable of neoplastic transformation in athymic nude mice (Haugen *et al.*, 1990). Maehle *et al.* (1992) demonstrated by DNA sequence analysis and immunocytochemical methods that *p53,* a tumor suppressor gene, contained a single base change (thymidine changed to cytosine at position 238) that resulted in greatly increased levels of expression in the immortalized and malignant cells. They proposed that nickel induced a mutation of *p53* that was responsible for the loss of senescence in the epithelial cells. The importance of such single-base-change events to the *in vivo* carcinogenesis process remains unclear, but a similar single base change in the K-*ras* gene was observed in a rat renal sarcoma induced by Ni_3S_2 (Higinbotham *et al.,* 1992).

Although tissue culture studies have shown that nickel compounds can lead to single-base-pair mutations in DNA as well as large chromosomal alterations, nickel is not a mutagen in bacteria (e.g., Biggart and Costa, 1986) and is only poorly mutagenic in animal cells [reviewed by Christie and Katsifis (1990)]. For example, Arrouijal *et al.* (1990) compared the mutagenicity of Ni_3S_2 in four assay systems. They found that the Ames test, using five *Salmonella typhimurium* strains, was uniformly negative at identifying mutagenic activity. A hypoxanthine-guanine phosphoribosyl transferase (HGPRT) point mutation test using V79 cells was similarly negative. In contrast, alterations were noted in metaphase chromosomes from nickel-treated human lymphocytes, and an increase in micronuclei was noted in erythrocytes from nickel-treated mice. Other investigators have noted various levels of positive mutagenic responses to variants of the HGPRT point mutation test [summarized by Christie and Katsifis (1990)]. In addition, mutagenicity of nickel compounds to tissue culture cells has been detected with several other types of assays. As one example, Biggart *et al.* (1987) observed that nickel chloride led to a 7- to 10-fold enhancement of expression of the conditionally defective v-*mos* transforming gene in rat kidney cells that were infected by a murine sarcoma virus mutant. The mutant virus possesses an out-of-frame feature that prevents v-*mos* expression at 39°C, but the defect can be suppressed by a splicing event at 33°C to generate the transformed phenotype. Nickel was shown to lead to a 70-base pair (bp) duplication that resulted in v-*mos* expression at 39°C (Chiocca *et al.,* 1991). Such a duplication was suggested to arise during DNA replication or improper repair synthesis after DNA breakage.

Additional cellular changes also have been noted in tissue culture cells upon addition of nickel compounds. For example, Holst and Nordlind (1988) found that treatment of peripheral blood T lymphocytes with nickel sulfate led to specific phosphorylation of certain nonhistone nuclear proteins. Such results are important because phosphorylation changes are known to be important in controlling cellular growth. In addition, Zhong *et al.* (1990) reported that nickel subsulfide and nickel disulfide induced 5.5-fold and 2-fold increases in hydrogen peroxide formation by resting polymorphonuclear leukocytes. Hydrogen peroxide can damage DNA; for example, it can oxidize thymidine to form 5-hydroxymethyl-2′-deoxyuridine. Consistent with a role for active oxygen molecules in carcinogenesis induced by nickel particles, Lin *et al.* (1991) found that vitamin E pretreatment of Chinese hamster ovary cells reduced the chromosome aberrations due to NiS particles but not the damage due to $NiCl_2$. In an unrelated experiment, Lin *et al.* (1992) provided Chinese hamster ovary cells with radioactively labeled amino acids and examined the specificity of nickel-induced protein–DNA cross-links. Although not present in the final preparation, nickel ion appeared to cause specific cross-link formation between DNA and histidine or cysteine residues (Lin *et al.,* 1992). Finally, many investigators have shown that nickel ion can inhibit various enzyme activities in tissue culture cells. For example, replication is greatly diminished by nickel chloride in a process that can be antagonized by addition of magnesium ion (Conway *et al.,* 1986).

Observations from tissue culture studies such as those described above or from animal studies described earlier allow one to begin to suggest mechanisms for how nickel can lead to toxic and carcinogenic effects. The chemistry and biochemistry of these mechanisms are described in more detail below.

9.4.6 Mechanisms of Nickel Damage

It is clear from the above discussions that nickel compounds can have a variety of toxic and carcinogenic effects. In the following paragraphs an attempt is made to summarize the potential chemical mechanisms by which this metal ion exerts its effects in animals by focusing on *in vitro* studies. Discussion of the possible mechanisms of nickel damage is divided according to the sites of action: effects at the membrane level due to lipid peroxidation, effects at the post-translational level on metabolic enzymes or structural proteins, effects at the transcription level by direct interaction with DNA or with regulatory proteins, and effects at the replication level by interaction with DNA polymerase or with DNA.

Nickel-induced effects at the membrane level are most consistent with peroxidative damage to membrane lipids. Here, I will describe the processes that can lead to lipid peroxidation and summarize mechanisms by which nickel can induce these effects. For an excellent general overview of the relationships between oxygen radicals, transition metals, oxygen toxicity, and disease, the reader is referred to the review by Halliwell and Gutteridge (1984).

Lipid peroxidation is initiated by abstraction of hydrogen atoms from polyunsaturated fatty acids by reactive species such as hydroxyl radicals (Eq. 9-1). The carbon-based lipid radicals can rearrange to yield conjugated dienes that subsequently can react with dioxygen to yield hydroperoxyl radicals (Eq. 9-2). The hydroperoxyl radicals can participate in peroxidative chain reactions by abstracting hydrogen atoms from other fatty acids (Eq. 9-3). The resulting lipid hydroperoxides are relatively stable, and these species are a normal component of lipids. The original hydrogen abstraction agent may arise from Fenton-type chemistry in which hydrogen peroxide reacts with ferrous ion to yield hydroxyl radical (Eq. 9-4). Alternatively, lipid hydroperoxides can react with reduced or oxidized iron species to form alkoxy radicals (Eq. 9-5) and peroxy radicals (Eq. 9-6). Either of these species can stimulate the peroxidative chain reaction. Thus, a number of mechanisms are available for lipid-damaging reactions.

$$\text{Lipid-H} + \cdot\text{OH} \rightarrow \text{Lipid}\cdot + H_2O \qquad (9\text{-}1)$$

$$\text{Lipid}\cdot + O_2 \rightarrow \text{Lipid-O-O}\cdot \qquad (9\text{-}2)$$

$$\text{Lipid-O-O}\cdot + \text{Lipid-H} \rightarrow \text{Lipid-O-O-H} + \text{Lipid}\cdot \qquad (9\text{-}3)$$

$$Fe^{2+} + H_2O_2 \rightarrow Fe^{3+} + OH^- + OH\cdot \qquad (9\text{-}4)$$

$$\text{Lipid-O-O-H} + Fe^{2+} \rightarrow \text{Lipid-O}\cdot + OH^- + Fe^{3+} \qquad (9\text{-}5)$$

$$\text{Lipid-O-O-H} + Fe^{3+} \rightarrow \text{Lipid-O-O}\cdot + H^+ + Fe^{2+} \qquad (9\text{-}6)$$

Turning now to the question of how nickel ion can stimulate lipid peroxidation, several hypotheses have been considered (Sunderman, 1987). First, nickel ion may displace ferrous ion (or cuprous ion) from cellular metalloproteins (Eq. 9-7) to allow Fenton chemistry to occur. Second, nickel ion may be capable of carrying out reactions (Eqs. 9-8, 9-9, and 9-10) that are analogous to the radical-generating reactions described above for iron. Such chemistry may be facilitated by intracellular complexing agents that aid in stabilizing the non-nickel(II) redox states. Finally, nickel ion may simply interfere with the normal cellular machinery that protects the cell against peroxidative damage that continuously occurs in our aerobic environment. Stinson et al. (1992) have provided evidence from temporal studies of changes in rat liver that events associated with lipid peroxidation are not simultaneously responsible for genetic damage. Nevertheless, similar metal-dependent oxidative reactions that cause DNA damage may be associated with carcinogenesis [see discussion below and Klein *et al.* (1991b) and Kasprzak (1991)].

$$Ni^{2+} + Protein\text{-}Fe^{2+} \rightarrow Fe^{2+} + Protein\text{-}Ni^{2+} \qquad (9\text{-}7)$$

$$Ni^{2+} + H_2O_2 \rightarrow Ni^{3+} + OH^- + OH\cdot \qquad (9\text{-}8)$$

$$Lipid\text{-}O\text{-}O\text{-}H + Ni^{2+} \rightarrow Lipid\text{-}O\cdot + OH^- + Ni^{3+} \qquad (9\text{-}9)$$

$$Lipid\text{-}O\text{-}O\text{-}H + Ni^{2+} \rightarrow Lipid\text{-}O\text{-}O\cdot + H^+ + Ni^{1+} \qquad (9\text{-}10)$$

Nickel-induced effects at the post-translational level include direct or indirect interactions with the function of metabolic enzymes or structural proteins. These interactions can lead to sometimes subtle but important cellular changes. There are many examples of enzymes that are inhibited or altered in their function by reversible binding of nickel ion (Nieboer *et al.,* 1984b). In various cases the nickel may displace the endogenous active cation, sterically hinder substrate binding, act as an allosteric effector, or function in some other manner. This type of alteration may be reflected in covalent changes in the same or other proteins. As a specific example, the presence of nickel has been shown to alter the specificity of protein kinases in the cell (Holst and Nordlind 1988). Similarly, developmentally important proteolytic processing events may be modified in the presence of nickel; for example, Beck *et al.* (1992) reported that *pNiXa,* a nickel-binding protease inhibitor from *Xenopus laevis,* may be associated with nickel-induced teratogenesis. Although an interaction between nickel and a protein can be established *in vitro,* it may be very difficult to establish whether such an interaction is important *in vivo.*

Nickel may also affect animal cells at the level of transcriptional regulation. One mechanism by which nickel ion can interfere with normal transcription is by direct interaction with DNA. Several investigators have used spectroscopic methods to study the ability of nickel to bind to DNA and induce a transition from the B- to the Z-form (Van de Sande *et al.,* 1982; Liquier *et al.,* 1984; Bourtayre *et al.,* 1984; Hacques and Marion, 1986). Because the nickel ion dissociation constant associated with this conformational change is so high (approximately 0.4 mM), however, it is unlikely that nickel induction of Z-DNA has a significant role in affecting transcriptional regulation unless certain sequences are especially prone to undergo this type of transformation. Rather, alterations in transcription regulation by nickel are liable to proceed via nickel ion binding to regulatory proteins. For example, Janknecht *et al.* (1991) have noted that many eucaryotic transcription factors possess (histidine-X)$_n$ repeats that could coordinate zinc or copper, and they termed this motif a histidine–metal zipper. It is reasonable to suppose that nickel could substitute for copper or zinc in these proteins and lead to alterations in mRNA synthesis. Nickel ion is also known to bind to zinc-finger proteins such as the transcription factor IIIA (TFIIIA) (Makowski *et al.,* 1991), and this type of interaction has been proposed as a potential site of nickel toxicity and carcinogenicity *in vivo* (Sunderman, 1990). The findings of more recent studies with purified TFIIIA protein (Makowski and Sunderman, 1992) or with synthetic zinc-finger peptides (Krizek and Berg, 1992), however, reduce the likelihood that interaction of nickel with zinc-finger proteins is significant *in vivo* because the affinity for nickel is greatly diminished in comparison to that for zinc. Nevertheless, nickel may bind tightly to other regulatory proteins.

Probably the most important aspect of nickel damage to animals involves genetic changes, especially those involved in carcinogenesis. Nickel-dependent effects at the DNA level include two general categories: (i) mutations that result in single-base-pair changes and (ii) larger types of DNA alterations such as rearrangements, deletions, and DNA–DNA or DNA–protein cross-links. Each of these topics is described below.

Nickel-dependent single-base-pair mutations can arise in DNA by protein inhibition, by oxidative mechanisms, and by nonoxidative processes. As an example of mutation by enzyme inhibition, Sirover and Loeb (1976) demonstrated that nickel chloride can interact with DNA polymerase to decrease the fidelity of DNA synthesis in an *in vitro* assay. Similarly, nickel can interfere with normal repair processes (e.g., Robison *et al.,* 1984), presumably by inhibiting repair enzymes [reviewed by Coogan *et al.* (1989) and Christie and Katsifis (1990)]. As an example of site-specific damage to DNA by an oxidative mechanism, Kasprzak and Hernandez (1989) have demonstrated that addition of hydrogen peroxide to Ni_3S_2 or $NiCl_2$ solutions containing 2'-deoxyguanosine or calf thymus DNA greatly enhanced the formation of 8-hydroxy-2'-deoxyguanosine and hydroxylated guanine residues. Furthermore, these investigators showed that hydrogen peroxide plus nickel ion appeared to be able to catalyze a deglycosylation of 2'-deoxyguanosine. Similar oxidative events are consistent with point mutations that are introduced *in vivo* (e.g., Higinbotham *et al.,* 1992). Perhaps related to how these reactions occur are the findings of Cotelle *et al.* (1992) and Inoue and Kawanishi (1989) that nickel ion bound to oligopeptides can give rise to active oxygen species (e.g., superoxide, hydroxyl radical, or singlet oxygen) in the presence of hydrogen peroxide. The mechanism for generating hydroxyl radical is likely to involve nickel-based Fenton chemistry (Eq. 9-8), where the metal ion is peptide bound. Superoxide is also likely to arise directly from nickel-peptide interaction with hydrogen peroxide, whereas singlet oxygen may be derived from a reaction of superoxide with hydroxyl radical or hydrogen peroxide (Inoue and

Kawanishi, 1989). Free-radical species also are able to be generated from lipid hydroperoxides by nickel ion in the presence of oligopeptides or thiol compounds, as shown by spectroscopic studies using a radical spin trap (Shi *et al.*, 1992, 1993). In order to better understand which active oxygen species is involved in the nickel-induced hydroxylation of 2'-deoxyguanosine, Kasprzak *et al.* (1992b) examined the effects of hydroxyl radical scavengers (ethanol, mannitol, and glutathione) and a singlet oxygen scavenger (histidine). They found that the effects of these and other compounds were inconsistent with the simple reactions of free hydroxyl radical or free singlet oxygen. Rather, they postulated the existence of a crypto-hydroxyl radical that remains metal associated prior to reaction with the nucleotide base. In contrast to these types of oxidative mechanisms, Schaaper *et al.* (1987) provided evidence that nickel-catalyzed hydrolytic depurination reactions can lead to point mutations in DNA. They found that nickel chloride induced a sixfold increase in depurination of [8-^{3}H]deoxyadenine-containing DNA. Thus, several mechanisms are available for nickel to induce single-base-pair changes in DNA.

Larger scale DNA alterations induced by nickel are known to include DNA–DNA and DNA–protein cross-links, as well as DNA strand breakage and rearrangements. Although the mechanisms of cross-linkage have not been well characterized, Lee *et al.* (1982) showed that the presence of denatured microsomes enhances the binding of nickel ion to calf thymus DNA by allowing the formation of stable complexes that contain protein, DNA, and nickel. Because nickel ion generally has a higher affinity for protein than DNA, such ternary complexes are likely to be important as the first step in any type of cross-linking reaction. In an extension of this type of study, Kasprzak and Bare (1989) examined the reactions of calf thymus DNA, a nucleohistone preparation, or free histone proteins to various treatments. They showed that DNA alone was unaffected by two-day incubation with nickel acetate, tetraglycine, or a mixture of these species. In contrast, the nucleohistone complex, comprised of DNA and histone proteins, exhibited clear increases in its melting temperature when provided with nickel acetate or nickel acetate plus tetraglycine, consistent with some type of nickel-mediated cross-linking event. Furthermore, in the latter treatment the solubility of histones was greatly reduced as if they were being cross-linked to the DNA or to each other. When histone proteins were incubated with nickel acetate and tetraglycine, polymerization of the histones was apparent by polyacrylamide gel analysis. Thus, oxidative processes associated with oligopeptide-bound nickel ion can participate in cross-linking reactions. However, it is unclear whether these specific reactions are significant to the *in vivo* cross-linking processes. For example, oxidation of tetraglycine may give rise to formaldehyde, which could subsequently lead to chemical cross-linking events.

In contrast to the lack of detailed studies concerning nickel ion-induced cross-linking reactions, numerous investigators have studied the mechanisms of nickel-dependent DNA cleavage. For example, Kawanishi *et al.* (1989) used [^{32}P]-5'-end-labeled fragments of human c-Ha-*ras*-1 DNA to demonstrate that nickel ions were able to cleave DNA in the presence of hydrogen peroxide. They suggested that nickel ion first binds to DNA, subsequent reaction with hydrogen peroxide generates an active oxygen species, and this species yields piperidine-labile sites in the DNA. On the basis of active oxygen scavenger studies, these investigators suggested that the species responsible for DNA damage was either hydroxyl radical that reacted with the DNA before being fully released into solution or the crypto-hydroxyl radical still bound to nickel ion, as described earlier. Consistent with the latter hypothesis, DNA cleavage was not totally random; rather, cytosine, thymine, and guanine residues were the preferred sites of cleavage, whereas adenine residues were resistant. Similarly, nickel(II) desferal in the presence of hydrogen peroxide was shown to be capable of specific DNA cleavage at CG sites (Joshi and Ganesh, 1992). In a variation of these types of experiments, Mack and Dervan (1990) synthetically attached a Gly-Gly-His sequence to the amino terminus of Hin recombinase and showed that the protein could serve as a sequence-specific oxidative endonuclease in the presence of nickel and monoperoxyphthalic acid (a peracid). The protein recognizes and binds to its normal recognition sequence, nickel binds to the tripeptide tail in a fashion analogous to that found in nickel interaction with serum albumin, and the peracid generates an active oxygen species that leads to specific DNA cleavage. Rather than production of hydroxyl radical, the investigators suggested that the mechanism involves the formation of a high-valent nickel-bound oxygen species that abstracts an hydrogen atom from the DNA. Some intrinsic DNA cleavage sequence specificity, especially for guanine residues in non-Watson–Crick duplex structures, has also been shown for simple nickel complexes in the presence of peracids (Chen *et al.*, 1991, 1992). By examining the effects of systematic variations in the complex structure, Muller *et al.* (1992; 1993) have shown that nickel ligands must possess strong in-plane donors and a flexible macrocyclic structure to accommodate the DNA reactivity. These investigators proposed the existence of a nickel(III) mechanistic intermediate containing a ligated guanine residue and the oxidant molecule. In addition, the same group has used oxygen-labile nickel(II) complexes to carry out DNA cleavage without the requirement for peracids or hydrogen peroxide. They found that simple addition of oxygen led to DNA strand breakage, again consistent with the presence of a nickel(III) intermediate (Cheng *et al.*, 1993). The relevance of these types of chemistry to *in vivo* nickel-induced DNA cleavage is unclear.

Although the discussion above has demonstrated that much is known about *in vitro* reactions of nickel, the relative importance of each of the above processes to the toxic and carcinogenic effects in animals remains to be established.

9.5 Perspective

Despite the importance of nickel to humans and other animals, many questions remain to be answered regarding the flux, the essentiality, and the mechanisms of toxicity and carcinogenicity of this metal ion. For example, little is known about the specificity and mechanism of the uptake of nickel ion into the body from the gastrointestinal tract. Once in the body, nickel is transported in the serum to various body tissues where it is accumulated to different levels by processes that are only partially understood. Similarly, the detailed mechanisms for excretion of the metal ion via the kidney are incompletely characterized. In addition, the cellular uptake of nickel ion is a relatively unexplored area. In terms of the essentiality of nickel, data for humans and most animals are lacking. Furthermore, in chickens, rats, pigs, sheep, goats, and steers, where various levels of support for a nickel requirement exist, the role(s) of the metal ion is not clearly established. Even in the case of a tunicate that is known to contain a novel nickel tetrapyrrole, it is not clear that nickel is required or what its functional role might be in the animal. Nickel-containing compounds can harm the body in many ways, each of which is only partially understood. Several types of immunological effects are attributed to nickel, and the metal can participate in additional harmful effects ranging from general toxicity to embryotoxicity to carcinogenicity. Although extensive studies have characterized the effects of nickel compounds on whole animals, tissue culture cells, and purified cellular components, many detailed mechanistic questions remain to be addressed in order to understand the effects of nickel at a molecular level. Clearly, further studies to characterize the types of nickel interactions with animals will continue to be an area of active research.

References

Abbracchio, M. P., Heck, J. D., and Costa, M., 1982. The phagocytosis and transforming activity of crystalline metal sulfide particles are related to their negative surface charge, *Carcinogenesis* 3:175–180.

Abdulwajid, A. W., and Sarkar, B., 1983. Nickel-sequestering renal glycoprotein, *Proc. Natl. Acad. Sci. USA* 80:4509–4512.

Anke, M., Grün, M., and Kronemann, H., 1980. Distribution of nickel in nickel-deficient goats and their offspring, in *Nickel Toxicology* (S. S. Brown and F. W. Sunderman, Jr., eds.), Academic Press, New York, pp. 69–72.

Anke, M., Grün, M., Groppel, B., and Kronemann, H., 1983. Nutritional requirements of nickel, in *Biological Aspects of Metals and Metal-Related Diseases* (B. Sarkar, ed.), Raven Press, New York, pp. 89–105.

Anke, M., Groppel, B., Kronemann, H., and Grün, M., 1984. Nickel—an essential element, in *Nickel in the Human Environment* (F. W. Sunderman, Jr. *et al.*, eds.), Oxford University Press, New York, pp. 339–365.

Arrouijal, F. Z., Hildebrand, H. F., Vophi, H., and Marzin, D., 1990. Genotoxic activity of nickel subsulfide α-Ni$_3$S$_2$, *Mutagenesis* 5:583–589.

Athar, M., Hasan, S. K., and Srivastava, R. C., 1987. Evidence for the involvement of hydroxyl radicals in nickel mediated enhancement of lipid peroxidation: Implications for nickel carcinogenesis, *Biochem. Biophys. Res. Commun.* 147:1276–1281.

Barnes, D. J., and Crossland, C. J., 1976. Urease activity in the staghorn coral, *Comp. Biochem. Physiol.* 55B:371–376.

Basrur, P. K., and Gilman, J. P. W., 1967. Morphological and synthetic response of normal and tumor muscle cultures to nickel subsulfide, *Cancer Res.* 27:1168–1177.

Beck, B. L., Henjum, D. C., Antonijczuk, K., Zaharia, O., Korza, G., Ozols, J., Hopfer, S. M., Barber, A. M., and Sunderman, F. W., Jr., 1992. *pNiXa*, a Ni^{2+}-binding protein in *Xenopus* oocytes and embryos, shows identity to *Ep*45, an estrogen-regulated hepatic serpin, *Res. Commun. Chem. Pathol. Pharmacol.* 77:3–16.

Benson, J. M., Burt, D. G., Cheng, Y. S., Hahn, F. F., Haley, P. J., Henderson, R. F., Hobbs, C. H., Pickrell, J. A., and Dunnick, J. K., 1989. Biochemical responses of rat and mouse lung to inhaled nickel compounds, *Toxicology* 57:255–266.

Bible, K. C., Buytendorp, M., Zierath, P. D., and Rinehart, K. L., 1988. Tunichlorin: A nickel chlorin isolated from the Caribbean tunicate *Trididemnum solidum, Proc. Natl. Acad. Sci. USA* 85:4582–4586.

Biggart, N. W., and Costa, M., 1986. Assessment of the uptake and mutagenicity of nickel chloride in *Salmonella* tester strains, *Mutat. Res.* 175:209–215.

Biggart, N. W., Gallick, G. E., and Murphy, E. C., Jr., 1987. Nickel-induced heritable alterations in retroviral transforming gene expression, *J. Virol.* 61:2378–2388.

Bourtayre, P., Pizzorni, L., Liquier, J., Tabourny, J., Taillandier, E., and Labarre, J. F., 1984. Z-form induction in DNA by carcinogenic nickel compounds: An optical spectroscopic study, in *Nickel and the Human Environment* (F. W. Sunderman, Jr. *et al.*, eds.), Oxford University Press, New York, pp. 227–234.

Brommundt, G., and Kavaler, F., 1987. La^{3+}, Mn^{2+}, and Ni^{2+} effects on Ca^{2+} pump and on Na$^+$–Ca^{2+} exchange in bullfrog ventricle, *Am. J. Physiol.* 253:C45–C51.

Brown, S. S., and Sunderman, F. W., Jr. (eds.), 1980. *Nickel Toxicology,* Academic Press, New York.

Brown, S. S., and Sunderman, F. W., Jr. (eds.), 1985. *Progress in Nickel Toxicology,* Blackwell Scientific, Oxford.

Campbell, J. A., 1943. Lung tumours in mice and man, *Br. Med. J.* 1:179–183.

Chen, X., Rokita, S. E., and Burrows, C. J., 1991. DNA modification: Intrinsic selectivity of nickel(II) complexes, *J. Am. Chem. Soc.* 113:5884–5886.

Chen, X., Burrows, C. J., and Rokita, S. E., 1992. Conformation-specific detection of guanine in DNA: Ends, mismatches, bulges, and loops, *J. Am. Chem. Soc.* 114:322–325.

Cheng, C. C., Rokita, S. E., and Burrows, C. J., 1993. Nickel(III)-promoted DNA cleavage with ambient oxygen, *Angew. Chem.* 32:277–278.

Chiocca, S. M., Sterner, D. A., Biggart, N. W., and Murphy, E. C., Jr., 1991. Nickel mutagenesis: Alteration of the MuSVts 110 thermosensitive splicing phenotype by nickel-induced duplication of the 3′ splice site, *Mol. Carcinogen.* 4:61–71.

Christie, N. T., and Katsifis, S. P., 1990. Nickel carcinogenesis, in *Biological Effects of Heavy Metals* (E. C. Foulkes, ed.), CRC Press, Boca Raton, Florida.

Christie, N. T., Sen, P., and Costa, M., 1988. Chromosomal alterations in cell lines derived from mouse rhabdomyosarcomas induced by crystalline nickel sulfide, *Biol. Metals* 1:43–50.

Ciccarelli, R. B., and Wetterhahn, K. E., 1982. Nickel distribution and DNA lesions induced in rat tissues by carcinogenic nickel carbonate, *Cancer Res.* 42:3544–3549.

Ciccarelli, R. B., and Wetterhahn, K. E., 1984. Molecular basis for the activity of nickel, in *Nickel in the Human Environment* (F. W. Sunderman, Jr. *et al.,* eds.), Oxford University Press, New York, pp. 201–213.

Ciccarelli, R. B., Hampton, T. H., and Jennette, K. W., 1981. Nickel carbonate induces DNA–protein crosslinks and DNA strand breaks in rat kidney, *Cancer Lett.* 12:349–354.

Conway, K., and Costa, M., 1989. Nonrandom chromosomal alterations in nickel-transformed Chinese hamster embryo cells, *Cancer Res.* 49:6032–6038.

Conway, K., Sen, P., and Costa, M., 1986. Antagonistic effect of magnesium chloride on the nickel chloride-induced inhibition of DNA replication in Chinese hamster ovary cells, *J. Biochem. Toxicol.* 1:11–26.

Coogan, T. P., Latta, D. M., Snow, E. T., and Costa, M., 1989. Toxicity and carcinogenicity of nickel compounds, *CRC Crit. Rev. Toxicol.* 19:341–384.

Cooley, L., Crawford, D. R., and Bishop, S. H., 1976. Urease from the lugworm, *Arenicola cristata, Biol. Bull.* 151:96–107.

Costa, M., 1991. Molecular mechanisms of nickel carcinogenesis, *Annu. Rev. Pharmacol. Toxicol.* 31:321–337.

Costa, M., and Mollenhauer, H. M., 1980a. Carcinogenic activity of particulate nickel compounds is proportional to their cellular uptake, *Science* 209:515–517.

Costa, M., and Mollenhauer, H. H., 1980b. Phagocytosis of nickel subsulfide particles during the early stages of neoplastic transformation in tissue culture, *Cancer Res.* 40:1688–1694.

Costa, M., Nye, J. S., Sunderman, F. W., Jr., Allpass, P. R., and Gondos, B., 1979. Induction of sarcomas in nude mice by implantation of Syrian hamster fetal cells exposed *in vitro* to nickel subsulfide, *Cancer Res.* 39:3591–3597.

Costa, M., Abbracchio, M. P., and Simmons-Hansen, J., 1981. Factors influencing the phagocytosis, neoplastic transformation, and cytotoxicity of particulate nickel compounds in tissue culture systems, *Toxicol. Appl. Pharmacol.* 60:313–323.

Cotelle, N., Trémolières, E., Bernier, J. L., Catteau, J. P., and Hénichart, J. P., 1992. Redox chemistry of complexes of nickel(II) with some biologically important peptides in the presence of reduced oxygen species: An ESR study, *J. Inorg. Biochem.* 46:7–15.

D'Alonzo, C. A., and Pell, S., 1963. A study of trace metals in myocardial infarction, *Arch. Environ. Health* 6:381–385.

Decock-Le Reverend, B., Perly, B., and Sarkar, B., 1987. Isolation and two-dimensional ^{1}H-NMR of peptide [1–24] of dog serum albumin and studies of its complexation with copper and nickel by NMR and CD spectroscopy, *Biochim. Biophys. Acta* 912:16–27.

DiPaolo, J. A., and Castro, B. C., 1979. Quantitative studies of *in vitro* morphological transformation of Syrian hamster cells by inorganic metal salts, *Cancer Res.* 39:1008–1023.

Diwan, B. A., Kasprzak, K. S., and Rice, J. M., 1992. Transplacental carcinogenic effects of nickel(II) acetate in the renal cortex, renal pelvis and adenohypophysis in F344/NCr rats, *Carcinogenesis* 13:1351–1357.

Doll, R., 1958. Cancer of the lung and nose in nickel workers, *Br. J. Ind. Med.* 15:217–223.

Dolovich, J., Evans, S. L., and Nieboer, E., 1984. Occupational asthma from nickel sensitivity. I. Human serum albumin in the antigenic determinant, *Br. J. Ind. Med.* 41:51–55.

Donskoy, E., Donskoy, M., Forouhar, F., Gillies, C. G., Marzouk, A., Reid, M. C., Zaharia, O., and Sunderman, F. W., Jr., 1986. Hepatic toxicity of nickel chloride in rats, *Ann. Clin. Lab. Sci.* 16:108–117.

Fishelson, Z., and Müller-Eberhard, H. J., 1982. C3 convertase of human complement: Enhanced formation and stability of the enzyme generated with nickel instead of magnesium, *J. Immunol.* 129:2603–2607.

Fishelson, Z., Pangburn, M. K., and Müller-Eberhard, H. J., 1983. C3 convertase of the alternate complement pathway. Demonstration of an active, stable C3b,Bb(Ni) complex, *J. Biol. Chem.* 258:7411–7415.

Foulkes, E. C., and McMullen, D. M., 1986. On the mechanism of nickel absorption in the rat jejunum, *Toxicology* 38:35–42.

Gitlitz, P. H., Sunderman, F. W., Jr., and Goldblatt, P. J., 1975. Aminoaciduria and proteinuria in rats after a single intraperitoneal injection of Ni(II), *Toxicol. Appl. Pharmacol.* 34:430–440.

Glennon, J. D., and Sarkar, B., 1982a. Nickel(II) transport in human blood serum. Studies of nickel(II) binding to human albumin and to native-sequence peptide, and ternary-complex formation with L-histidine, *Biochem. J.* 203:15–23.

Glennon, J. D., and Sarkar, B., 1982b. The non-specificity of dog serum albumin and the *N*-terminal model peptide glycylglycyl-L-tyrosine *N*-methylamide for nickel is due to the lack of histidine in the third position, *Biochem. J.* 203:25–31.

Hacques, M. F., and Marion, C., 1986. DNA polymorphism: Spectroscopic and electro-optical characterizations of Z-DNA and other types of left-handed helical structures induced by Ni^{2+}, *Biopolymers* 25:2281–2293.

Halliwell, B., and Gutteridge, J. M. C., 1984. Oxygen toxicity, oxygen radicals, transition metals and disease, *Biochem. J.* 219:1–14.

Hansen, K., and Stern, R. M., 1983. *In vitro* toxicity and transformation potency of nickel compounds, *Environ. Health Perspect.* 51:223–226.

Hasan, H., and Ali, S. F., 1981. Effects of thallium, nickel, and cobalt administration on the lipid peroxidation in different regions of the rat brain, *Toxicol. Appl. Pharmacol.* 57:8–13.

Haugen, A., Ryberg, D., Hansteen, I.-L., and Amstad, P., 1990. Neoplastic transformation of a human kidney epithelial cell line transfected with v-Ha-*ras* oncogene, *Int. J. Cancer* 45:572–577.

Heck, J. D., and Costa, M., 1983. Influence of surface charge and dissolution on the selective phagocytosis of potentially carcinogenic particulate metal compounds, *Cancer Res.* 43:5652–5656.

Higinbotham, K. G., Rice, J. M., Diwan, B. A., Kasprzak, K. S., Reed, C. D., and Perantoni, A. O., 1992. GGT to GTT transversions in codon 12 of the K-*ras* oncogene in rat renal sarcomas induced with nickel subsulfide or nickel subsulfide/iron are consistent with oxidative damage to DNA, *Cancer Res.* 52:4747–4751.

Hildebrand, H. F., Decaestecker, A.-M., Arrouijal, F.-Z., and Marinez, R., 1991. *In vitro* and *in vivo* uptake of nickel sulfides by rat lymphocytes, *Arch. Toxicol.* 65:324–329.

Himmelhoch, S. R., Sober, H. A., Valle, B. L., Peterson, E. A., and Fuwa, K., 1966. Spectrographic and chromatographic resolution of metalloproteins in human serum, *Biochemistry* 5:2523–2532.

Holst, M., and Nordlind, K., 1988. Phosphorylation of nuclear proteins of peripheral blood T lymphocytes activated by nickel sulfate and mercuric chloride, *Int. Arch. Allergy Appl. Immunol.* 85:337–340.

Hopfer, S. M., Plowman, M. C., Sweeney, K. R., Bantle, J. A., and Sunderman, F. W., Jr., 1991. Teratogenicity of Ni^{2+} in *Xenopus laevis,* assayed by the FETAX procedure, *Biol. Trace Elem. Res.* 29:203–216.

Inoue, S., and Kawanishi, S., 1989. ESR evidence for superoxide, hydroxyl radicals and singlet oxygen produced from hydrogen peroxide and nickel(II) complex of glycylglycyl-L-histidine, *Biochem. Biophys. Res. Commun.* 159:445–451.

Janknecht, R., Sander, C., and Pongs, O., 1991. (HX)$_n$ repeats: A pH-controlled protein–protein interaction motif of eukaryotic transcription factors? *FEBS Lett.* 295:1–2.

Jones, D. C., May, P. M., and Williams, D. R., 1980. Computer simulation models of low-molecular-weight nickel(II) complexes and therapeuticals *in vivo*, in *Nickel Toxicology* (S. S. Brown and F. W. Sunderman, Jr., eds.), Academic Press, New York, pp. 73–80.

Joshi, R. R., and Ganesh, K. N., 1992. Specific cleavage of DNA at CG sites by Co(III) and Ni(II) desferal complexes, *FEBS Lett.* 313:303–306.

Kapsenberg, M. L., Wierenga, E. A., Stiekema, F. E. M., Tiggelman, A. M. B. C., and Bos, J. D., 1992. Th1 lymphokine production profiles of nickel-specific CD4$^+$ T-lymphocyte clones from nickel contact allergic and non-allergic individuals, *J. Invest. Dermatol.* 98:59–63.

Kasprzak, K. S., 1991. The role of oxidative damage in metal carcinogenesis, *Chem. Res. Toxicol.* 4:604–615.

Kasprzak, K. S., and Bare, R. M., 1989. *In vitro* polymerization of histones by carcinogenic nickel compounds, *Carcinogenesis* 10:621–624.

Kasprzak, K. S., and Hernandez, L., 1989. Enhancement of hydroxylation and deglycosylation of 2′-deoxyguanosine by carcinogenic nickel compounds, *Cancer Res.* 49:5964–5968.

Kasprzak, K. S., Gabryel, P., and Jarczewska, K., 1983. Carcinogenicity of nickel(II)hydroxides and nickel(II)sulfate in Wistar rats and its relation to the *in vitro* dissolution rates, *Carcinogenesis* 4:275–279.

Kasprzak, K. S., Diwan, B. A., Konishi, N., Misra, M., and Rice, J. M., 1990. Initiation by nickel acetate and promotion by sodium barbital of renal cortical epithelial tumors in male F344 rats, *Carcinogenesis* 11:647–652.

Kasprzak, K. S., Diwan, B. A., Rice, J. M., Misra, M., Riggs, C. W., Olinski, R., and Dizdaroglu, M., 1992a. Nickel(II)-mediated oxidative DNA base damage in renal and hepatic chromatin of pregnant rats and their fetuses. Possible relevance to carcinogenesis, *Chem. Res. Toxicol.* 5:809–815.

Kasprzak, K. S., North, S. L., and Hernandez, L., 1992b. Reversal by nickel(II) of inhibitory effects of some scavengers of active oxygen species upon hydroxylation of 2′-deoxyguanosine *in vitro*, *Chem.-Biol. Interact.* 84:11–19.

Kawanishi, S., Inoue, S., and Yamamoto, K., 1989. Site-specific DNA damage induced by nickel(II) ion in the presence of hydrogen peroxide, *Carcinogenesis* 10:2231–2235.

King, M. M., and Huang, C. Y., 1983. Activation of calcineurin by nickel ions, *Biochem. Biophys. Res. Commun.* 114:955–961.

Kirchgessner, M., and Schnegg, 1976. Malate dehydrogenase and glucose-6-phosphatase activity in livers of Ni-deficient rats, *Bioinorg. Chem.* 6:155–161.

Kirchgessner, M., and Schnegg, A., 1980. Biochemical and physiological effects of nickel deficiency, in *Nickel in the Environment* (J. O. Nriagu, ed.), John Wiley & Sons, New York, pp. 635–652.

Klein, C. B., Conway, K., Wang, X. W., Bhamra, R. K., Lin, X., Cohen, M. D., Annab, L., Barrett, J. C., and Costa, M., 1991a. Senescence of nickel-transformed cells by an X chromosome: Possible epigenetic control, *Science* 251:796–799.

Klein, C. B., Frenkel, K., and Costa, M., 1991b. The role of oxidative processes in metal carcinogenesis, *Chem. Res. Toxicol.* 4:592–604.

Knight, J. A., Plowman, M. R., Hopfer, S. M., and Sunderman, F. W., Jr., 1991. Pathological reactions in lung, liver, thymus, and spleen of rats after subacute parenteral administration of nickel sulfate, *Ann. Clin. Lab. Sci.* 21:275–283.

Krizek, B. A., and Berg, J. M., 1992. Complexes of zinc finger proteins with Ni^{2+} and Fe^{2+}, *Inorg. Chem.* 31:2984-2986.

Kuehn, K., Frase, C. B., and Sunderman, F. W., Jr., 1982. Phagocytosis of particulate nickel compounds by rat peritoneal macrophages *in vitro, Carcinogenesis* 3:321-326.

Langård, S., and Stern, R. M., 1984. Nickel in welding fumes—a cancer hazard to welders? A review of epidemiological studies on cancer in welders, in *Nickel in the Human Environment* (F. W. Sunderman, Jr. *et al.,* eds.), Oxford University Press, New York, pp. 95-103.

Laurie, S. H., and Pratt, D. E., 1986. A spectroscopic study of nickel(II)–bovine serum albumin binding and reactivity, *J. Inorg. Biochem.* 28:431-439.

Laussac, J.-P., and Sarkar, B., 1984. Characterization of the copper(II)- and nickel(II)-transport site of human serum albumin. Studies of copper(II) and nickel(II) binding to peptide 1-24 of human serum albumin by ^{13}C and ^{1}H NMR spectroscopy, *Biochemistry* 23:2832-2838.

Lechner, J. F., Tokiwan, T., McClendon, I. A., and Haugen, A., 1984. Effects of nickel sulfate on growth and differentiation of normal human bronchial epithelial cells, *Carcinogenesis* 5: 1697-1703.

Lee, J. E., Ciccarelli, R. B., and Jennette, K. W., 1982. Solubilization of the carcinogen nickel subsulfide and its interaction with deoxyribonucleic acid and protein, *Biochemistry* 21:771-778.

Leonard, A., and Jacquet, P., 1984. Embryotoxicity and genotoxicity of nickel, in *Nickel in the Human Environment* (F. W. Sunderman, Jr. *et al.,* eds.), Oxford University Press, New York, pp. 277-291.

Lin, S.-M., Hopfer, S. M., Brennan, S. M., and Sunderman, F. W., Jr., 1989. Protein blotting method for detection of nickel-binding proteins, *Res. Commun. Chem. Pathol. Pharmacol.* 65:275-288.

Lin, X., Sugyama, M., and Costa, M., 1991. Differences in the effect of vitamin E on nickel sulfide or nickel chloride-induced chromosomal aberrations in mammalian cells, *Mutat. Res.* 260:159-164.

Lin, X., Zhuang, Z., and Costa, M., 1992. Analysis of residual amino acid–DNA crosslinks induced in intact cells by nickel and chromium compounds, *Carcinogenesis* 13:1763-1768.

Liquier, J., Bourtayre, P., Pizzorni, L., Sournies, F., Labarre, J. F., and Taillandier, E., 1984. Spectroscopic studies of conformational transitions in double stranded DNAs in the presence of carcinogenic nickel compounds and antitumoral drug (SOAZ), *Anticancer Res.* 4:41-44.

Lu, C. C., Matsumoto, M., and Iijima, S., 1979. Teratogenic effects of nickel chloride on embryonic mice and its transfer to embryonic mice, *Teratology* 19:137-142.

Lucassen, M., and Sarkar, B., 1979. Nickel(II)-binding constituents of human blood serum, *J. Toxicol. Environ. Health* 5:897-905.

Lumb, G., and Sunderman, F. W., Jr., 1988. The mechanism of malignant tumor induction by nickel subsulfide, *Ann. Clin. Lab. Sci.* 18:353-366.

Luo, S.-Q., Plowman, M. C., Hopfer, S. M., and Sunderman, F. W., Jr., 1993. Mg^{2+}-deprivation enhances and Mg^{2+}-supplementation diminishes the embryotoxic and teratogenic effects of Ni^{2+}, Co^{2+}, Zn^{2+}, and Cd^{2+} for frog embryos in the *FETAX* assay, *Ann. Clin. Lab. Sci.* 23: 121-129.

Mack, D. P., and Dervan, P. B., 1990. Nickel-mediated sequence specific oxidative cleavage of DNA by a designed metalloprotein, *J. Am. Chem. Soc.* 112:4604-4606.

Maehle, L., Metcalf, R. A., Ryberg, D., Bennett, W. P., Harris, C. C., and Haugen, A., 1992. Altered *p53* gene structure and expression in human epithelial cells after exposure to nickel, *Cancer Res.* 52:218-221.

Makowski, G. S., and Sunderman, F. W., Jr., 1992. The interactions of zinc, nickel, and cadmium with *Xenopus* transcription factor IIIA, assessed by equilibrium dialysis, *J. Inorg. Biochem.* 48:107-119.

Makowski, G. S., Lin, S. M., Brennan, S. M., Smialowitz, H. M., Hopfer, S. M., and Sunderman, F. W., Jr., 1991. Detection of two Zn-finger proteins of *Xenopus laevis*, TFIIIA, and p43, by probing Western blots of ovary cytosol with $^{65}Zn^{2+}$, $^{63}Ni^{2+}$, or $^{109}Cd^{2+}$, *Biol. Trace Elem. Res.* 29:93–109.

Mas, A., Holt, D., and Webb, M., 1985. The acute toxicity and teratogenicity of nickel in pregnant rats, *Toxicology* 35:47–57.

McDonald, J. A., Vorhaben, J. E., and Campbell, J. W., 1980. Invertebrate urease: Purification and properties of the enzyme from a land snail, *Otala lactea, Comp. Biochem. Physiol.* 66B: 223–231.

McNeeley, M. D., Sunderman, W. F., Jr., Nechay, M. W., and Levine, H., 1971. Abnormal concentrations of nickel in serum in cases of myocardial infarction, stroke, burns, hepatic cirrhosis, and uremia, *Clin. Chem.* 17:1123–1128.

Menné, T., 1992. Nickel contact hypersensitivity, in *Nickel and Human Health: Current Perspectives* (E. Nieboer and J. O. Nriagu, eds.), John Wiley & Sons, New York, pp. 193–200.

Menon, C. R., and Nieboer, E., 1986. Uptake of nickel(II) by human peripheral mononuclear leukocytes, *J. Inorg. Biochem.* 28:217–225.

Misra, M., Olinski, R., Dizdaroglu, M., and Kasprzak, K. S., 1993. Enhancement by L-histidine of nickel(II)-induced DNA–protein cross-linking and oxidative DNA base damage in the rat kidney, *Chem. Res. Toxicol.* 6:33–37.

Morgan, W. T., 1981. Interactions of histidine-rich glycoprotein of serum with metals, *Biochemistry* 20:1054–1061.

Muller, J. G., Chen, X., Dadiz, A. C., Rokita, S. E., and Burrows, C. H., 1992. Ligand effects associated with the intrinsic selectivity of DNA oxidation promoted by nickel(II) macrocyclic complexes, *J. Am. Chem. Soc.* 114:6407–6411.

Muller, J. G., Chen, X., Dadiz, A. C., Rokita, S. E., and Burrows, C. H., 1993. Macrocyclic nickel complexes in DNA recognition and oxidation, *Pure Appl. Chem.* 65:545–550.

National Research Council, 1975. *Nickel*, National Academy of Sciences, Washington, D.C.

Nicklin, S., and Nielsen, G. T., 1992. Nickel and the immune system: Current concepts, in *Nickel and Human Health: Current Perspectives* (E. Nieboer and J. O. Nriagu, eds.), John Wiley & Sons, New York, pp. 239–259.

Nieboer, E., and Nriagu, J. O. (eds.), 1992. *Nickel and Human Health: Current Perspectives,* John Wiley & Sons, New York.

Nieboer, E., Stafford, A. R., Evans, S. L., and Dolovich, J., 1984a. Cellular binding and/or uptake of nickel(II) ions, in *Nickel in the Human Environment* (F. W. Sunderman, Jr. *et al.,* eds.), Oxford University Press, New York, pp. 321–331.

Nieboer, E., Maxwell, R. I., and Stafford, A. R., 1984b. Chemical and biological reactivity of insoluble nickel compounds and the bioinorganic chemistry of nickel, in *Nickel in the Human Environment* (F. W. Sunderman, Jr. *et al.,* eds.), Oxford University Press, New York, pp. 321–331.

Nieboer, E., Tom, R. T., and Sanford, W. E., 1988a. Nickel metabolism in man and animals, in *Nickel and Its Role in Biology* (H. Sigel and A. Sigel, eds.), *Metal Ions in Biological Systems,* Vol. 23, Marcel Dekker, New York, pp. 91–121.

Nieboer, E., Rossetto, F. E., and Menon, C. R., 1988b. Toxicology of nickel compounds, in *Nickel and Its Role in Biology* (H. Sigel and A. Sigel, eds.), *Metal Ions in Biological Systems,* Vol. 23, Marcel Dekker, New York, pp. 359–402.

Nieboer, E., Sanford, W. E., and Stace, B. C., 1992. Absorption, distribution, and excretion of nickel, in *Nickel and Human Health: Current Perspectives* (E. Nieboer and J. O. Nriagu, eds.), John Wiley & Sons, New York, pp. 49–68.

Nielsen, F. H., 1971. Studies on the essentiality of nickel, in *Newer Trace Elements in Nutrition* (W. Mertz and W. E. Cornatzer, eds.), Marcel Dekker, New York, pp. 215–253.

Nielsen, F. H., 1984. Nickel, in *Biochemistry of the Essential Ultratrace Elements* (E. Frieden, ed.), Plenum Press, New York, pp. 293–308.

Nielsen, F. H., 1986. Nickel, in *Trace Elements in Human and Animal Nutrition,* 5th ed. (W. Mertz, ed.), Academic Press, New York, pp. 245–273.

Nielsen, F. H., 1991. Nutritional requirements for boron, silicon, vanadium, nickel, and arsenic: Current knowledge and speculation, *FASEB J.* 5:2661–2667.

Nielsen, F. H., and Ollerich, D. A., 1974. Nickel: A new essential trace element, *Fed. Proc.* 33: 1767–1772.

Nielsen, F. H., and Sauberlich, H. E., 1970. Evidence of a possible requirement for nickel by the chick, *Proc. Soc. Exp. Biol. Med.* 134:845–849.

Nielsen, F. H., Myron, D. R., Givand, S. H., and Ollerich, D. A., 1975a. Nickel deficiency and nickel–rhodium interaction in chicks, *J. Nutr.* 105:1607–1619.

Nielsen, F. H., Myron, D. R., Givand, S. H., Zimmerman, T. J., and Ollerich, D. A., 1975b. Nickel deficiency in rats, *J. Nutr.* 105:1620–1630.

Nielsen, F. H., Shuler, T. R., McLeod, T. G., and Zimmerman, T. J., 1984. Nickel influences iron metabolism through physiologic, pharmacologic and toxicologic mechanisms in the rat, *J. Nutr.* 114:1280–1288.

Nishimura, M., and Umeda, M., 1979. Induction of chromosomal aberrations in cultured mammalian cells by nickel compounds, *Mutat. Res.* 68:337–349.

Nomoto, S., and Sunderman, F. W., Jr., 1988. Presence of nickel in alpha-2-macroglobulin isolated from human serum by high performance liquid chromatography, *Ann. Clin. Lab. Sci.* 18: 78–84.

Nomoto, S., McNeeley, M. D., and Sunderman, F. W., Jr., 1971. Isolation of a nickel α_2-macroglobulin from rabbit serum, *Biochemistry* 10:1647–1651.

Okamoto, M., 1987. Induction of ocular tumor by nickel subsulfide in the Japanese common newt, *Cynops pyrrhogaster, Cancer Res.* 47:5213–5217.

Onkelinx, C., and Sunderman, F. W., Jr., 1980. Modeling of nickel metabolism, in *Nickel in the Environment* (J. O. Nriagu, ed.), John Wiley & Sons, New York, pp. 525–545.

Parker, K., and Sunderman, F. W., Jr., 1974. Distribution of ^{63}Ni in rabbit tissues following intravenous injection of $^{63}NiCl_2$, *Res. Commun. Chem. Pathol. Pharmacol.* 7:755–762.

Patierno, S. R., and Costa, M., 1985. DNA–protein cross-links induced by nickel compounds in intact cultured mammalian cells, *Chem.-Biol. Interact.* 55:75–91.

Pienta, R. J., Poley, J. A., and Lebherz, W. B., III, 1977. Morphological transformation of early passage golden Syrian hamster embryo cells derived from cryopreserved primary cultures as a reliable *in vitro* bioassay for identifying diverse carcinogens, *Int. J. Cancer* 19:642–655.

Predki, P. F., Whitfield, D. M., and Sarkar, B., 1992a. Characterization and cellular distribution of acidic peptide and oligosaccharide metal-binding compounds from kidneys, *Biochem. J.* 281:835–841.

Predki, P. F., Harford, C., Brar, P., and Sarkar, B., 1992b. Further characterization of the *N*-terminal copper(II)- and nickel(II)-binding motif of proteins. Studies of metal binding to chicken serum albumin and the native sequence peptide, *Biochem. J.* 287:211–215.

Pullarkat, R. K., Sbaschnig-Agler, M., and Reha, H., 1981. Biosynthesis of phosphatidylserine in rat brain microsomes, *Biochim. Biophys. Acta* 664:117–123.

Rayner-Canham, G. W., Van Roode, M., and Burke, J., 1985. Nickel and cobalt concentrations in the tunicate *Halocynthia pyriformis:* Evidence for essentiality of the two metals, *Inorg. Chim. Acta* 106:L37–L38.

Rezuke, W. N., Knight, J. A., and Sunderman, F. W., Jr., 1987. Reference values for nickel concentrations in human tissues and bile, *Am. J. Ind. Med.* 11:419–426.

Robison, S. H., Cantoni, O., and Costa, M., 1984. Analysis of metal-induced DNA lesions and DNA-repair replication in mammalian cells, *Mutat. Res.* 131:173–181.

Romani, A., and Scarpa, A., 1992. Regulation of cell magnesium, *Arch. Biochem. Biophys.* 298: 1–12.

Sadler, T. W., 1980. Effects of maternal diabetes on early embryogenesis. I. The teratogenic potential of diabetic serum, *Teratology* 21:339–347.

Saillenfait, A. M., Sabate, J. P., Langonne, I., and de Ceaurriz, J., 1991. Nickel chloride teratogenesis in cultured rat embryos, *Toxicol. In Vitro* 5:83–89.

Sarkar, B., 1980. Nickel in blood and kidney, in *Nickel Toxicology* (S. S. Brown and F. W. Sunderman, eds.), Academic Press, New York, pp. 81–84.

Sarkar, B., 1984. Nickel metabolism, in *Nickel in the Human Environment* (F. W. Sunderman, Jr. *et al.,* eds.), Oxford University Press, New York, pp. 367–384.

Schaaper, R. M., Koplitz, R. M., Tkeshelashvili, L. K., and Loeb, L. A., 1987. Metal-induced lethality and mutagenesis: Possible role of apurinic intermediates, *Mutat. Res.* 177:179–188.

Schiffer, R. B., Sunderman, F. W., Jr., Baggs, R., and Moynihan, J. A., 1991. The effects of exposure to dietary nickel and zinc upon humoral and cellular immunity in SJL mice, *J. Neuroimmunol.* 34:229–239.

Schroeder, H. A., and Mitchner, M., 1971. Toxic effects of trace elements on the reproduction of mice and rats, *Arch. Environ. Health* 23:102–106.

Schroeder, H. A., Balassa, J. J., and Tipton, I. H., 1962. Abnormal trace metals in man—nickel, *J. Chronic Dis.* 15:51–65.

Sen, P., and Costa, M., 1985. Induction of chromosomal damage in Chinese hamster ovary cells by soluble and particulate nickel compounds: Preferential fragmentation of the heterochromatic long arm of the X-chromosome by carcinogenic crystalline NiS particles, *Cancer Res.* 45:2320–2325.

Shi, X., Dalal, N. S., and Kasprzak, K. S., 1992. Generation of free radicals from lipid hydroperoxides by Ni^{2+} in the presence of oligopeptides, *Arch. Biochem. Biophys.* 299:154–162.

Shi, X., Dalal, N. S., and Kasprzak, K. S., 1993. Generation of free radicals in reactions of Ni(II)-thiol complexes with molecular oxygen and model lipid hydroperoxides, *J. Inorg. Biochem.* 15:211–225.

Shirakawa, T., Kusaka, Y., and Morimoto, K., 1992. Specific IgE antibodies to nickel in workers with known reactivity to cobalt, *Clin. Exp. Allergy* 22:213–218.

Shirali, P., Decaestecker, A. M., Marez, T., Hildebrand, H. F., Bailly, C., and Martinez, R., 1991. Ni_3S_2 uptake by lung cells and its interaction with plasma membrane, *J. Appl. Toxicol.* 11: 279–288.

Silvennoinen-Kassinen, S., Poikonen, K., and Ikäheimo, I., 1991. Characterization of nickel-specific T cell clones, *Scand. J. Immunol.* 33:429–434.

Simmons, J. E., Jr., 1961. Urease activity in trypanorhynch cestodes, *Biol. Bull.* 121:535–546.

Sinigaglia, F., Scheidegger, D., Garotta, G., Scheper, R., Pletscher, M., and Lanzaveccgia, A., 1985. Isolation and characterization of Ni-specific T cell clones from patients with Ni-contact dermatitis, *J. Immunol.* 135:3929–3932.

Sirover, M. A., and Loeb, L. A., 1976. Infidelity of DNA synthesis *in vitro:* Screening for potential metal mutagens or carcinogens, *Science* 194:1434–1436.

Smialowicz, R. J., Rogers, R. R., Riddle, M. M., and Stott, G. A., 1984. Immunological effects of nickel: I. Suppression of cellular and humoral immunity, *Environ. Res.* 33:413–427.

Smialowicz, R. J., Rogers, R. R., Rowe, D. G., Riddle, M. M., and Luebke, R. W., 1987. The effects of nickel on immune function in the rat, *Toxicology* 44:271–281.

Smith, J. C., and Hackley, B., 1968. Distribution and excretion of nickel-63 administered intravenously to rats, *J. Nutr.* 95:541–546.

Snow, E. T., 1992. Metal carcinogenesis: Mechanistic implications, *Pharmacol. Ther.* 53:31–65.

Spears, J. W., 1984. Nickel as a "newer trace element" in the nutrition of domestic animals, *J. Anim. Sci.* 59:823–835.

Spears, J. W., and Hatfield, E. E., 1978. Nickel for ruminants. I. Influence of dietary nickel on ruminal urease activity, *J. Anim. Sci.* 47:1345–1350.

Spears, J. W., Smith, C. J., and Hatfield, E. E., 1977. Rumen bacterial urease requirement for nickel, *J. Dairy Sci.* 7:1073–1076.

Spears, J. W., Hatfield, E. E., Forbes, R. M., and Koenig, S. E., 1978. Studies on the role of nickel in the ruminant, *J. Nutr.* 108:313–320.

Spears, J. W., Hatfield, E. E., and Forbes, R. M., 1979. Nickel for ruminants. II. Influence of dietary nickel on performance and metabolic parameters, *J. Anim. Sci.* 48:649–657.

Stinson, T. J., Jaw, S., Jeffrey, E. H., and Plewa, M. J., 1992. The relationship between nickel chloride-induced peroxidation and DNA strand breakage in rat liver, *Toxicol. Appl. Pharmacol.* 117:98–103.

Sunderman, F. W., Jr., 1977. A review of the metabolism and toxicology of nickel, *Ann. Clin. Lab. Sci.* 7:377–398.

Sunderman, F. W., Jr., 1987. Lipid peroxidation as a mechanism of acute nickel toxicity, *Toxicol. Environ. Chem.* 15:59–69.

Sunderman, F. W., Jr., 1988. Nickel, in *Handbook on Toxicology of Inorganic Compounds* (H. G. Siler and H. Sigel, eds.), Marcel Dekker, New York, pp. 454–468.

Sunderman, F. W., Jr., 1989a. Mechanisms of nickel carcinogenesis, *Scand. J. Work Environ. Health* 15:1–12.

Sunderman, F. W., 1989b. A pilgrimage into the archives of nickel toxicology, *Ann. Clin. Lab. Sci.* 19:1–16.

Sunderman, F. W., Jr., 1990. Regulation of gene expression by metals: Zinc finger-loop domains in transcription factors, hormone receptors, and proteins encoded by oncogenes, in *Metal Ions in Biology and Medicine* (P. Collery, L. A. Poirier, M. Manfait, and J. C. Etienne, eds.), John Libby Eurotext, Paris, pp. 549–554.

Sunderman, F. W., Jr., 1992. Toxicokinetics of nickel in humans, in *Nickel and Human Health: Current Perspectives* (E. Nieboer and J. O. Nriagu, eds.), John Wiley & Sons, New York, pp. 69–76.

Sunderman, F. W., Jr., Nomoto, S., Pradhan, A. M., Levine, H., Bernstein, S. H., and Hirsch, R., 1970. Increased concentrations of serum nickel after acute myocardial infarction, *N. Engl. J. Med.* 283:896–899.

Sunderman, F. W., Jr., Nomoto, S., Morang, R., Nechay, M. W., Burke, C. N., and Nielsen, S. W., 1972. Nickel deprivation in chicks, *J. Nutr.* 102:259–268.

Sunderman, F. W., Jr., Shen, S. K., Mitchell, J. M., Allpass, P. R., and Damjanov, I., 1978. Embryotoxicity and fetal toxicity of nickel in rats, *Toxicol. Appl. Pharmacol.* 43:381–390.

Sunderman, F. W., Jr., Costa, E. R., Fraser, C., Hui, G., Levine, J. J., and Tse, T. P. H., 1981. [63]Nickel-constituents in renal cytosol of rats after injection of [63]nickel chloride, *Ann. Clin. Lab. Sci.* 11:488–496.

Sunderman, F. W., Jr., Mangold, B. L. K., Wong, S. H. Y., Shen, S. K., Reid, M. C., and Jansson, I., 1983. High-performance size-exclusion chromatography of [63]Ni-constituents in renal cytosol and microsomes from [63]NiCl$_2$-treated rats, *Res. Commun. Chem. Pathol. Pharmacol.* 39: 477–492.

Sunderman, F. W., Jr., Aitio, A., Berlin, A., Bishop, C., Buringh, E., Davis, W., Gounar, M., Jacquignon, P. C., Mastromatteo, E., Rigaut, J. P., Rosenfeld, C., Saracci, R., and Sors, A. (eds.), 1984. *Nickel in the Human Environment,* Oxford University Press, New York.

Sunderman, F. W., Jr., Marzouk, A., Hopfer, S. M., Zaharia, O., and Reid, M. C., 1985. Increased lipid peroxidation in tissues of nickel chloride-treated rats, *Ann. Clin. Lab. Sci.* 15:229–236.

Suzuki, K., Bennon, Y., Mitsuoka, T., Takebe, S., Kobashi, K., and Hase, J., 1979. Urease-producing species of intestinal anaerobes and their activities, *Appl. Environ. Microbiol.* 37: 379–382.

Swierenga, S. H. H., and Basrur, P. K., 1968. Effect of nickel on cultured rat embryo muscle cells, *Lab. Invest.* 19:663–674.

Tabata, M., and Sarkar, B., 1992. Specific nickel(II)-transfer process between the native sequence peptide representing the nickel(II)-transport site of human serum albumin and L-histidine, *J. Inorg. Biochem.* 45:93–104.

Templeton, D. M., 1987. Interaction of toxic cations with the glomerulus: Binding of Ni to purified glomerular basement membrane, *Toxicology* 43:1–15.

Templeton, D. M., 1992. Nickel at the renal glomerulus: Molecular and cellular interactions, in *Nickel and Human Health: Current Perspectives* (E. Nieboer and J. O. Nriagu, eds.), John Wiley & Sons, New York, pp. 135–170.

Templeton, D. M., and Sarkar, B., 1985. Peptide and carbohydrate complexes of nickel in human kidney, *Biochem. J.* 230:35–42.

Templeton, D. M., and Sarkar, B., 1986. Nickel binding to the C-terminal tryptic fragment of a peptide from human kidney, *Biochim. Biophys. Acta* 884:383–386.

Tsangaris, J. M., Chang, J. W., and Martin, R. B., 1969. Cupric and nickel ion interactions with proteins as studied by circular dichroism, *Arch. Biochem. Biophys.* 130:53–58.

Tveito, G., Hansteen, I.-L., Dalen, H., and Haugen, A., 1989. Immortalization of normal human kidney epithelial cells by nickel(II), *Cancer Res.* 49:1829–1835.

Umeda, M., and Nishimura, M., 1979. Inducibility of chromosomal aberrations by metal compounds in cultured mammalian cells, *Mutat. Res.* 67:221–229.

Van de Sande, J. H., McIntosh, L. P., and Jovin, T. M., 1982. Mn^{2+} and other transition metals at low concentration induce the right-to-left helical transformation of poly[d(G-C)], *EMBO J.* 1:777–782.

Wang, X. W., and Costa, M., 1991. Changes in protein phosphorylation in wild-type and nickel-resistant cells and their involvement in morphological elongation, *Biol. Metals* 4:201–206.

Wang, X. W., Imbra, R. J., and Costa, M., 1988. Characterization of mouse cell lines resistant to nickel(II) ions, *Cancer Res.* 48:6850–6854.

Wase, A. W., Goss, D. M., and Boyd, M. J., 1954. The metabolism of nickel. I. Spatial and temporal distribution of ^{63}Ni in the mouse, *Arch. Biochem. Biophys.* 51:1–4.

Wozney, M. A., Bryant, M. P., Holdeman, L. V., and Moore, W. E. C., 1977. Urease assay and urease-producing species of anaerobes in the bovine and human feces, *Appl. Environ. Microbiol.* 33:1097–1104.

Zhong, Z., Troll, W., Koenig, K. L., and Frenkel, K., 1990. Carcinogenic sulfide salts of nickel and cadmium induce H_2O_2 formation by human polymorphonuclear leukocytes, *Cancer Res.* 50:7564–7570.

Index

DATE DUE

NOV 1 4 1997	
FEB 1 3 1998	
APR 1 8 1998	
MAY 1 8 1998	
APR 1 9 1999	
OCT 2 9 2000	
DEC 1 8 2000	
APR 2 4 2006	
SEP 1 4 2007	
NOV 1 3 2014	